19th Asia Pacific Bioinformatics Conference (APBC 2021)

Online

3 - 5 February 2021

ISBN: 978-1-7138-5563-7

Printed from e-media with permission by:

Curran Associates, Inc.
57 Morehouse Lane
Red Hook, NY 12571

Some format issues inherent in the e-media version may also appear in this print version.

Printed by Curran Associates, Inc. (2022)

For further information, please contact BioMed Central Ltd at the address below. The collection is available online at https://bmcbioinformatics.biomedcentral.com/articles/supplements/volume-22-supplement-10

BioMed Central
The Campus, 4 Crinan Street
London N1 9XW
United Kingdom

Phone: +44 (0) 20 3192 2009

Email: ORSupport@spingernature.com
Web: www.biomedcentral.com

Additional copies of this publication are available from:

Curran Associates, Inc.
57 Morehouse Lane
Red Hook, NY 12571 USA
Phone: 845-758-0400
Fax: 845-758-2633
Email: curran@proceedings.com
Web: www.proceedings.com

TABLE OF CONTENTS

Dessie *et al. BMC Bioinformatics* (2021) 22:270
https://doi.org/10.1186/s12859-021-04189-2

BMC Bioinformatics

METHODOLOGY Open Access

A novel miRNA-based classification model of risks and stages for clear cell renal cell carcinoma patients

Eskezeia Y. Dessie[1,4], Jeffrey J. P. Tsai[1], Jan-Gowth Chang[2*] and Ka-Lok Ng[1,3,4*]

From The 19th Asia Pacific Bioinformatics Conference (APBC 2021) Tainan, Taiwan. 3-5 February 2021

*Correspondence:
d6781@mail.cmuh.org.tw;
ppiddi@gmail.com
[1] Department of Bioinformatics and Medical Engineering, Asia University, Taichung, Taiwan
[2] Department of Laboratory Medicine, China Medical University, Taichung, Taiwan
Full list of author information is available at the end of the article

Abstract

Background: Clear cell renal cell carcinoma (ccRCC) is the most common subtype of renal carcinoma and patients at advanced stage showed poor survival rate. Despite microRNAs (miRNAs) are used as potential biomarkers in many cancers, miRNA biomarkers for predicting the tumor stage of ccRCC are still limitedly identified. Therefore, we proposed a new integrated machine learning (ML) strategy to identify a novel miRNA signature related to tumor stage and prognosis of ccRCC patients using miRNA expression profiles. A multivariate Cox regression model with three hybrid penalties including Least absolute shrinkage and selection operator (Lasso), Adaptive lasso and Elastic net algorithms was used to screen relevant prognostic related miRNAs. The best subset regression (BSR) model was used to identify optimal prognostic model. Five ML algorithms were used to develop stage classification models. The biological significance of the miRNA signature was analyzed by utilizing DIANA-mirPath.

Results: A four-miRNA signature associated with survival was identified and the expression of this signature was strongly correlated with high risk patients. The high risk patients had unfavorable overall survival compared with the low risk group (HR = 4.523, *P*-value = 2.86e−08). Univariate and multivariate analyses confirmed independent and translational value of this predictive model. A combined ML algorithm identified six miRNA signatures for cancer staging prediction. After using the data balancing algorithm SMOTE, the Support Vector Machine (SVM) algorithm achieved the best classification performance (accuracy = 0.923, sensitivity = 0.927, specificity = 0.919, MCC = 0.843) when compared with other classifiers. Furthermore, enrichment analysis indicated that the identified miRNA signature involved in cancer-associated pathways.

Conclusions: A novel miRNA classification model using the identified prognostic and tumor stage associated miRNA signature will be useful for risk and stage stratification for clinical practice, and the identified miRNA signature can provide promising insight to understand the progression mechanism of ccRCC.

Keywords: Clear cell renal cell carcinoma, Biomarkers, MicroRNAs, Survival analysis, Regularized cox model, Machine learning classifiers

Background

Renal cell carcinoma (RCC) is one of the top ten cancer diagnoses, and it accounts for 3–5% of all new cases in females and males [1]. Recently, there are more than 140,000 death per year associated with RCC [2]. Clear cell renal cell carcinoma (ccRCC) is the most common RCC subtype and it represents 70–80%, of all renal malignant tumors [3]. Despite many advances in effective therapeutic and diagnostic strategies in ccRCC, and the overall survival rate is still poor, particularly for advanced-stage ccRCC patients[4]. CcRCC has poor prognosis due to the resistance to chemotherapy and radiotherapy[5]. Late tumor staging is the main risk factor of ccRCC patients [6] and detection of ccRCC patients at early-stage is crucial for better diagnosis and treatment options.

Currently, the development of next generation sequencing (NGS) technology has enabled researchers to explore genetic alterations in tumorigenesis and discovering molecular biomarkers for many cancers[7]. NGS allows to examine the possible contributions of the upstream molecular regulators of gene expression such as miRNAs. MiRNA play critical roles in regulating various physiological and pathological processes, including regulation of cell division, apoptosis, cell maturation, angiogenesis, metastasis, migration, invasion, differentiation of cells, metabolism, and proliferation by negative regulation of gene expression [8–10]. Moreover, in various cancers, dysregulated miRNAs can be used as biomarkers [11–13]

Statistical and machine learning approaches have been used to predict gene sets as biomarkers for patients with ccRCC [14]. Ng and Taguchi employed the tensor decomposition method to identify miRNA signature in ccRCC [15]. Previously, studies were used miRNA expression profiles of liver and breast cancer patients, followed by a support vector machine (SVM) with genetic algorithm, to predict the early and advanced stages [16, 17]. Recently, miRNA profiles were used to detect lung cancer subtypes [18]. Several studies have reported miRNA biomarkers in ccRCC. For example, a three-miRNA signature including miR-21, miR-155 and miR-584 is associated with survival in ccRCC [11]. Zhao and Bai identified 13-miRNA signature associated with overall survival in ccRCC [19]. However, the study of multi-miRNA signature models to predict the risks and tumor stages of ccRCC patients are still limited and hence, we aimed to select a small set of miRNAs as signature that can predict risk as well as tumor stages in ccRCC patients using genomic profiles, so that identified miRNA signature can provide promising insight to understand the progression and development mechanism of ccRCC.

In this study, we proposed a computational method for identifying prognostic-associated miRNA signature as well as predicting the early and late tumor stages of ccRCC using miRNA expression profiles. We identified a four miRNA signature associated with the prognosis of ccRCC from high-dimensional miRNA expression profiles using multivariate Cox regression with Elastic-net, Lasso and Adaptive lasso penalties followed by best subset regression analysis. The prognostic risk model involving four miRNA signature effectively classified ccRCC patients into high and low risk groups; prognosis was significantly worse in high-risk group when compared with low-risk groups. Furthermore, we extracted significant miRNAs that can distinguish early and late tumor stages using various machine learning approaches. We identified a six miRNA signature strongly related to tumor stages of ccRCC patients. The five machine learning algorithms were used to evaluate classification performance of a six miRNA signature using independent testing set. Finally, the SVM algorithm achieved the best classification performance when compared with other classifiers.

Results

We proposed a computational method including penalized Cox models and machine learning approach to identify miRNA signature for risk and tumor stage prediction using miRNA profiles, which consists of several steps as described in detail in the "Methods" section. To develop optimal prognostic predictive model for ccRCC patients, combined penalized Cox models (including Elastic-net, Lasso and Adaptive lasso), best subset regression and risk score model were used. Furthermore, a combined machine learning approach was used to prioritize and identify miRNA signatures associated with early and late tumor stages in ccRCC patients. The systematic pipeline of the overall process is shown in Additional file 1: Fig. S1.

Identification of dysregulated miRNAs in ccRCC patients

After TCGA-ccRCC data quality assessment, preprocessing and normalization, a total of 1046 miRNA expression profiles were used for differentially analysis based on "limma" package in R. We identified 124 differentially expressed miRNAs (DEMs), of which 80 downregulated and 44 upregulated miRNAs in 254 primary tumor tissue samples compared with 71 normal samples, using the criteria of absolute value log2foldchange >1 and Benjamin-Hochberg (BH) adjusted P-value <0.05 (Additional file 1: Table S1). These abnormal miRNAs were used for subsequent survival and stage classification model development.

Identification of prognostic-associated miRNAs and development of risk classification model

DEMs that altered in tumor samples are potential prognostic and diagnostic signatures. To identify significant prognostic DEMs, the TCGA cohort of ccRCC patient (n = 252) having survival information were used. Subsequently, three regularized survival methods (including Elastic-net, Lasso, and Adaptive lasso) with ten-fold cross validation were implemented to obtain the optimal lambda (λ) values that obtained from the smallest partial likelihood deviances. The estimated optimal penalty parameter λ values for the three algorithms were $\lambda_{opt}^{Enet}=0.093$, $\lambda_{opt}^{Lasso}=0.056$ and $\lambda_{opt}^{Ad.lasso}=0.015$ and these optimal tuning parameters were used to choose informative features (miRNAs) that were associated with patient survival. Elastic-net, Lasso, and Adaptive lasso algorithms identified 13 miRNAs, 11 miRNAs and 6 miRNAs respectively (Table 1). A union of candidate miRNAs selected by the three methods including: miR-21, miR-223, miR-146b, miR-30b, miR-3613, miR-187, miR-203, miR-514-3, miR-129-2, miR-200a, miR-508, miR-1.2 and miR-934 were used for BSR analysis. Then, all subset prognostic models created by the identified 13 miRNAs were assessed using BSR analysis based on the "glmulti" package in R and finally we obtained an optimal miRNA prognostic model with four-miRNA signatures (including miR-30b, miR-21, miR-187 and miR-150 200a) having the smallest AIC value (Additional file 1: Fig. S2). Then, using the regression coefficient obtained from the result of multivariate Cox regression analysis of the four-miRNA signatures (Additional file 1: Table S2), we developed a risk score (RS) model, which is given by,

Table 1 The penalized Cox regression coefficients of selected miRNAs from three methods

Gene name	Elastic-net	Lasso	Adaptive lasso
miR-21	0.513	0.705	1.049
miR-223	0.117	0.129	0.117
miR-146b	0.083	0.017	–
miR-30b	0.068	0.057	0.061
miR-3613	0.058	0.034	–
miR-187	0.045	0.053	–
miR-203	−0.011	–	–
miR-514-3	−0.019	–	–
miR-129-2	−0.037	−0.003	–
miR-200a	−0.071	−0.060	–
miR-508	−0.130	−0.151	−0.349
miR-1.2	−0.146	−0.171	−0.271
miR-934	−0.268	−0.205	−1.248

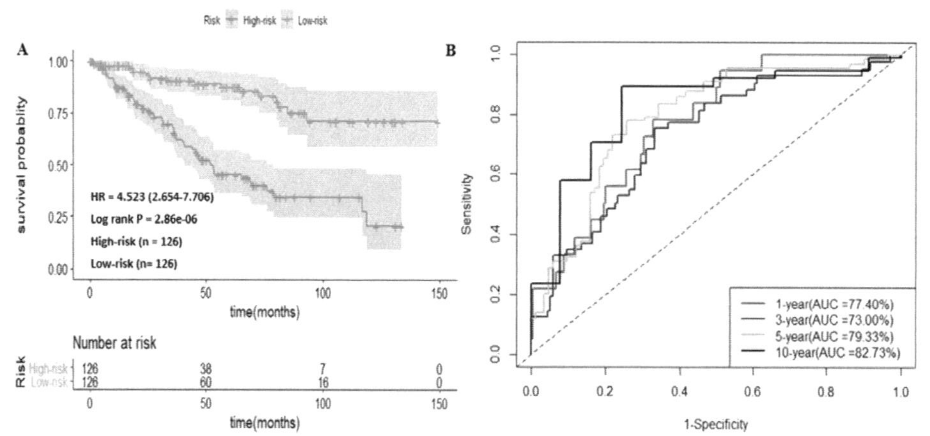

Fig. 1 Prognostic classification of ccRCC patients using the risk score model. **a** Kaplan–Meier (KM) survival analysis for high risk and low risk subgroup stratification based on the risk score model. **b** Validation of risk score classification performance by time dependent ROC analysis

RS = (0.525 × miR-30b expression) + (1.485 × miR-21 expression) + (0.485 × miR-187 expression) − (0.320 × miR-200a expression). Based on this risk score definition, we stratified ccRCC patients into high risk and low risk groups. The Kaplan–Meier curve shows that high risk group related to poor overall survival time relative to low risk group (Fig. 1a). Furthermore, the AUC values of the time dependent ROC curve were 77.40%, 73.00%, 79.33% and 82.73% for a 1-year, 3-year, 5-year and 10-year survival (Fig. 1b). These results demonstrated that the time dependent risk prediction based on a four-miRNA signature can be used for risk assessment for ccRCC patients.

Univariate and multivariate analysis for risk prediction evaluation

In order to further examine whether the identified miRNA signature can be used as independent predictor, we performed univariate and multivariate Cox analysis (Table 2). The results of univariate Cox analysis indicated that age, pathological stages and the miRNA signature are significantly correlated with overall survival. Meanwhile,

Table 2 Univariate and multivariate Cox analyses showed risk score is an independent risk factor

variables	Univariate Cox analysis		Multivariate Cox analysis	
	P-value	HR (95% CI)	*P*-value	HR (95% CI)
Age	0.001*	1.808 (1.324–2.47)	0.061	1.598 (0.978–2.610)
Gender	0.757	0.953 (0.702–1.294)	0.078	0.610 (0.352–1.058)
Pathological stage	2e−16*	3.912 (2.857–5.362)	0.048*	3.303 (1.028–10.949)
Pathological T stage	3.8e−16*	3.204 (2.369–4.331)	0.272	0.573 (0.212–1.546)
Pathological M stage	2e−16*	4.254 (3.117–5.805)	0.035*	1.957 (1.048–3.655)
Pathological Grade	3.42e−08*	2.598 (1.851–3.646)	0.217	1.462 (0.804–2.633)
Four-miRNA signature	2.14e−08*	4.628 (2.707–7.911)	4.23e−05*	4.324 (2.385–7.839)

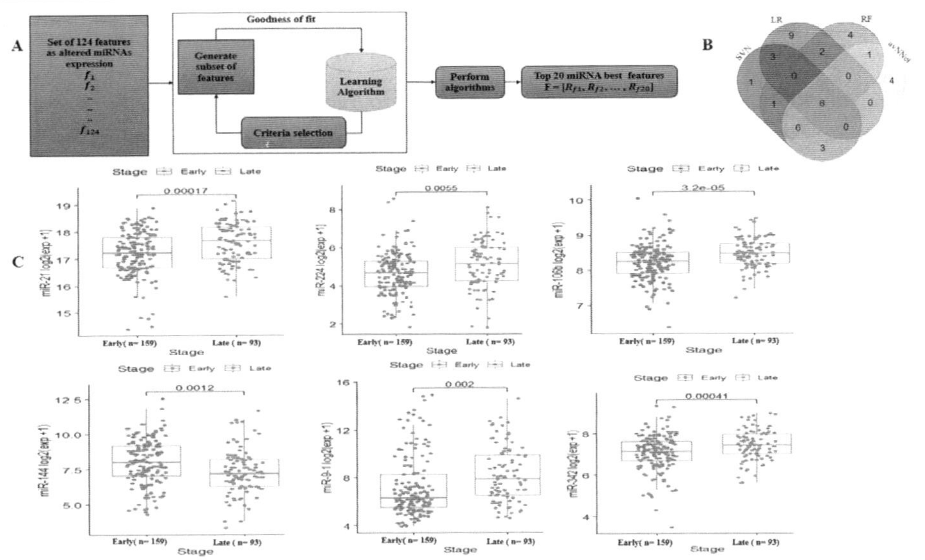

Fig. 2 Stage associated features (miRNAs). **a** Implementation of feature selection using various machine learning algorithms. **b** Common features among top 20 ranked miRNAs selected by four ML learning algorithms. **c** Results of six differentially expressed miRNAs between early and late stage of ccRCC patients

multivariate analysis showed that pathological stage (*P*-value = 0.048), pathological M stage (*P*-value = 0.035) and four miRNA signature (*P*-value = 4.23e−05) are significantly associated with unfavorable prognosis.

Identification of miRNA signature for classification of tumor stages

In order to identify tumor stage associated miRNA signatures, 124 DEMs expression profiles and the corresponding ccRCC patients with stage information (n = 252) were used. We applied ensemble learning feature selection algorithms including logistic regression (LR), random forest (RF), support vector machine with radial kernel (SVMR) and average neural network (avNNet) model to identify optimal classifier of tumor stages. The maximum Relevance Minimum Redundancy (mRMR) algorithms were used to identify miRNA features having the most correlation with tumor stage and the least correlation with miRNA features themselves. The detail of mRMR algorithm is described in Fig. 2a. The utility of mRMR based ensemble ML algorithm is enhancing informative feature selection by minimizing the bias that might be introduced by single algorithm. We identified top ranked features of 20 miRNAs using four ML methods and the ranking of miRNAs and their corresponding features importance relevance are shown (Additional file 1: Table S3). To determine the common number of miRNA features for tumor stage classification, we performed an overlapping analysis of selected miRNAs by these algorithms and six common miRNAs including miR-106b, miR-144, miR-224, miR-9-1, miR-21, miR-342 (Fig. 2b) were identified. These identified miRNAs were used to develop a stage prediction model. We also further verified that the identified 6 miRNA features were differentially expressed between early and late stage groups. The box-plot shows for each miRNA, there was a significant expression difference in early and late stage subgroups (Fig. 2c). Interestingly, the identified 6 stage-associated miRNAs are significantly associated with the overall survival of patients with ccRCC (Additional file 1: Fig. S3).

SMOTE sampling data and stage classification performance

In this study, SMOTE sampling was used to solve the class imbalance problem. Balanced dataset obtained from the SMOTE algorithm was then divided into training set and test set (80:20 ratio). The identified six miRNA signature was used to build prediction model based on five ML algorithms—SVMR, LR, Naïve Bayes, avNNet and KNN. The classification performance evaluated after data balancing is shown in Table 3. The accuracy of all models were found to be in the range of 0.62–0.923 with maximum accuracy of 0.923 for SVMR. The highest sensitivity and specificity, MCC and precision were observed for SVMR. To check whether SMOTE data sampling resulted improvement prediction, we used the original data (without SMOTE balancing) and constructed the training set and test set in a 80:20 ratio. Stage prediction and their classification performance delivered by the five ML methods were explored. The best accuracy and sensitively was observed in Naïve Bayes with an accuracy 0.775 and sensitivity 0.778. The details of model performance comparison are shown in (Additional file 1: Table S4). It is noteworthy that SMOTE data balancing algorithm can improve the prediction accuracy of all ML models except Naïve Bayes. The overall model prediction results indicated that the identified

Table 3 The performance of ML based models constructed by identified six miRNA signatures on balanced training set and test set using the SMOTE algorithm

Algorithms	Methods	Performance measures					
		ACC	Sensitivity	Specificity	MCC	F-score	Precision
SVMR	tenfold	0.990	0.987	0.993	0.981	0.989	0.991
	Test	0.923	0.927	0.919	0.843	0.911	0.895
LR	tenfold	0.688	0.649	0.713	0.357	0.620	0.594
	Test	0.612	0.473	0.716	0.194	0.510	0.553
Naïve Bayes	tenfold	0.761	0.737	0.776	0.508	0.711	0.688
	Test	0.721	0.709	0.730	0.436	0.684	0.661
avNNet	tenfold	0.902	0.918	0.892	0.801	0.882	0.848
	Test	0.783	0.709	0.838	0.553	0.736	0.453
KNN	tenfold	0.889	0.874	0.900	0.773	0.870	0.866
	Test	0.775	0.782	0.770	0.547	0.748	0.717

six-signature miRNAs can classify tumor stages of ccRCC patients with reasonable prediction performance using miRNAs expression profiles.

Enrichment analysis and biological roles of identified miRNA signatures

The identified miRNA signatures are linked with the development and progression of various cancer types. Overexpression of miR-21 leads to amplified cell proliferation and reduced apoptosis [20] and high expression of miR-21 associated with poor survival in lung cancer and ccRCC [11, 21]. Downregulated miR-30b-5p act as a tumor suppressor to regulate renal cell carcinoma in cell proliferation, metastasis and epithelial-to-mesenchymal transition by targeting G-protein subunit α-13 [22]. Decreased miR-187 in clear cell renal cell carcinoma inhibits cell growth, migration though targeting B7-H3 and correlated with lower survival [23]. MiR-106p-5p upregulation targets several negative regulators of the Wnt/β-catenin pathway [24]. MiR-144 promotes RCC development by hampering mTOR expression [25]. Fujii N et al. reported that higher expression of miR-224 associated with poor progression-free survival and overall survival in ccRCC [26]. MiR-200a regulates epithelial to mesenchymal transition-associated with gene expression and regulates prognosis in colorectal cancer [27]. In addition, miR-200a consistently decreased in RCC and serve as diagnostic biomarker for the early detection of RCC [28].

Biological roles of the identified miRNAs were assessed by using KEEG pathways and GO annotation analyses via DIANA-mirPath. The enriched biological pathways of the upregulated and downregulated identified miRNA signatures are presented in Fig. 3a, b. The upregulated miRNA signatures are enriched in Prion diseases, Lysine degradation, ECM-receptor interaction, Proteoglycans in cancer, Fatty acid elongation,Pathways in cancer, Cell cycle, FoxO signaling pathway, p53 signaling pathway, TGF-beta signaling pathway, Biosynthesis of unsaturated fatty acids, Viral carcinogenesis, signaling pathways regulating pluripotency of stem cells, Renal cell carcinoma and other biological pathways. The detail biological pathways and number of target genes for upregulated miRNA signatures are described in Additional file 1: Table S5. Similarly, downregulated

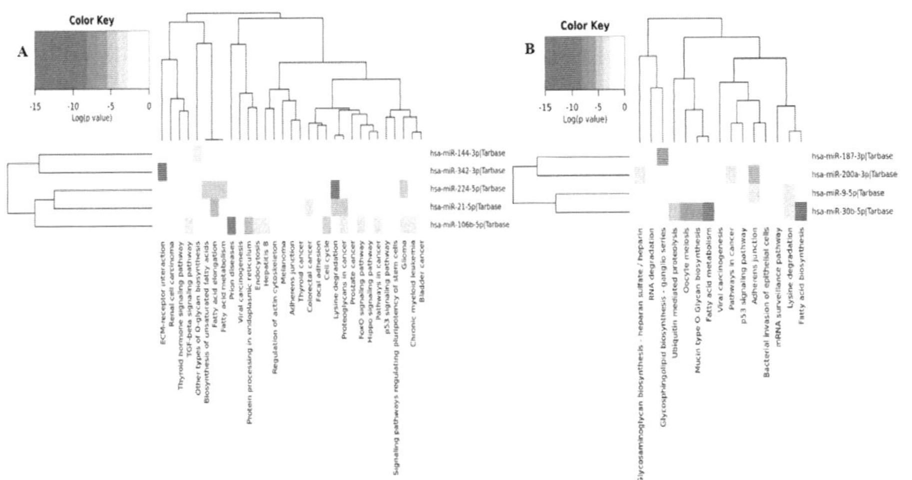

Fig. 3 KEGG pathway analysis for up and down regulated miRNA signatures. **a** Enriched KEGG pathways for upregulated miRNA signatures. **b** Enriched KEGG pathways for downregulated miRNA signatures

miRNA signatures involved in Fatty acid biosynthesis, Adherens junction, Fatty acid metabolism, Lysine degradation, Pathways in cancer, Viral carcinogenesis, p53 signaling pathway, mRNA surveillance pathway, RNA degradation other KEGG pathways. The detail summary of downregulated miRNA signatures enriched pathways and number of target genes are indicated in Additional file 1: Table S6.

GO analysis also showed that upregulated miRNA signatures participated in cellular nitrogen compound metabolic process, biosynthetic process, mitotic cell cycle, cell death, DNA metabolic process, innate immune response, cell cycle, cell proliferation and others. The detail biological processes and target genes of upregulated miRNA signatures are shown in Additional file 1: Table S7. Furthermore, the biological processes of downregulated miRNAs are enriched in nucleobase-containing compound catabolic process, macromolecular complex assembly, mitotic cell cycle, cellular protein modification process, catabolic process, biosynthetic process, gene expression, viral process, cellular component assembly, cellular protein metabolic process, small molecule metabolic process, Fc-epsilon receptor signaling pathway, response to stress, cell death and other biological process. The detail biological processes and target genes for downregulated miRNA signatures are shown in Additional file 1: Table S8.

Discussion

In this study, we showed an efficient strategy to identify miRNA signatures that can stratify high risk and low risk patients as well as classify early and late tumor stages. Several computational and machine leering algorithms have been developed to explore miRNA-associated diseases [17, 29]. Previous studies also used miRNA profiles to identify biomarkers for risk stratification [11, 30]; however, only a few research works have been conducted to explore miRNA signatures for early tumor stage of ccRCC. Therefore, in our study, we proposed a novel hybrid ML strategy to identify miRNA signatures associated with overall survival and tumor stages classification for ccRCC patients. The major findings and contributions of our work are outlined as follow. Identification of risk and stage predictive miRNA signatures with good predictive performance. The use of multivariate Cox regression with Elastic-net, Lasso and adaptive Lasso penalties followed by optimal subset prognostic model selection strategy identified a four novel-miRNA signature, namely miR-30b, miR-21, miR-187 and miR-200a. This signature can effectively stratify high risk and low risk subgroups with good classification performance (time dependent ROC); hence, the four-miRNA signature may be use as a prognostic biomarker. Furthermore, the proposed combined ML algorithms identified six top ranked miRNAs (miR-106b, miR-144, miR-224, miR-9-1, miR-21 and miR-342) based on their relative importance and their classification performance were evaluated by five ML methods. The classification performance of the six-stage associated miRNAs revealed effective stratifying ability early stage and late stage. The stage classification performance can be further improved if we applied SMOTE algorithm to prepare balanced early and late tumor stage data. The consistency and validation of the predictive models were assessed using an independent test set. The biological significance of the identified miRNA signatures in ccRCC development and progression were discussed and confirmed by the existed literature. The biological roles of the identified miRNA signatures were examined using enrichment analyses.

Future works: we will attempt to identify more robust features and classification models and apply different balanced data algorithms to improve the efficiency of the classification performance of early and late stage other than LR, Naive Bayes, KNN, and SVM classifiers.

Conclusions

In summary, we have attempted to identify potential miRNA signatures for stratification of risk using integrated statistical approach. An optimal four-miRNA based prognostic model with the smallest AIC criterion was identified and the four-miRNA signature can effectively classify ccRCC patients into high and low risk groups. Moreover, using state-of-the-art ML algorithms, we identified six top ranked miRNA signatures that can classify early and late stages for patients with ccRCC. These six miRNA signature classified early stage and late stage with reasonable good classification performance. Finally, we explored the biological roles of the identified risk- and stage-associated miRNA signatures and these signatures involved important biological pathways. Overall, we expect that our findings provide promising insight to explore the role of miRNAs in ccRCC patients and could help risk and stage classifications.

Methods

Input data

The RNAseq expression profiles and their corresponding clinical data (including TNM stage, survival time, sex, age) for ccRCC were retrieved from the Firebrowse database (http://firebrowse.org/). The clinical information consists of 325 samples (254 tumors and 71 normal tissues), and clinical data consists of 131, 28, 48, and 45 samples of stage I, stage II, stage III and stage IV respectively. Finally, we considered both early stage (stage I and stage II, n = 159) and late stage (stage III and stage IV, n = 93) samples of patients for classification purpose.

Preprocessing and identification of differentially expressed miRNAs

Normalization of the miRNA profiles were carried out by using the 'edgeR' package [19]. Differential analysis of miRNAs was conducted by utilizing the 'LIMMA' package in R [19] and miRNAs that satisfy the criterion of the absolute value of log2fold change > 1 and Benjamin-Hochberg (BH) adjusted P-value < 0.05 were considered as differentially expressed miRNAs (DEMs).

Data standardization for ML modeling

In this calculation, z-scaling was used to normalized the count per million (CPM) of miRNA expression values, which is defined by the following equation.

$$\begin{aligned} x &= log_2(CPM + 1), \\ z &= \frac{x - \overline{x}}{\sigma}. \end{aligned} \tag{1}$$

where x is the expression value of miRNA, $\overline{x}$ is the mean expression values of miRNA of the samples and σ is the standard deviation of expression values of miRNA of the samples and z represents normalized miRNA expression that follow the normal distribution with zero mean and unit standard deviation.

Identification of miRNA signature associated patient survival and construction of risk classification model

Feature selection (FS) is an important process to improve classification performance by avoiding irrelevant/noise features. To select survival-associated features (miRNAs), Cox model regression model was proposed. Cox regression is defined as:

$$h(t, X_{i\beta}) = h_0(t)\exp(X_{i\beta}) \tag{2}$$

Here, i denotes for ccRCC patients, and the Xs are the covariates (miRNAs). $X_{i\beta} = \beta_1 x_{i1} + \cdots + \beta_k x_{ik}$, $h_0(t)$ is the baseline hazard function at time t, $\beta = (\beta_1, \beta_2, \ldots, \beta_k)$ is the vector of regression coefficients, and β denotes the k-dimension regression coefficient vector of covariates. However, when we deal with high-dimension data such as genomic data, number of penalized Cox models have been proposed (including: Lasso, Adaptive lasso and Elastic-net) to solve the overfitting problem [31–33]. For each model, the estimated $\hat{\beta}$ values are obtained by minimizing the negative log-likelihood function with different penalty functions as follows:

$$\hat{\beta}_{\mathrm{Lasso}} = \arg\min_{\beta}\left\{-\sum_{i=1}^{n}\delta_i\left(X_{i\beta} - \log\left(\sum_{i=1}^{n}\exp\left(X_i\beta\right)\right)\right) + \lambda\sum_{i=1}^{K}|\beta_i|\right\} \tag{3}$$

$$\hat{\beta}_{\mathrm{Ad.\,lasso}} = \arg\min_{\beta}\left\{-\sum_{i=1}^{n}\delta_i\left(X_{i\beta} - \log\left(\sum_{i=1}^{n}\exp\left(X_i\beta\right)\right)\right) + 2\lambda\sum_{i=1}^{K}w_{nj}|\beta_i|\right\} \tag{4}$$

$$\hat{\beta}_{\mathrm{Enet}} = \arg\min_{\beta}\left\{-\sum_{i=1}^{n}\delta_i\left(X_{i\beta} - \log\left(\sum_{i=1}^{n}\exp\left(X_i\beta\right)\right)\right) + \lambda\sum_{i=1}^{K}|\beta_i| + (1-\alpha)\sum_{i=1}^{K}\beta_i^2\right\} \tag{5}$$

Here δ_i is an indicator for the uncensored observation, λ is called penalty and $w_{nj} = \left|\hat{\beta}_{nj}\right|^{-1}$ is calculated from the initial estimator $\hat{\beta}_n$.

Lasso Cox and adaptive Lasso Cox were used to identify relevant miRNAs associated with survival time by the shrinkage of some of the irrelevant miRNAs regression coefficients to zero. Adaptive Lasso Cox imposes an adaptive weighted penalty term in comparison with the Lasso, model, which further reduce the number of less-relevant miRNAs in such a way that the resulting coefficient estimates are sparse. Elastic-net is suitable for screening relevant miRNAs when there is a multicollinearity problem in the genomic data [34]. We used (ten-fold) cross-validation of Lasso, Adaptive lasso and Elastic-net algorithms to obtain the predicted optimal λ value for each method. The optimal λ value that minimize the estimated mean-squared prediction error and this optimal λ was used to select candidate miRNAs. All these methods were implemented using the 'glmnet' [35] package in *R*. Then, to make use of the strength of each method, we proposed a combined feature selection approach in the study of miRNA-disease association. The union of candidate miRNAs identified by the three algorithms were used to identify the best miRNAs combination that could predict prognosis of ccRCC patients more effectively. More detailed description of selecting the best subset prognostic model is described below:

Let $l = 1, 2, \ldots, k$, where k is the total number of candidate miRNAs identified by Lasso, Adaptive lasso and Elastic-net algorithms.

a) Construct all possible combinatorial subset model having l miRNA candidates
b) Compare all possible models with the Akaike information criterion (AIC)
c) Select the best subset prognostic model, having the smallest AIC, with l miRNAs using the 'glmulti' package [36] in *R*.

After identifying the best subset prognostic model, we developed risk score (a linear combination of best subset miRNA expression and coefficients of miRNAs obtained from multivariate analysis) to evaluate risk prediction performance. The risk score (RS) computed as follow,

$$\text{Risk Score} = \sum_{i=1}^{k} x_i \beta_i \quad (6)$$

Then, RS was used to classified ccRCC patients into high risk and low risk group using median RS as cutoff. The time dependent receiver operating characteristics (ROC) was used to estimate the survival time difference between high risk and low risk group using the 'survivalROC' package in R.

Identification of miRNA signature associated with tumor stages for early and late tumor stage classification

To select best subset features (miRNAs) that associated with tumor stages, we proposed to use hybrid feature selection methods. Feature selection methods were conducted based on ten-fold cross-validation. The four popular ML algorithms proposed in this study include: LR, RF, SVMR and avNNet, and top ranked miRNAs based on their importance as the best features were identified from high-dimensional data. We utilized the "caret" package in *R* that consists of several complex ML algorithms for classification and prediction problems.

The processing of feature selection using four ML algorithms are briefly discussed as follows.

Ensemble logistic regression (LR) model is a ML model used as a classification model in feature selection to identify features that can distinguish binary samples of patients. Let $\boldsymbol{x} \in \boldsymbol{R}^k$ denotes an observation consists of k feature miRNA values and let $y \in \{-1, +1\}$ represents the corresponding binary outcome; such as, early and late stage. A LR model is a condition probability distribution (CPD) of the class level y given the feature (miRNAs) vector $\boldsymbol{x}$ is defined by

$$p(y/\boldsymbol{x}) = \frac{1}{1 + \exp\left(-y\left(w^T x + v\right)\right)} \quad (7)$$

where $w \in R^k$ is the weight vector and $v \in \boldsymbol{R}$ are parameters of LR. Ensemble LR for relevant feature selection is stable with respect to variation of the learning samples, since it uses t-test to rank features, which does not consider dependence between features. LR transfers the strongly correlated features to the LR gradient decent optimization algorithm to increase classification performance [37].

The random forest (RF) algorithm is a non-linear regression model, which has been used in several regression and classification problems in a variety of computational studies. RF learning is flexible algorithm that minimizes prediction errors and screen the most important features related to class level when feature size is large. It consists of constructing trees by splitting random sample of r features from a set of n features using bootstrapped training data.

$$IG(n) = 1 - \sum_{i=1}^{j} (ki)^2 \tag{8}$$

where IG denotes impurity of a node n, ki denotes the features (miRNAs), and j is the fraction of each ki feature (miRNAs) [38].

Support vector machine (SVM) algorithm has been applied in many biological problems, such as biomarkers selection and classification problems [16, 39]. SVM is non-probabilistic classifier which maximize the margin of the decision boundary to classify two classes using support vectors to achieve the best classification. The general formulation of the SVM classifier is defined as

$$minimize \frac{1}{2} w^2 + \varphi \sum_{i=1}^{n} x_i \tag{9}$$

Here, w denotes the normal vector of the hyperplane, φ is the classifier parameter, x_i are the variables and n is the number of vectors in the training dataset.

The avNNet model is a type of neural network (NN), that learns nearly infinite number of mapping functions and works like natural human neurons. The inputs (features) connect to class label (staging), the connections are called edges. The input feature connects either forward/backward propagation hidden nodes to compute neurons. The number of hidden layers determines the depth of NN. A feedforward NN having more than one hidden layer is called deep network [40, 41]. The NN is commonly applied to the discovery of biomarkers in cancer studies [42].

$$\beta_j^r = \vartheta\left(\sum_K w_{jk}^r \beta_k^{r-1} + \alpha_j^r\right) \tag{10}$$

where β_j^r represents the jth neuron in the rth layer is associated with activation in the $(r-1)$th layer. The ϑ is a vectorising function parameter. The w_{jk}^r denotes a rth weight matrix for each layer r of the jth row and kth column, and α_j^r denotes the j bias for each layer r.

Stage classification methods

SVMR, NB, avNNet, KNN and LR were utilized for building predictive and classification models. Each model was constructed by ten-fold cross validation to avoid over/under fitting. The cost function was optimized [100–1000 iterations with 100 steps per iteration] to attain accurate classification.

Data balancing

The proportion of early stage patients is approximately twice of late stage patients; this creates the data imbalance problem which leads to biased prediction. Therefore, we performed the data balancing procedure by using the Synthetic Minority Oversampling Technique (SMOTE) algorithm [43] (included in the DMwR package). The SMOTE is a common method to solve data imbalance problem more effectively, prior to applying the ML classifier [42]. The dataset was divided randomly into training set (80%) and the remaining 20% as independent test set. The training set was employed to train ML algorithms in classifying early and late stage of patients based on ten-fold cross validation. The test set was used as independent test set and used to assess the classification performance of five ML methods.

Classification performance evaluation metrics

To evaluate the performance of the classification models, we used the following measures: specificity, sensitivity, accuracy (ACC), precision, and Matthews correlation coefficient (MCC). The mathematical formulas of the measures are given below:

$$Sensitivity = \frac{TP}{TP + FN} \tag{11}$$

$$Specificity = \frac{TN}{TN + FP} \tag{12}$$

$$ACC = \frac{TP + TN}{TP + TN + FP + FN} \tag{13}$$

$$Precision = \frac{TP}{TP + FP} \tag{14}$$

$$\mathrm{MCC} = \frac{TP \times TN - FP \times FN}{\sqrt{(TP + FP)(TP + FN)(TN + FP)(TN + FN)}} \tag{15}$$

where TP, TN, FP, and FN denote true positive, true negative, false positive and false negative respectively.

Functional enrichment analysis

We assessed the biological relevance of the identified prognostic and diagnostic miRNA signatures using the DIANA-mirPath [44].

Abbreviations
RCC: Renal cell carcinoma; ccRCC: Clear cell renal cell carcinoma; ML: Machine learning; TCGA: The Cancer Genome Atlas; SMOTE: Synthetic Minority Oversampling Technique; LASSO: Least absolute shrinkage and selection operator; HR: Hazard ratio; SVM: Support Vector Machine; SVMR: Support Vector Machine with radial kernel; MRMR: Maximum Relevance Minimum Redundancy; LR: Logistic regression; KNN: K-nearest neighbors algorithm; AVNNET: Averaging Neural network; GO: Gene Ontology; KEGG: Kyoto Encyclopedia of Genes and Genomes.

Supplementary Information

The online version contains supplementary material available at https://doi.org/10.1186/s12859-021-04189-2.

Additional file 1. Lists of supplementary materials, including workflow, differentially expressed miRNAs, KM plots of ccRCC patients, Cox regression results, and the results of enrichment analysis of the miRNA signatures.

Acknowledgements
We would like to thank the peer reviewers for their constructive comments, which have helped improve the manuscript. Our gratitude also goes to Michael Burton, Asia University, proofread the manuscript.

About this supplement
This article has been published as part of BMC Bioinformatics Volume 22 Supplement 10 2021: Selected articles from the 19th Asia Pacific Bioinformatics Conference (APBC 2021): bioinformatics. The full contents of the supplement are available at https://bmcbioinformatics.biomedcentral.com/articles/supplements/volume-22-supplement-10.

Authors' contributions
E-YD conceptualized, draft and reviewed and edited the manuscript. J-JPT managed the project and secured the necessary acquisition funding. J-GC contributed to guide data analysis, and interpretation and manuscript writing. K-LN foresee the research, managed the project, reviewed and edited the manuscript. All authors read and approved the final manuscript.

Funding
Dr. Ka-Lok Ng and Mr. Eskezeia Y. Dessie works are supported by the Taiwan Ministry of Science and Technology (MOST) (Grant Numbers: MOST 109-2221-E-468-013 and MOST 108-2221-E-468-020). Dr. Ka-Lok Ng work is also supported by Asia University (grant numbers: 106-asia-06, 106-asia-09, 107-asia-02 and 107-asia-09). Dr. Jeffrey J.P. Tsai work is supproted by MOST (Grant Number: MOST 109-2321-B-468-001). Dr. Jan-Gowth Chang is supported by China Medical University Hospital (DMR-107-209). The publication costs are funded by MOST (Grant Numbers: MOST 109-2221-E-468-013 and MOST 108-2221-E-468-020). The funding bodies were not involved in the design of the study, in the collection, analysis of the data, or in writing of the manuscript.

Availability of data and materials
The datasets analyzed during the current study are available in the Firebrowse database, https://gdac.broadinstitute.org/.

Declarations

Ethics approval and consent to participate
Not applicable.

Consent for publication
Not applicable.

Competing interests
None to declare.

Author details
[1]Department of Bioinformatics and Medical Engineering, Asia University, Taichung, Taiwan. [2]Department of Laboratory Medicine, China Medical University, Taichung, Taiwan. [3]Department of Medical Research, China Medical University Hospital, China Medical University, Taichung, Taiwan. [4]Center for Artificial Intelligence and Precision Medicine Research, Asia University, Taichung, Taiwan.

Received: 18 April 2021 Accepted: 12 May 2021
Published online: 25 May 2021

References

1. Siegel RL, Miller KD, Jemal A. Cancer statistics, 2018. CA Cancer J Clin. 2018;68(1):7–30.
2. Ferlay J, et al. Cancer incidence and mortality worldwide: sources, methods and major patterns in GLOBOCAN 2012. Int J Cancer. 2015;136(5):E359–86.
3. Zhang L, et al. The complex roles of microRNAs in the metastasis of renal cell carcinoma. J Nanosci Nanotechnol. 2013;13(5):3195–203.
4. Siegel RL, Miller KD, Jemal A. Cancer statistics, 2016. CA Cancer J Clin. 2016;66(1):7–30.

5. Wood CG. Multimodal approaches in the management of locally advanced and metastatic renal cell carcinoma: combining surgery and systemic therapies to improve patient outcome. Clin Cancer Res. 2007;13(2 Pt 2):697s–702s.
6. Amin MB, et al. The Eighth Edition AJCC Cancer Staging Manual: continuing to build a bridge from a population-based to a more "personalized" approach to cancer staging. CA Cancer J Clin. 2017;67(2):93–9.
7. Li, Y., et al., Transcriptomic and functional network features of lung squamous cell carcinoma through integrative analysis of GEO and TCGA data.
8. Chang RK, et al. MicroRNA expression profiles in nonepithelial ovarian tumors. Int J Oncol. 2018;52(1):55–66.
9. Lujambio A, et al. Genetic unmasking of an epigenetically silenced microRNA in human cancer cells. Cancer Res. 2007;67(4):1424–9.
10. Macfarlane LA, Murphy PR. MicroRNA: biogenesis, function and role in cancer. Curr Genomics. 2010;11(7):537–61.
11. Liang B, Zhao J, Wang X. A three-microRNA signature as a diagnostic and prognostic marker in clear cell renal cancer: an in silico analysis. PLoS ONE. 2017;12(6):e0180660.
12. Jacob H, et al. A four-microRNA classifier as a novel prognostic marker for tumor recurrence in stage II colon cancer. Sci Rep. 2018;8(1):6157.
13. Asakura K, et al. A miRNA-based diagnostic model predicts resectable lung cancer in humans with high accuracy. Commun Biol. 2020;3(1):134.
14. Li F, et al. An improved clear cell renal cell carcinoma stage prediction model based on gene sets. BMC Bioinform. 2020;21(1):232.
15. Ng K-L, Taguchi YH. Identification of miRNA signatures for kidney renal clear cell carcinoma using the tensor-decomposition method. Sci Rep. 2020;10(1):15149.
16. Yerukala Sathipati S, Ho S-Y. Identifying a miRNA signature for predicting the stage of breast cancer. Sci Rep. 2018;8(1):16138.
17. Yerukala Sathipati S, Ho S-Y. Novel miRNA signature for predicting the stage of hepatocellular carcinoma. Sci Rep. 2020;10(1):14452.
18. Sherafatian M, Arjmand F. Decision tree-based classifiers for lung cancer diagnosis and subtyping using TCGA miRNA expression data. Oncol Lett. 2019;18(2):2125–31.
19. Zhao E, Bai X. Nomogram based on microRNA signature contributes to improve survival prediction of clear cell renal cell carcinoma. Biomed Res Int. 2020;2020:7434737.
20. Oue N, et al. High miR-21 expression from FFPE tissues is associated with poor survival and response to adjuvant chemotherapy in colon cancer. Int J Cancer. 2014;134(8):1926–34.
21. Li C, et al. Non-small cell lung cancer associated microRNA expression signature: integrated bioinformatics analysis, validation and clinical significance. Oncotarget. 2017;8(15):24564–78.
22. Liu W, et al. MiR-30b-5p functions as a tumor suppressor in cell proliferation, metastasis and epithelial-to-mesenchymal transition by targeting G-protein subunit α-13 in renal cell carcinoma. Gene. 2017;626:275–81.
23. Zhao J, et al. MicroRNA-187, down-regulated in clear cell renal cell carcinoma and associated with lower survival, inhibits cell growth and migration though targeting B7–H3. Biochem Biophys Res Commun. 2013;438(2):439–44.
24. Lu J, et al. miR-106b-5p promotes renal cell carcinoma aggressiveness and stem-cell-like phenotype by activating Wnt/β-catenin signalling. Oncotarget. 2017;8(13):21461–71.
25. Xiang C, Cui SP, Ke Y. MiR-144 inhibits cell proliferation of renal cell carcinoma by targeting MTOR. J Huazhong Univ Sci Technol Med Sci. 2016;36(2):186–92.
26. Fujii N, et al. Extracellular miR-224 as a prognostic marker for clear cell renal cell carcinoma. Oncotarget. 2017;8(66):109877–88.
27. Pichler M, et al. MiR-200a regulates epithelial to mesenchymal transition-related gene expression and determines prognosis in colorectal cancer patients. Br J Cancer. 2014;110(6):1614–21.
28. Wang C, et al. Circulating miR-200a is a novel molecular biomarker for early-stage renal cell carcinoma. ExRNA. 2019;1(1):25.
29. You ZH, et al. PBMDA: A novel and effective path-based computational model for miRNA-disease association prediction. PLoS Comput Biol. 2017;13(3):e1005455.
30. Tang J, et al. Identification of miRNA-based signature as a novel potential prognostic biomarker in patients with breast cancer. Dis Markers. 2019;2019:3815952.
31. Tibshirani R. The lasso method for variable selection in the Cox model. Stat Med. 1997;16(4):385–95.
32. Zou H. The adaptive lasso and its oracle properties. J Am Stat Assoc. 2006;101(476):1418–29.
33. Zou H, Hastie T. Regularization and variable selection via the Elastic Net. J R Stat Soc Ser B (Stat Methodol). 2005;67(2):301–20.
34. Lee S, Lim H. Review of statistical methods for survival analysis using genomic data. Genomics Inform. 2019;17(4):e41.
35. Kuhn, M., et al., Package 'caret'. 2020.
36. Calcagno V, de Mazancourt C. glmulti: An R Package for Easy Automated Model Selection with (Generalized) Linear Models. J Stat Softw. 2010;34(12):29.
37. Zakharov R, Dupont P. Ensemble logistic regression for feature selection. Berlin: Springer; 2011.
38. Liaw A, Wiener M. Classification and Regression by randomForest. R news. 2007;2:18–22.
39. Chu F, Wang L. Applications of support vector machines to cancer classification with microarray data. Int J Neural Syst. 2005;15(6):475–84.
40. Kriegeskorte N, Golan T. Neural network models and deep learning. Curr Biol. 2019;29(7):R231–6.
41. Guang-Bin H, Qin-Yu Z, Chee-Kheong S. Extreme learning machine: a new learning scheme of feedforward neural networks. In: IEEE international joint conference on neural networks (IEEE Cat. No.04CH37541). 2004.
42. Mirza B, et al. Machine learning and integrative analysis of biomedical big data. Genes (Basel). 2019;10(2):87.
43. Chawla NV, et al. SMOTE: synthetic minority over-sampling technique. J Artif Int Res. 2002;16(1):321–57.
44. Vlachos IS, et al. DIANA-miRPath v3.0: deciphering microRNA function with experimental support. Nucleic Acids Res. 2015;43(W1):W460–6.

Publisher's Note

Wu et al. BMC Bioinformatics (2021) 22:271
https://doi.org/10.1186/s12859-021-04192-7

BMC Bioinformatics

RESEARCH Open Access

A tool for analyzing and visualizing ribo-seq data at the isoform level

Wei-Sheng Wu[1*], Yi-Hong Tsao[2], Sheng-Cian Shiue[1], Ting-Yu Chen[1], Yan-Yuan Tseng[3] and Joseph T. Tseng[2*]

From The 19th Asia Pacific Bioinformatics Conference (APBC 2021) Tainan, Taiwan. 3-5 February 2021

*Correspondence: wessonwu@mail.ncku.edu.tw; tctseng@mail.ncku.edu.tw
[1] Department of Electrical Engineering, National Cheng Kung University, Tainan 701, Taiwan
[2] Department of Biotechnology and Bioindustry Sciences, National Cheng Kung University, Tainan 701, Taiwan
Full list of author information is available at the end of the article

Abstract

Background: Translational regulation is one important aspect of gene expression regulation. Dysregulation of translation results in abnormal cell physiology and leads to diseases. Ribosome profiling (RP), also called ribo-seq, is a powerful experimental technique to study translational regulation. It can capture a snapshot of translation by deep sequencing of ribosome-protected mRNA fragments. Many ribosome profiling data processing tools have been developed. However, almost all tools analyze ribosome profiling data at the gene level. Since different isoforms of a gene may produce different proteins with distinct biological functions, it is advantageous to analyze ribosome profiling data at the isoform level. To meet this need, previously we developed a pipeline to analyze 610 public human ribosome profiling data at the isoform level and constructed HRPDviewer database.

Results: To allow other researchers to use our pipeline as well, here we implement our pipeline as an easy-to-use software tool called RPiso. Compared to Ribomap (a widely used tool which provides isoform-level ribosome profiling analyses), our RPiso (1) estimates isoform abundance more accurately, (2) supports analyses on more species, and (3) provides a web-based viewer for interactively visualizing ribosome profiling data on the selected mRNA isoforms.

Conclusions: In this study, we developed RPiso software tool (http://cosbi7.ee.ncku.edu.tw/RPiso/) to provide isoform-level ribosome profiling analyses. RPiso is very easy to install and execute. RPiso also provides a web-based viewer for interactively visualizing ribosome profiling data on the selected mRNA isoforms. We believe that RPiso is a useful tool for researchers to analyze and visualize their own ribosome profiling data at the isoform level.

Keywords: Ribo-seq, Ribosome profiling, Pipeline, Tool, Software, Isoform-level, Visualization

Background

Translational regulation, which governs the mRNA translational efficiencies and protein degradation rates, is a key mechanism for protein synthesis [1, 2]. Translational regulation enables a cell to change its proteome to maintain cellular homeostasis in response to internal and external stimuli [1, 2]. Aberrations in the translational control leads to human diseases such as cancer [3]. Therefore, detailed knowledge of the molecular mechanisms of translational regulation is essential in understanding cellular homeostasis and disease.

Ribosome profiling, also called ribo-seq, is a powerful experimental technique to study translational regulation [4]. By deep sequencing of ribosome-protected mRNA fragments, ribo-seq captures a snapshot of translation at a specific time and in a specific physiological condition. Ribo-seq has been used to examine many aspects of translation such as alternative initiation, frameshift, and the dynamics of elongation and termination [5–7]. With the continuous advance of the ribo-seq technique, more biological insights into the mechanisms of eukaryotic translation will be revealed in the near future [7].

Bioinformatics tools are needed to make sense of ribo-seq data. Nowadays many software tools have been developed to analyze users' ribo-seq data. For example, RiboProfiling [8] is a Bioconductor package that provides quality assessment and quantification of Ribo-seq data. RiboTools [9] is a Galaxy toolbox for the qualitative analysis (e.g. identification of translational ambiguities and stop codon readthrough events) on ribo-seq data. RiboGalaxy [10] provides a set of online tools (e.g. ribocount, riboplot, and riboseqR) for the analysis and visualization of ribo-seq data. Descriptions of many other tools could be found in two review papers [11, 12].

Although various existing tools have fulfilled most of the needs for ribo-seq data processing, one challenge remains to be addressed. Almost all existing tools analyze ribo-seq data at the gene level rather than the isoform level. In higher eukaryotes, many genes produce multiple mRNA isoforms [13, 14]. Different mRNA isoforms of a gene may produce a variety of proteins with distinct biological functions. Therefore, it is advantageous to analyze ribo-seq data at the isoform level so that more biological insights can be extracted from ribo-seq data.

Ribomap [15] is a widely used software tool dedicated to quantify isoform-level ribosome profiles. Ribomap assigns ribo-seq reads to different mRNA isoforms based on the estimated mRNA isoform abundance from RNA-seq. Our group previously developed a pipeline to analyze 610 human ribo-seq datasets at the isoform level and constructed HRPDviewer [16] database to provide the ribosome profiling results on each isoform in these ribo-seq datasets. To allow other researchers to do the same analyses on their ribo-seq data from a species of interest, here we implement our pipeline as an easy-to-use software tool called RPiso. RPiso is dedicated to quantify isoform-level ribosome profiles. RPiso incorporates Bowtie [17] for transcriptome mapping and RSEM [18] for isoform abundance estimation. The goals of HRPDviewer and RPiso are very different. HRPDviewer is a database which allows users to view the analyzing results of the 610 public human ribo-seq data at the isoform level. HRPDviewer does not allow users to analyze their ribo-seq data. On the contrary, RPiso is developed as an easy-to-use software tool to analyze users' ribo-seq data. Using RPiso allows users to calculate and view the translational level of each isoform/gene from their ribo-seq data.

Compared to Ribomap, our RPiso has two unique features. First, RPiso supports analysis on multiple species. RPiso precompiled the reference mRNA transcriptomes of five species (human, mus, rat, yeast and zebrafish). On the contrary, Ribomap only

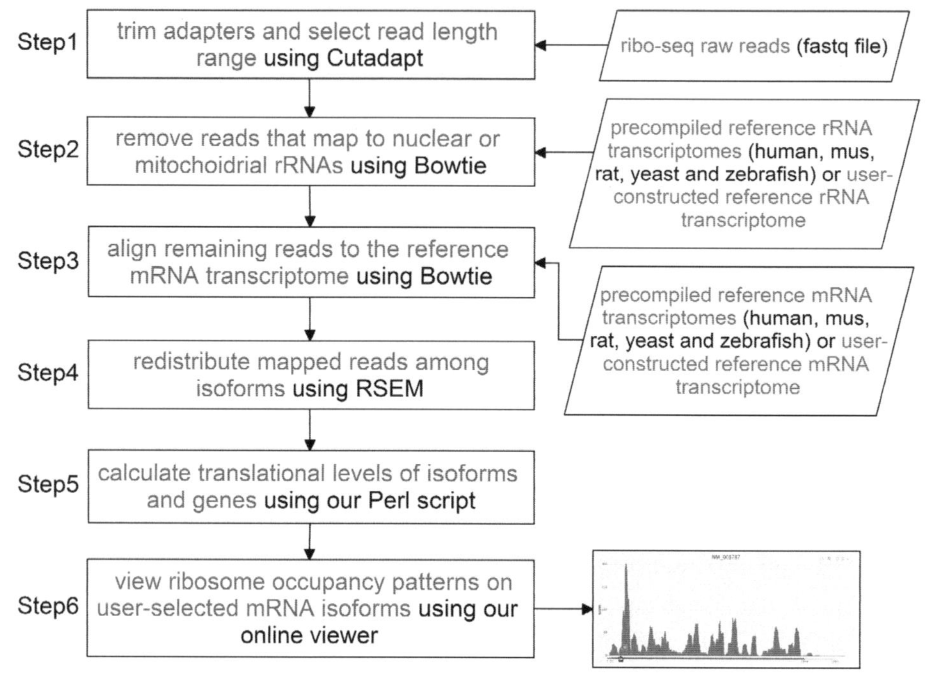

Fig. 1 RPiso software workflow. RPiso consists of six processing steps

precompiled the human reference mRNA transcriptome. RPiso also provides step-by-step instructions on how to prepare the reference mRNA transcriptomes of other species of interest, but Ribomap does not provide such information. Second, RPiso has a web-based viewer for interactively visualizing ribosome profiling data on the selected isoforms while Ribomap does not provide any visualization. RPiso's online viewer helps users to find out novel biological insights. By viewing the ribo-seq data mapped on different mRNA isoforms of a gene, users can know which mRNA isoforms are highly or lowly translated in the physiological condition under study and gain an accurate understanding of differential translational regulation of different isoforms of a gene.

Here we give an example. Human cyclin G1 (encoded by gene CCNG1) plays important roles during the DNA damage response and its dysfunctions lead to cancers. CCNG1 has two mRNA isoforms (NM_004060 and NM_199246). Using RPiso to analyze ribo-seq data from a cell cycle study in Hela cell [19], we have shown in our HRPDviewer study [16] that the cyclin G1 mRNA isoform NM_004060 is constitutively translated throughout the cell cycle with peak levels at early G1 phase but the other cyclin G1 mRNA isoform NM_199246 is only translated in G1 phase, indicating that the two mRNA isoforms of the gene CCNG1 may be under different translational regulations in the cell cycle process.

Implementation

RPiso software workflow

RPiso consists of several processing steps (Fig. 1). First, the adaptor sequences of the ribo-seq raw reads were trimmed and the reads within a certain length range (default 27 ~ 40) were kept both using Cutadapt [20]. Second, the contaminating reads (i.e. the reads mapped to the nuclear rRNAs or mitochondrial rRNAs) were removed using Bowtie [17]. The reference rRNA transcriptomes of five species (human, mus, rat, yeast, and zebrafish) were already pre-complied. Users have to construct the reference rRNA transcriptome if their ribo-seq reads come from other species. Third, the remaining reads were aligned to the reference mRNA transcriptome using Bowtie [17]. The reference mRNA transcriptomes of five species (human, mus, rat, yeast, and zebrafish) were already pre-complied. Users have to construct the reference mRNA transcriptome if their ribo-seq reads come from other species. Our software manual provides step-by-step instructions on the reference rRNA and mRNA transcriptomes preparation (Additional file 1). Fourth, the redistribution of the mapped reads among isoforms was accomplished using RSEM [18]. RSEM uses a generative statistical model which handles read mapping uncertainty in a statistically rigorous manner [18, 21]. Although RSEM was originally designed for RNA-seq, we have shown in our HRPDviewer database paper [16] that RSEM can also be used to handle read mapping uncertainty for Ribo-seq with high accuracy. Fifth, the translational levels of each mRNA isoform and each gene were calculated using our own Perl script (RPisoCalculation.pl). The translational level of an mRNA isoform is defined as the average Normalized Reads Per Kilobase per Million mapped reads (NRPKM) of its coding region. The translational level of a gene is defined as the sum of the translational levels of all its mRNA isoforms. The detailed mathematical formula could be found in our HRPDviewer database paper [16]. Finally, the ribosome occupancy patterns on the user-selected mRNA isoforms could be seen using our web-based viewer, which was developed in Python using the Django MTV framework. The ribosome occupancy patterns were plotted by a feature-rich JavaScript library called Plotly.js. If users do not want to use our web-based viewer, we also provide an.html file which contains all the figures of the ribosome occupancy patterns on the user-selected mRNA isoforms.

Configuration of RPiso software

The configuration of RPiso is shown in Fig. 2. The first layer is the "RPiso" directory. The second layer consists of five directories ("Data", "References", "Scripts", "Programs", and "XXX"). The "Data" directory stores a user's ribo-seq fastq file. The "References" directory contains two sub-directories. The "NCBI" subdirectory contains the reference transcriptome files for both the mRNAs and rRNAs of five species (human, mus, rat, yeast, and zebrafish) retrieved from NCBI. The "Gene_list" subdirectory contains lists of user-given gene names whose Ribo-seq profiles could be visualized by our web-based viewer. The "Scripts" directory contains all the scripts of RPiso. Users have to execute RPiso in this directory. The "Programs" directory contains two state-of-the-art read-processing tools (Bowtie-1.2.2-linux-x86_64 and RSEM-1.3.1) used in our RPiso software. The "XXX" directory contains all the output files of our RPiso software after analyzing users' ribo-seq.fastq file. XXX stands for the user-defined output folder name. The six output files in the "XXX" directory are introduced as follows. First, the "XXX.genes.results" file contains the translational levels of all genes. Second, the "XXX.isoforms.results" file contains the translational levels of all isoforms. Third, "XXX.normalized.readdepth" file contains the normalized reads per million mapped reads (NRPM) of all the positions on each isoform. Fourth, XXX_summary file summarizes the mapping rate

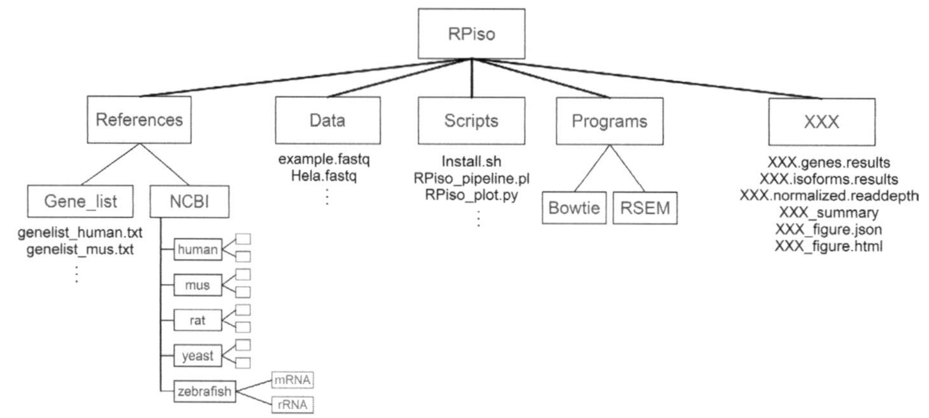

Fig. 2 Configuration of RPiso software. The first layer is the "RPiso" directory. The second layer consists of five directories ("Data", "References", "Scripts", "Programs", and "XXX"). XXX stands for the user-defined output folder name

of each processing step of our RPiso software tool. Fifth, XXX_figure.json file contains the ribosome occupancy patterns on all isoforms of the user-selected genes (given in the "Gene_list" folder) that can be interactively visualized by our online viewer (http://cosbi7.ee.ncku.edu.tw/RPiso/). Sixth, XXX_figure.html file contains all the figures of the ribosome occupancy patterns on the user-selected mRNA isoforms. This alternative is for those users who do not want to use our online viewer.

Results and discussion

The usage of RPiso software

First, download RPiso.tar.gz from our website (http://cosbi7.ee.ncku.edu.tw/RPiso/). Second, decompress RPiso.tar.gz in a Linux system and users will have the following four folders: "Data", "References", "Scripts", and "Programs". Third, run Install.sh in the "Scripts" folder. This shell script will install three programs (Cutadapt 1.18, Bowtie-1.2.2-linux-x86_64 and RSEM-1.3.1) and construct the rRNA and mRNA transcriptome reference indices of five pre-compiled species (human, mus, rat, yeast, and zebrafish). Users need to do extra steps to construct the rRNA and mRNA transcriptome reference indices of the species of interest other than the five pre-complied species. The detailed instructions can be found in our RPiso manual (Additional file 1). To serve users who are familiar with Docker, we also provide a docker image of RPiso at https://hub.docker.com/r/n26091225/rpiso. Fourth, put users' ribo-seq data in the "Data" folder. Here we use a part of the ribo-seq data of human Hela cell with RPL19 (Ribosomal Protein L19) knockdown from our lab as a sample data (named example.fastq). Fifth, run our RPiso pipeline (RPiso_pipeline.pl) in the "Scripts" folder as follows: "nohup perl RPiso_pipeline.pl -adapter CTGTAGGCACCATCAAT -species human -output ExOut example.fastq &". The first parameter "-adapter" specifies the adapter sequence (e.g. CTGTAGGCACCATCAAT). The second parameter "-species" specifies the species being analyzed (e.g. human). The third parameter "-output" specifies the output folder name (e.g. ExOut). The final parameter specifies the user's ribo-seq file name (e.g. example.fastq). After running RPiso_pipeline.pl, users will find an output folder (e.g. ExOut) with six files (ExOut.

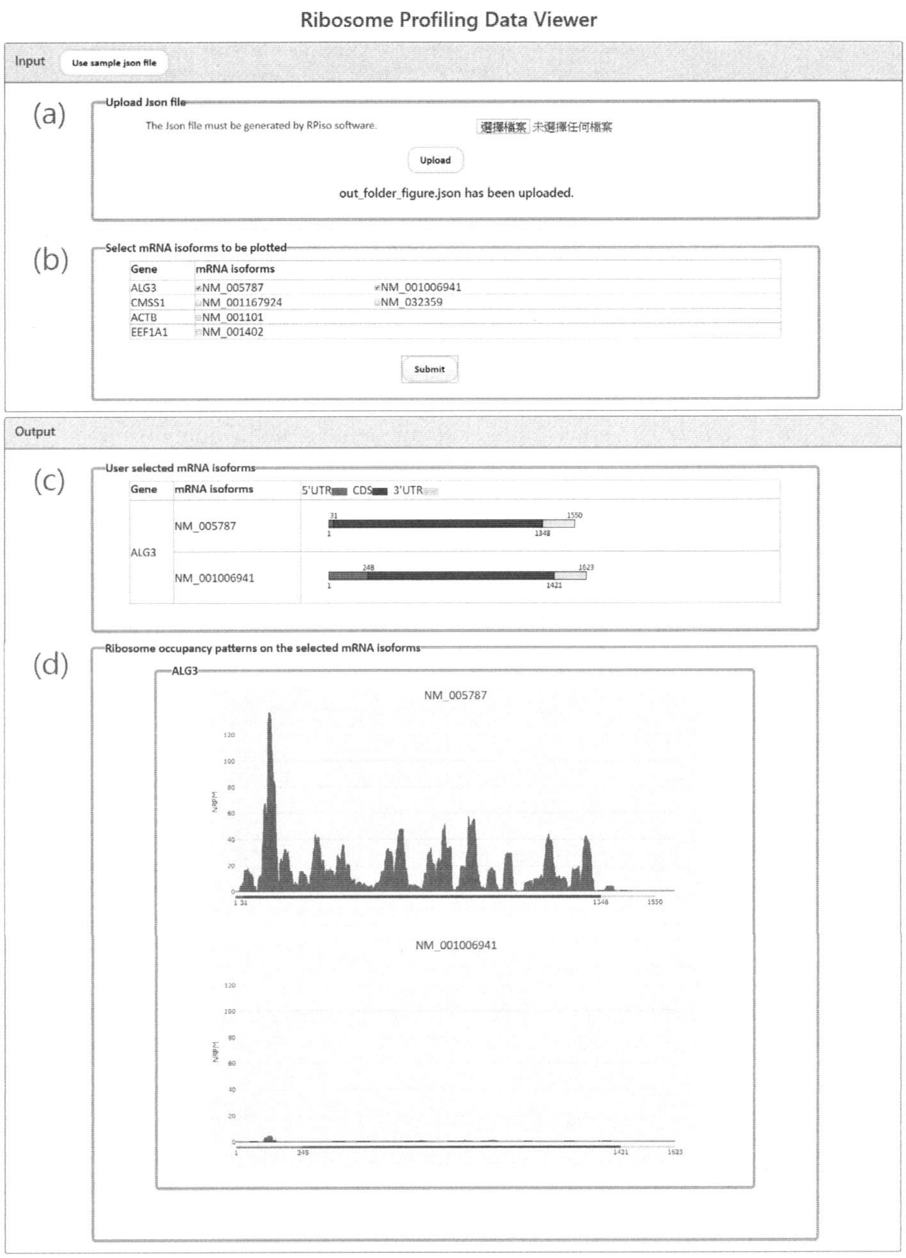

Fig. 3 RPiso's online viewer. To use the online viewer, users have to (**a**) upload the Json file generated by RPiso and (**b**) select the mRNA isoforms to be plotted. After submission, users will see (**c**) the information of 5'UTR, CDS, and 3'UTR for all selected mRNA isoforms and (**d**) the ribosome occupancy patterns on all the selected mRNA isoforms. The value on y-axis represents the normalized reads per million mapped reads (NRPM)

genes.results, ExOut.isoforms.results, ExOut.normalized.readdepth, ExOut_summary, ExOut_figure.json and ExOut_figure.html). The descriptions of these six files have been mentioned in the previous subsection "Configuration of RPiso software". Finally, upload ExOut_figure.json into our online viewer. Users will see the ribosome occupancy patterns on all positions of all the isoforms of the user-selected genes (Fig. 3). If users do not want to use our web-based viewer, they can just open ExOut_figure.html to see all the figures of the ribosome occupancy patterns on the user-selected mRNA isoforms.

Our RPiso estimates isoform abundance more accurately than Ribomap does

Ribomap [15] is a widely used software tool dedicated to quantify isoform-level ribosome profiles. Since our RPiso software tool also aims to provide isoform-level analysis on the ribo-seq data, the performance comparison between RPiso and Ribomap is a necessity. Following the authors of Ribomap, the human Hela cell ribo-seq data (GSM546920) in Guo et al. [22] was used as a testing dataset to evaluate the performance.

As ribo-seq footprints primarily originate from CDS regions, accurate attribution of footprints to a particular isoform over others would rely on that isoform's unique differences in CDS exonic sequence composition. (Note that this rationale has been used in HRPDviewer [16] for performance evaluation.) Assume that a gene of interest has two isoforms (isoform A and isoform B). If the unique exon of isoform A has more ribo-seq footprints than the unique exon of isoform B has, then a good isoform-level software tool should redistribute more footprints to isoform A than to isoform B. That is, the translational level of isoform A should be higher than that of isoform B. Here we used this assertion as the performance index to compare the performance of Ribomap [15] and our RPiso.

To make it easy to apply the above rationale, here we only consider genes with two isoforms. From all genes in the human genome, we selected 106 human genes with two isoforms. The gene collection criteria are as follows: (1) each gene must have exactly two isoforms, (2) each isoform must have exactly one unique exon, and (3) at least one of the two unique exons must have ribo-seq footprints. By checking the outputs of both software on these 106 genes, we found that RPiso and Ribomap have the same assertions on 77 genes but have opposite assertions on the other 29 genes. Therefore, the results on these 29 genes can be used to compare the performance of RPiso and Ribomap. Take the gene ALG3 as an example. ALG3 has two isoforms (NM_005787 and NM_001006941). Each isoform has its unique exon. Using Integrative Genomics Viewer [23] to visualize the ribo-seq footprints (from the BAM file generated by Bowtie) shows that the unique exon of NM_005787 has much more ribo-seq footprints (91 vs. 2 uniquely mapped reads) than the unique exon of NM_001006941 has (Fig. 4). Therefore, the translational level of NM_005787 should be higher than that of NM_001006941. Our RPiso supports this assertion (0.9 vs. 0.01) while Ribomap contradicts this assertion (110 vs. 274). Therefore, our RPiso outperformed Ribomap in this case. In total, our RPiso outperformed Ribomap in 86% (25/29) cases (Table 1), suggesting that our RPiso estimates isoform abundance more accurately than Ribomap does.

Conclusions

In this study, we developed RPiso software tool to provide isoform-level ribosome profiling analyses. RPiso is very easy to install and execute. Compared to Ribomap (a widely used software tool which provides isoform-level ribosome profiling analyses), our RPiso has four advantages. First, while Ribomap needs RNA-seq to assign ribo-seq reads to

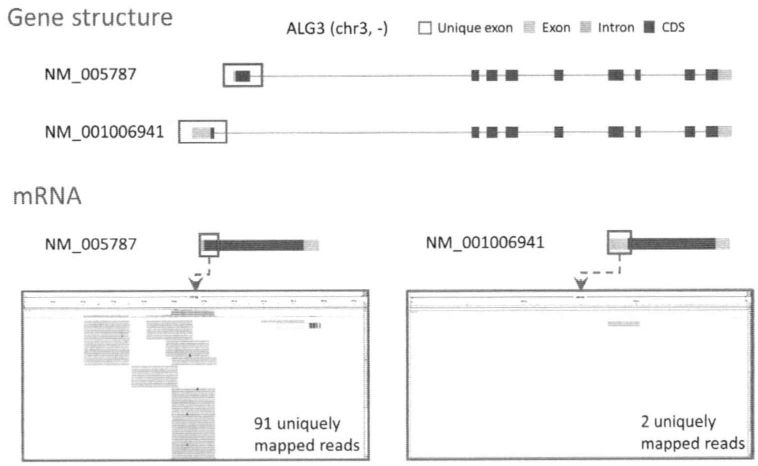

	NM_005787	NM_001006941
TL calculated by RPiso	0.9	0.01
TL calculated by Ribomap	110	274

Fig. 4 Performance comparison of RPiso and Ribomap using the gene ALG3. ALG3 has two isoforms (NM_005787 and NM_001006941). Each isoform has its unique exon. Using Integrative Genomics Viewer [23] to visualize the ribo-seq footprints (from the BAM file generated by Bowtie [17]) shows that the unique exon of NM_005787 has much more ribo-seq footprints (91 vs. 2 uniquely mapped reads) than the unique exon of NM_001006941 has. Therefore, a good isoform-level software tool should redistribute more footprints to NM_005787 than to NM_001006941. That is, the translational level (TL) of NM_005787 should be higher than that of NM_001006941. Our RPiso supports this assertion (0.9 vs. 0.01) while Ribomap contradicts this assertion (110 vs. 274). Therefore, our RPiso outperformed Ribomap in this case

different mRNA isoforms, our RPiso can do the same task only based on ribo-seq reads alone. That is, Ribomap needs both RNA-seq and ribo-seq as inputs while RPiso can analyze ribo-seq alone. Second, RPiso estimates isoform abundance more accurately than Ribomap does. RPiso outperforms Ribomap in 86% (25/29) of the case studies (Table 1). Third, RPiso supports analysis on multiple species while Ribomap supports only human. RPiso precompiled the reference mRNA transcriptomes of five species (human, mus, rat, yeast, and zebrafish). For each species, the following data are provided: rRNA sequences, mRNA sequences, mRNA annotations (mRNA length, 5'UTR length, CDS length, and 3'UTR length, and mRNA-gene mapping). From mRNA-gene mapping information, users can know all the mRNA isoforms of a gene of interest. Moreover, we have run Bowtie to generate the rRNA and mRNA transcriptome reference indices of each of the five species. Therefore, if the user's ribo-seq data belong to these five species, users then can run RPiso without doing extra steps to construct the transcriptome reference indices first. RPiso also provides step-by-step instructions on how to prepare the reference mRNA transcriptomes on other species of interest, but Ribomap does not provide such information. Fourth, RPiso has a web-based viewer for interactively visualizing ribosome occupancy patterns on the selected isoforms while Ribomap does not give any visualization. RPiso's online viewer helps users to find out novel biological insights. For example, by viewing the ribo-seq data mapped on two isoforms (NM_005787 and NM_001006941) of ALG3, users can know the translational level of NM_005787 is much higher than that of NM_001006941 (Fig. 3d) in the physiological condition under study

Table 1 The 29 genes (with two isoforms) used to compare the performance of RPiso and Ribomap

Gene name	isoform A versus isoform B	Ribo-seq reads uniquely mapped to the unique exon of each isoform (by Bowtie): A versus B	TL(A) versus TL(B) calculated by RPiso	TL(A) versus TL(B) calculated by by Ribomap
ALG3	NM_005787 versus NM_001006941	91 versus 2 (A > B)	0.9 versus 0.01 (A > B)	110 versus 274 (A < B)
CNRIP1	NM_015463 versus NM_001111101	1 versus 0 (A > B)	0.01 versus 0 (A > B)	1 versus 2 (A < B)
COL5A1	NM_000093 versus NM_001278074	11 versus 0 (A > B)	0.69 versus 0 (A > B)	92 versus 1104 (A < B)
COQ6	NM_182476 versus NM_182480	9 versus 4 (A > B)	0.3 versus 0.11 (A > B)	12 versus 162 (A < B)
CRY2	NM_021117 versus NM_001127457	4 versus 0 (A > B)	0.06 versus 0 (A > B)	4 versus 30 (A < B)
EEF1E1	NM_001135650 versus NM_004280	124 versus 0 (A > B)	1.76 versus 0.6 (A > B)	135 versus 232 (A < B)
EGLN3	NM_022073 versus NM_001308103	30 versus 0 (A > B)	0.55 versus 0.02 (A > B)	22 versus 98 (A < B)
GALNT2	NM_004481 versus NM_001291866	181 versus 0 (A > B)	4.21 versus 0 (A > B)	234 versus 1926 (A < B)
KLF10	NM_005655 versus NM_001032282	53 versus 0 (A > B)	1.7 versus 0 (A > B)	89 versus 681 (A < B)
LCORL	NM_001166139 versus NM_153686	22 versus 13 (A > B)	0.17 versus 0.15 (A > B)	31 versus 98 (A < B)
LIG3	NM_013975 versus NM_002311	57 versus 0 (A > B)	0.4 versus 0.01 (A > B)	65 versus 300 (A < B)
LRIG3	NM_153377 versus NM_001136051	9 versus 0 (A > B)	0.09 versus 0.04 (A > B)	17 versus 128 (A < B)
LRP5	NM_002335 versus NM_001291902	3 versus 0 (A > B)	0.46 versus 0 (A > B)	321 versus 338 (A < B)
LYSMD1	NM_001136543 versus NM_212551	7 versus 0 (A > B)	0.18 versus 0 (A > B)	7 versus 28 (A < B)
MAP1S	NM_018174 versus NM_001308363	5 versus 0 (A > B)	0.15 versus 0 (A > B)	10 versus 135 (A < B)
MAT2B	NM_013283 versus NM_182796	23 versus 0 (A > B)	3.61 versus 0 (A > B)	58 versus 1068 (A < B)
NAF1	NM_138386 versus NM_001128931	127 versus 0 (A > B)	1.2 versus 0 (A > B)	133 versus 406 (A < B)
NDUFA11	NM_175614 versus NM_001193375	28 versus 0 (A > B)	1.92 versus 0 (A > B)	29 versus 355 (A < B)
NSMAF	NM_001144772 versus NM_003580	17 versus 13 (A > B)	0.11 versus 0.09 (A > B)	25 versus 167 (A < B)
NT5DC2	NM_001134231 versus NM_022908	62 versus 0 (A > B)	0.67 versus 0 (A > B)	62 versus 296 (A < B)
PPIL3	NM_130906 versus NM_032472	18 versus 0 (A > B)	0.59 versus 0 (A > B)	61 versus 75 (A < B)
PRKG1	NM_006258 versus NM_001098512	42 versus 0 (A > B)	0.2 versus 0 (A > B)	44 versus 77 (A < B)
RASA1	NM_002890 versus NM_022650	176 versus 0 (A > B)	0.45 versus 0 (A > B)	192 versus 321 (A < B)
RFC3	NM_002915 versus NM_181558	240 versus 0 (A > B)	2.01 versus 1.43 (A > B)	279 versus 808 (A < B)
RPS15	NM_001018 versus NM_001308226	22 versus 0 (A > B)	18.06 versus 0 (A > B)	24 versus 2482 (A < B)
CPD	NM_001304 versus NM_001199775	9 versus 0 (A > B)	0.11 versus 0.52 (A < B)	640 versus 33 (A > B)

Table 1 (continued)

Gene name	isoform A versus isoform B	Ribo-seq reads uniquely mapped to the unique exon of each isoform (by Bowtie): A versus B	TL(A) versus TL(B) calculated by RPiso	TL(A) versus TL(B) calculated by by Ribomap
PPP2R2A	NM_002717 versus NM_001177591	12 versus 0 (A > B)	0.44 versus 1.77 (A < B)	859 versus 39 (A > B)
RPL14	NM_001034996 versus NM_003973	3 versus 0 (A > B)	0.2 versus 65.57 (A < B)	12,815 versus 83 (A > B)
SUCLG2	NM_003848 versus NM_001177599	54 versus 0 (A > B)	0.41 versus 1.98 (A < B)	911 versus 16 (A > B)

Take the gene ALG3 as an example. ALG3 has two isoforms (NM_005787 and NM_001006941). Each isoform has its unique exon. The unique exon of NM_005787 has much more Ribo-seq footprints (91 vs. 2 uniquely mapped reads) than the unique exon of NM_001006941 has. Therefore, the translational level (TL) of NM_005787 should be higher than that of NM_001006941. Our RPiso supports this assertion (0.9 vs. 0.01) while Ribomap contradicts this assertion (110 vs. 274). Therefore, our RPiso outperformed Ribomap in this case. In total, our RPiso outperformed Ribomap in 86% (25/29) of the cases (bold-faced names), suggesting that our RPiso estimates isoform abundance more accurately than Ribomap does

(i.e. human Hela cell with RPL19 (Ribosomal Protein L19) knockdown), indicating that the two isoforms of ALG3 may be under different translational regulation. We believe that RPiso is a useful software tool for researchers to analyze and visualize their own ribo-seq data at the isoform level.

Availability and requirements

Project name: RPiso.

Project home page: [main site] http://cosbi7.ee.ncku.edu.tw/RPiso/, [backup site 1] http://cosbi4.ee.ncku.edu.tw/RPiso/, and [backup site 2] http://cosbi6.ee.ncku.edu.tw/RPiso/.

Operating system(s): Linux ubuntu 14.04 LTS (or 16.04 LTS).

Programming language: Perl 5.22.1 and Python 2.7.12 (or 3.5.2).

Other requirements: Cutadapt 1.18, Bowtie-1.2.2-linux-x86_64 and RSEM-1.3.1.

License: none required.

Any restrictions to use by non-academics: no restriction.

Abbreviations

NRPKM: Normalized Reads Per Kilobase per Million mapped reads; NRPM: Normalized Reads Per Million mapped reads; RP: Ribosome profiling.

Supplementary Information

The online version contains supplementary material available at https://doi.org/10.1186/s12859-021-04192-7.

Additional file 1: RPiso manual.

Acknowledgements

We thank National Cheng Kung University and Ministry of Science and Technology of Taiwan for their support.

About this supplement

This article has been published as part of BMC Bioinformatics Volume 22 Supplement 10 2021: Selected articles from the 19th Asia Pacific Bioinformatics Conference (APBC 2021): bioinformatics. The full contents of the supplement are available at https://bmcbioinformatics.biomedcentral.com/articles/supplements/volume-22-supplement-10.

Authors' contributions
WSW, YYT, and JTT conceived the research topic and provided essential guidance. WSW, YHT and TYC developed the RPiso pipeline. WSW, TYC and SCS developed the online viewer. WSW wrote the manuscript. SCS prepared all the figures in the manuscript. All authors read, edited and approved the final manuscript.

Funding
Publication costs are funded by Ministry of Science and Technology of Taiwan [107-2221-E-006-225-MY3 and 108-2628-E-006-004-MY3]. YYT was supported by NCI RO1 CA204962. The funders had no role in study design, data collection and analysis, decision to publish, or preparation of the manuscript.

Availability of data and materials
All the data in RPiso are available at http://cosbi7.ee.ncku.edu.tw/RPiso/.

Declarations

Ethics approval and consent to participate
Not applicable.

Consent for publication
Not applicable.

Competing interests
The authors declare that they have no competing interests.

Author details
[1]Department of Electrical Engineering, National Cheng Kung University, Tainan 701, Taiwan. [2]Department of Biotechnology and Bioindustry Sciences, National Cheng Kung University, Tainan 701, Taiwan. [3]Center for Molecular Medicine and Genetics, School of Medicine, Wayne State University, Detroit, MI 48201, USA.

Received: 18 April 2021 Revised: 18 April 2021 Accepted: 12 May 2021
Published online: 25 May 2021

References

1 Kelen K, Van Der Beyaert R, Inzé D, Veylder De L. Translational control of eukaryotic gene expression. Crit Rev Biochem Mol Biol. 2009; 44(4):143–68.
2 Hershey JW, Sonenberg N, Mathews MB. Principles of translational control: an overview. Cold Spring Harb Perspect Biol. 2012; 4(12):a011528.
3 Ruggero D. Translational control in cancer etiology. Cold Spring Harb Perspect Biol. 2013; 5(2):a012336.
4 Ingolia NT, Ghaemmaghami S, Newman JR, Weissman JS. Genome-wide analysis in vivo of translation with nucleotide resolution using ribosome profiling. Science. 2009; 324(5924):218–23.
5 Jackson RJ, Hellen CU, Pestova TV. The mechanism of eukaryotic translation initiation and principles of its regulation. Nat Rev Mol Cell Biol. 2010; 11(2):113–27.
6 Michel AM, Choudhury KR, Firth AE, Ingolia NT, Atkins JF, Baranov PV. Observation of dually decoded regions of the human genome using ribosome profiling data. Genome Res. 2012; 22(11):2219–29.
7 Andreev DE, O'Connor PB, Loughran G, Dmitriev SE, Baranov PV, Shatsky IN. Insights into the mechanisms of eukaryotic translation gained with ribosome profiling. Nucl Acids Res. 2017; 45(2):513–26.
8 Popa A, Lebrigand K, Paquet A, Nottet N, Robbe-Sermesant K, Waldmann R, Barbry P. RiboProfiling: a bioconductor package for standard Ribo-seq pipeline processing. F1000Res. 2016;5:1309.
9 Popa A, Lebrigand K, Paquet A, Nottet N, Robbe-Sermesant K, Waldmann R, Barbry P. RiboTools: a Galaxy toolbox for qualitative ribosome profiling analysis. Bioinformatics. 2015;31(15):2586–8.
10 Michel AM, Mullan JP, Velayudhan V, O'Connor PB, Donohue CA, Baranov PV. RiboGalaxy: a browser based platform for the alignment, analysis and visualization of ribosome profiling data. RNA Biol. 2016;13(3):316–9.
11 Wang H, Wang Y, Xie Z. Computational resources for ribosome profiling: from database to web server and software. Brief Bioinform. 2017;20(1):144–55.
12 Kiniry SJ, Michel AM, Baranov PV. Computational methods for ribosome profiling data analysis. Wiley Interdiscip Rev RNA. 2019;2019:e1577.
13 Breitbart RE, Andreadis A, Nadal-Ginard B. Alternative splicing: a ubiquitous mechanism for the generation of multiple protein isoforms from single genes. Annu Rev Biochem. 1987;56:467–95.
14 Andreadis A, Gallego ME, Nadal-Ginard B. Generation of protein isoform diversity by alternative splicing: mechanistic and biological implications. Annu Rev Cell Biol. 1987;3:207–42.
15 Wang H, McManus J, Kingsford C. Isoform-level ribosome occupancy estimation guided by transcript abundance with Ribomap. Bioinformatics. 2016;32(12):1880–2.
16 Wu WS, Jiang YX, Chang JW, Chu YH, Chiu YH, Tsao YH, Nordling TEM, Tseng YY, Tseng JT. HRPDviewer: human ribosome profiling data viewer. Database (Oxford). 2018;2018:bay074.
17 Langmead B, Trapnell C, Pop M, Salzberg SL. Ultrafast and memory-efficient alignment of short DNA sequences to the human genome. Genome Biol. 2009;10(3):R25.
18 Li B, Dewey CN. RSEM: accurate transcript quantification from RNA-Seq data with or without a reference genome. BMC Bioinform. 2011;12;323.

19 Stumpf CR, Moreno MV, Olshen AB, Taylor BS, Ruggero D. The translational landscape of the mammalian cell cycle. Mol Cell. 2013;52:574–82.
20 Martin M. Cutadapt removes adapter sequences from high-throughput sequencing reads. EMBnet J. 2011;17:1–10.
21 Li B, Ruotti V, Stewart RM, Thomson JA, Dewey CN. RNA-Seq gene expression estimation with read mapping uncertainty. Bioinformatics. 2010;26(4):493–500.
22 Guo H, Ingolia NT, Weissman JS, Bartel DP. Mammalian microRNAs predominantly act to decrease target mRNA levels. Nature. 2010;466(7308):835–40.
23 Robinson JT, Thorvaldsdóttir H, Winckler W, Guttman M, Lander ES, Getz G, Mesirov JP. Integrative genomics viewer. Nat Biotechnol. 2011;29:24–6.

BMC Bioinformatics

RESEARCH

Open Access

MitoTox: a comprehensive mitochondrial toxicity database

Yu-Te Lin[1], Ko-Hong Lin[2], Chi-Jung Huang[1] and An-Chi Wei[1,2*]

From The 19th Asia Pacific Bioinformatics Conference (APBC 2021) Tainan, Taiwan. 3-5 February 2021

*Correspondence:
acwei86@ntu.edu.tw
[1] Graduate Institute of Biomedical Electronics and Bioinformatics, National Taiwan University, Taipei, Taiwan
Full list of author information is available at the end of the article

Abstract

Background: Mitochondria play essential roles in regulating cellular functions. Some drug treatments and molecular interventions have been reported to have off-target effects damaging mitochondria and causing severe side effects. The development of a database for the management of mitochondrial toxicity-related molecules and their targets is important for further analyses.

Results: To correlate chemical, biological and mechanistic information on clinically relevant mitochondria-related toxicity, a comprehensive mitochondrial toxicity database (MitoTox) was developed. MitoTox is an electronic repository that integrates comprehensive information about mitochondria-related toxins and their targets. Information and data related to mitochondrial toxicity originate from various sources, including scientific journals and other electronic databases. These resources were manually verified and extracted into MitoTox. The database currently contains over 1400 small-molecule compounds, 870 mitochondrial targets, and more than 4100 mitochondrial toxin-target associations. Each MitoTox data record contains over 30 fields, including biochemical properties, therapeutic classification, target proteins, toxicological data, mechanistic information, clinical side effects, and references.

Conclusions: MitoTox provides a fully searchable database with links to references and other databases. Potential applications of MitoTox include toxicity classification, prediction, reference and education. MitoTox is available online at http://www.mitotox.org.

Keywords: Database, Mitochondria, Mitochondrial toxicity, Toxin-target association

Background

Mitochondria, the so-called "energy powerhouse" of a cell, generate approximately 95% of cellular adenosine triphosphate (ATP) through oxidative phosphorylation. Mitochondria are also essential for life-supporting functions and cell fate decisions. Mitochondria play multiple roles in the regulation of cellular physiological and pathological mechanisms, including ATP generation, metabolic control, signal transduction, immune response, and apoptosis [1, 2]. Mitochondrial function and dynamics may change due to pathological responses during disease development. In many cases of mitochondrial dysfunction, oxidative phosphorylation is inhibited, thus inducing an elevated level of glycolysis, which leads to fatal lactate accumulation in the serum [3]. Other potential syndromes include cardiomyopathy, rhabdomyolysis, peripheral neuropathy, optic neuropathy, and hepatic steatosis, which are often observed in mitochondrial diseases [4]. Mitochondrial dysfunction has been linked to common diseases, such as cardiovascular diseases, neurodegenerative diseases, diabetes, and cancer [5].

Moreover, mitochondria have been recognized as unintended drug targets of many pharmaceutical and therapeutic agents that can damage mitochondria and lead to changes in mitochondrial morphology and function. Among these, drug-induced mitochondrial toxicity was found to be responsible for cardiotoxicity, hepatotoxicity, neurotoxicity, nephrotoxicity, ototoxicity, and many other organs-related toxicities [6–8]. The heart, brain, and liver, which rely heavily on oxidative phosphorylation or serve as the main organ of drug metabolism, appear to be the main targets of mitochondrial toxicity [9]. Since the late 1990s, the US Food and Drug Administration has withdrawn dozens of drugs from the market due to their hepatotoxicity or cardiotoxicity, and many of these drugs have been linked to mitochondrial dysfunction [10]. For example, troglitazone, a member of the thiazolidinedione class of antidiabetic drugs, was withdrawn in 2000 with reported lethal hepatotoxicity caused by the off-target effect on the mitochondrial electron transport chain (complex I) [7, 11]. Cerivastatin, a lipid-lowering drug withdrawn in 2001 due to rhabdomyolysis, has been linked to mitochondrial toxicity by affecting complex III-related respiration [12, 13]. Table 1 lists additional examples of pharmaceutical drugs that are related to mitochondrial toxicity. Attention was directed to the mitochondria because of these severe side effects, and drug development has begun to consider mitochondria-related toxicity as well as mitochondrial-targeted therapeutics [10]. For cardiologists, this is appealing because mitochondria are the major source of ATP and reactive oxygen species (ROS) in the myocardium and because mitochondrial dysfunction plays an important role in heart failure and arrhythmias [14]. Several drugs used in clinical settings have shown cardiac toxicity due to their direct effects on cardiac mitochondria [15]. Understanding how these molecules act on mitochondria can provide a mechanistic explanation for their toxicological or pharmacological effects.

In addition to pharmaceutical exposures, environmental factors and pollutants have been identified as previously unrecognized mitochondrial toxicants, including neurotoxin 1-methyl-4phenyl-1, 2, 3, 6-tetrahydropyridine (MPTP), pesticides (e.g., paraquat and rotenone), and heavy metals (e.g., mercury and cadmium); all of which are now widely recognized mitochondrial toxins [16, 17]. Mitochondrial damage has a broader impact on health than is commonly recognized.

Drug-induced mitochondrial toxicity has been studied for more than fifty years. Multiple mechanisms of mitochondrial toxicity have been reported, including inhibition of oxidative phosphorylation, uncoupling, oxidative stress, irreversible opening of the mitochondrial permeability transition pore, inhibition of fatty acid oxidation or TCA cycle to reduce NADH and $FADH_2$ production, and impairment of mitochondrial DNA (mtDNA) replication or mtDNA-encoded protein synthesis [11, 18]. Although large-scale screening studies reporting drug-induced mitochondrial membrane potential loss

Table 1 Examples of mitochondrial toxicity-related drugs in MitoTox

Indication	Drug Name	Mechanism of mitochondrial toxicity	Status	Toxicity	$t_{1/2}$ (h)[a]	Protein binding (%)[a]	LD_{50} (rat)[a]
Antidiabetic	Troglitazone	Inhibition of ETC	Withdrawn, 2000	Hepatotoxicity	16–34	> 99%	1.9768 mol/kg
	Rosiglitazone	Inhibition of ETC	Approved, investigational	Cardiotoxicity	3–4	99.8%	2.4515 mol/kg
	Pioglitazone	Inhibition of ETC	Approved, investigational	Cardiotoxicity	3–7/16–24	> 99%	2.0115 mol/kg
	Ciglitazone	Inhibition of ETC	Discontinued	NA	NA	NA	NA
	Darglitazone	Inhibition of ETC	Discontinued	NA	NA	NA	NA
	Muraglitazar	Inhibition of ETC	Discontinued	NA	NA	NA	NA
	Metformin	Inhibition of complex I; uncoupling; impaired TCA cycle	Approved	Lactic acidosis	6.2	> 90%	1000 mg/kg
Anticancer	Doxorubicin	Increased ROS; mtDNA adduct; iron overload	Approved	Cardiotoxicity	20–48	74–76%	21.8 mg/kg
	Cisplatin	Inhibition of Complex I	Approved	Nephrotoxicity	0.3–0.7	> 90%	2.7612 mol/kg
Hyperlipidemia	Fenofibrate	Inhibition of complex I	Approved	Hepatotoxicity	20	~ 99%	> 2000 mg/kg
	Clofibrate	Inhibition of complex I	Approved	Hepatotoxicity	18–22	95–97%	940 mg/kg
	Ciprofibrate	Inhibition of complex I	Approved	Hepatotoxicity	NA	NA	NA
Psychotropic	Valproic acid	Inhibition of TCA cycle	Approved	Coma and respiratory depression	9–16	90%	670 mg/kg
	Clozapine	Inhibition of the ETC	Approved	Metabolic syndrome	8	97%	3.0838 mol/kg
	Fluoxetine	Uncoupler	Approved	CNS, GI effects	1–3 days	94.5%	2.6048 mol/kg
	Nefazodone	Inhibition of Complex I and complex IV	Withdrawn	Hepatotoxicity	2–4	> 99%	2.9067 mol/kg
Analgesic	Acetaminophen	Oxidative stress	Approved	Hepatotoxicity	1–4	25%	1944 mg/kg
	Aspirin	mPTP potentiation	Approved	Cardiotoxicity, GI effects	0.25	99.5%	920–1480 mg/kg
	Diclofenac	Inhibition of ETC and ATP synthase	Approved	Nephrotoxicity	2	> 99%	3.6447 mol/kg
Antibacterial	Imipenem	Oxidative stress	Approved	Nephrotoxicity	1.3–5.1	20%	1.8089 mol/kg
	Chloramphenicol	Depletion of iron	Approved	Aplastic anemia	1.5–3.5	50–60%	2500 mg/kg
Antiviral	Zidovudine	mtDNA replication (inhibit mtDNA polymerase-γ)	Approved	Myopathy	0.5–2.9	30–38%	NA

Table 1 (continued)

[a] From DrugBank (https://www.drugbank.ca/)

[19–21], the mechanisms of mitochondrial toxicity and dysfunction are incompletely understood.

A number of public toxicant databases, such as Aggregated Computational Toxicology Resource (ACToR) [22], Toxin and Toxin Target Database (T3DB) [23], and the Comparative Toxicogenomics Database (CTD) [24], provide comprehensive information about the bioassay data, chemical structures of the toxicants, toxin targets, and chemical-gene-disease relationships. The NIH Tox21 project has developed a library of ~ 10,000 chemicals and drugs (Tox21 10 K), and has generated > 50 million quantitative high-throughput screening (qHTS) data points using over 70 cell-based assays covering a broad range of biological pathways. A few studies have described the screening of the effects on mitochondrial membrane potential and other mitochondrial functional indicators using the Tox21 library [19, 21, 25]; however, there is a lack of currently available database specifically dedicated to mitochondrial toxicity or recording of multiple aspects of the toxin-mitochondria relationships. A keyword search of DrugBank, a public database for drug information [26], demonstrated that "mitochondrial" appears in the list of keywords of only 81 drugs. A PubMed search using the keyword "mitochondrial" resulted in approximately 350,000 hits, and the number of hits for a search for "mitochondrial toxicity" was approximately 31,000 (as of August 2020). These statistics imply that mitochondria are an important research topic; however, the attention to mitochondrial toxicity is insufficient.

Limited information about drug-induced mitochondrial toxicity is available in the literature, and there is no public database specifically for mitochondrial toxicity. With the increasing demands for integrating basic and clinical research, the current paper aims to develop an integrative database for mitochondrial toxicity and identify the toxicological mechanism and corresponding mitochondrial targets. Understanding how these molecules alter mitochondrial phenotype can provide a mechanistic explanation for their toxicological or pharmacological effects on the mitochondria, which will facilitate the development of safer and more effective drugs. This open-access database will help develop and improve detection and screening procedures for mitochondrial toxicity and further facilitate the identification and analysis of mitochondrial functional and pathological changes.

Methods

Database construction

Many mechanisms cause mitochondrial toxicity, as mentioned previously. Drugs and compounds that have been reported to exhibit mitochondrial toxicity in the literature were first collected into the mitochondrial toxicity database MitoTox. Through literature searches of journal articles and review papers, drugs and compounds with validated evidence of drug-induced mitochondrial toxicity were recorded. The keywords for mitochondrial toxicity were used to search the collected texts to characterize the mechanisms of mitochondrial toxicity of each drug. Chemical properties, bioactivity, and clinical applications, toxicity and side effect information of the compounds are automatically curated in the MitoTox database from the DrugBank, PubChem, UniProt, and SIDER databases. Additional information is curated manually from the KEGG, BRENDA, IPA [27], LiverTox (http://livertox.nih.gov/; search term: "mitochondria"), and MedicinesComplete (AHFS Drug Information, [online] London: Pharmaceutical Press http://www.new.medicinescomplete.com/, search term: "mitochondrial") databases.

For example, to acquire clinical information on these drugs, side effects, frequency of side effects, indications, and ATC code, files were downloaded and analyzed from the SIDER Database (http://sideeffects.embl.de/) and DrugBank (https://go.drugbank.com/atc). The drug synonym names and drug targets from the "external Drug Links" and "target Drug-UniProt Links" were obtained from the DrugBank website. To further expand our drug list, the complete DrugBank database XML file (version 5.1.4) was downloaded and manually curated from the DrugBank database (https://www.drugbank.ca/releases/latest) to perform a further search of the text relevant to mitochondria. The Ingenuity Pathway Analysis (IPA; QIAGEN Inc., https://www.qiagenbioinformatics.com/products/ingenuity-pathway-analysis) database (as of Novmember 2020; version 57662101) was also used to find the correlation of mitochondrial function and target molecules.

Database implementation

MitoTox is an open platform for developing web-based toxicological data discovery and reporting applications. The database is implemented in Python3 under the web framework with Django 3.0.3, the relational database with PostgreSQL, the webserver in Nginx (connected to the web framework by the Gunicorn library), and the Ubuntu system on the backend. The front end is constructed using HTML5, CSS3, and JavaScript (Fig. 1). Additional libraries, jQuery, D3.js, and Bootstrap were utilized to help us visualize the data and complete the web design. A GUI and searchable drop-down menus for the database were created by Django's admin library and the Django-selectable library. The Django web framework basic security measures were provided by the Django web framework to restrict access to the modification of the entries of all database tables.

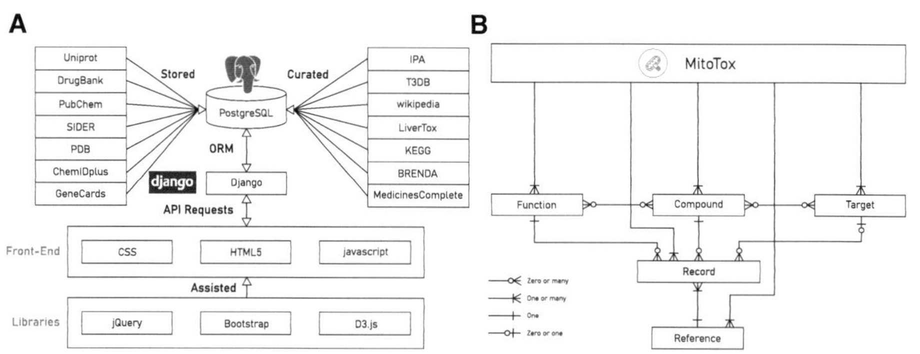

Fig. 1 MitoTox database workflow and structure. **a** MitoTox is an open-access database constructed using the Django framework with a PostgreSQL database as the backend. **b** The MitoTox database structure includes entries of compounds, targets, functions and references

Database description

MitoTox includes over 1400 small compound entries corresponding to more than 34,000 different synonyms. These compounds are linked to over 870 mitochondrial-related protein, protein subunit and RNA targets through toxin and toxin-target bonds. Figure 2 shows a summary and statistics of the MitoTox database. The toxicities are classified into 8 major groups according to their function and toxic mechanisms, into 225 functional subgroups according to their toxicity mechanisms, and into over 650 target subgroups according to their molecular targets. All the entries and relationships recorded in the database are supported by references; currently, there are approximately 300 references. One important feature of MitoTox is the storage of experimental results of in vitro and in vivo assays for the detection of drug-induced mitochondrial toxicity. The experimental results from the mitochondrial functional and toxicity assays were recorded with the dosage, corresponding experimental methods, model system, and results including IC50 (half-maximal inhibitory concentration), EC50 (half-maximal effective concentration), or Cmax (the maximum or peak serum concentration).

Results

To better understand the physiological impact of each drug or compound on mitochondria, we constructed a drug database combining general pharmaceutical information and experimental data on the drugs. We searched for mitochondria-related keywords in the complete DrugBank file and found 1011 items that were possibly related to drug-induced mitochondrial toxicity. Through manual curation, we identified 180 true positive items of drug-induced mitochondrial toxicity. We applied the same pipeline to obtain drug targets, synonym names, and clinical information for these drugs. From the keyword search in PubMed, over 300 review articles and research papers reporting drugs and molecules were manually recorded in the MitoTox database. Collectively, our database currently contains 1400 drugs or compounds that have been tested for or show evidence of mitochondrial toxicity, which are classified into 77 categories of mitochondrial toxicity mechanisms (Fig. 3).

Mitochondrial dysfunction could stem from structural, enzymatic, and biochemical changes [28]. Mitochondrial toxicities arise from different mechanisms, including

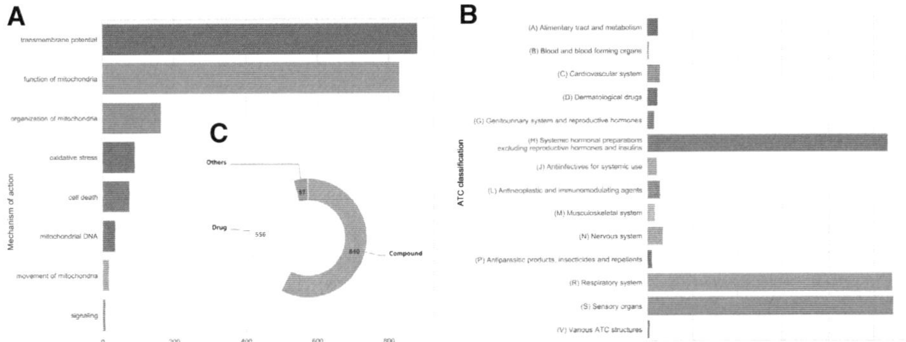

Fig. 2 Summary and statistics of the MitoTox database. The mitochondrial toxicity-related molecules are classified according to **a** the affected mitochondrial functions, **b** the affected organ systems according to The Anatomical Therapeutic Chemical (ATC) Classification, and **c** types

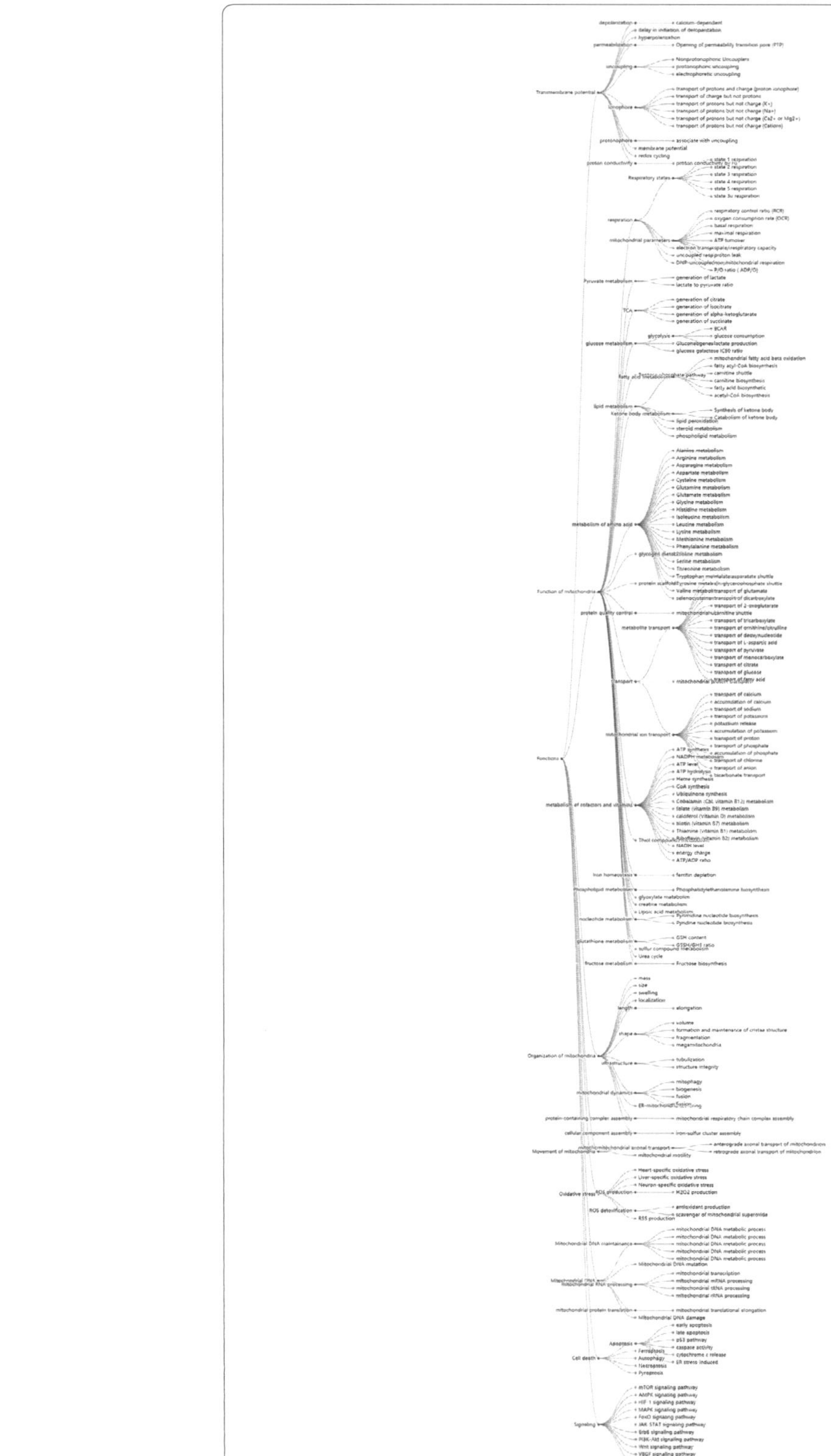

Fig. 3 Classification of mitochondrial toxicity mechanisms according to functionality. The mechanisms of toxicity are classified into multiple layers. The first classification layer includes eight main categories in: transmembrane potential, functions of mitochondria, organization of mitochondria, movement of mitochondria, oxidative stress, mitochondrial DNA, cell death, and signaling. Each category is further classified into subclasses according to their mechanism of action

interference with the mitochondrial respiratory chain, the tricarboxylic acid (TCA) cycle, and mitochondrial DNA replication, leading to loss of mitochondrial membrane potential and increased mitochondrial oxidative stress, and eventually cell death. Based on these results, mitochondrial toxicity mechanisms are classified into eight main categories: alternation of transmembrane potential, function of mitochondria, organization of mitochondria, movement of mitochondria, oxidative stress, cell death, mitochondrial DNA, and metabolic-related signaling pathways (Fig. 4a). Each category is further classified into subgroups according to the corresponding toxicity mechanism. For example, in the category of transmembrane potential, inhibiting ETC or uncoupling or modulating transport of ions could be mechanisms that affect mitochondrial membrane potential and therefore were classified under the mitochondrial membrane potential category.

Mitochondrial toxicity-related targets are classified into mitochondrial enzymes in the matrix, transporters or carriers on the membranes, mitochondrial DNA replication machinery, RNAs, protein synthesis and assembly machinery, small molecules and metabolites, etc. The targets can also be classified based on their locations, such as the mitochondrial matrix, inner mitochondrial membrane (IMM), outer mitochondrial membrane (OMM), and intermembrane space (IMS). In each category, there are molecular targets corresponding to altering mitochondrial function and morphology.

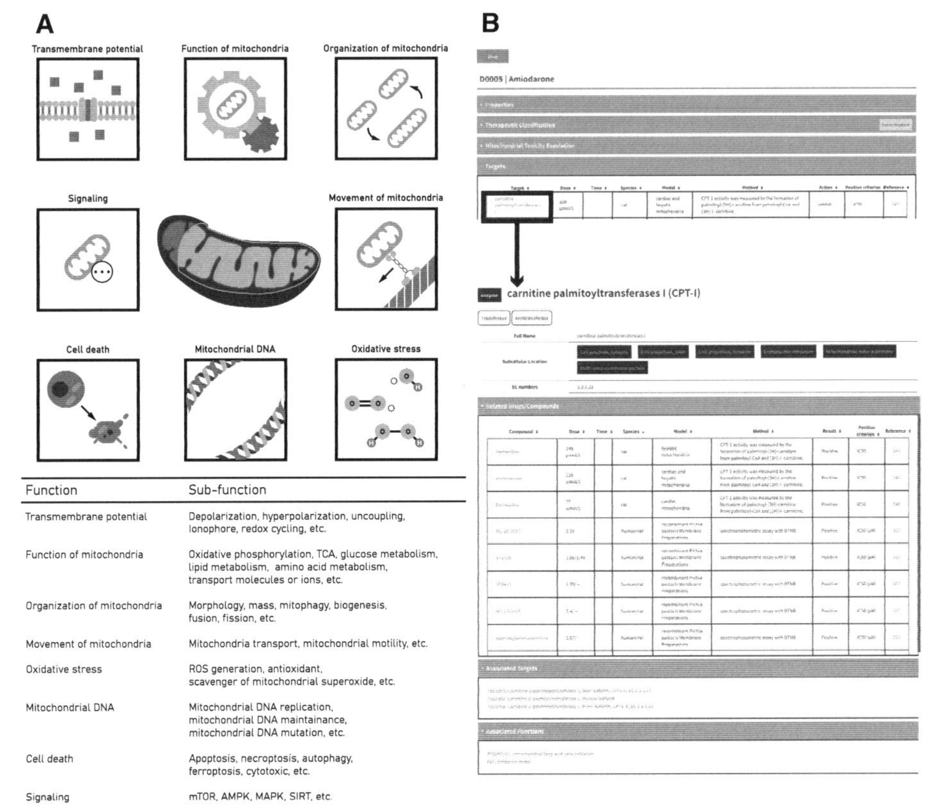

Fig. 4 Classification of molecular targets of drugs that induce mitochondrial toxicity. **a** The molecular targets are classified into eight main categories according to the action of the targets. **b** A screenshot of the mitochondrial toxin target record using "Amiodarone" and one of its targets, "carnitine palmitoyltransferase I (CPT-1)," as an example

For example, in the ETC category, there are complex I, complex II, complex III, complex IV, and ATP synthase proteins. In each protein complex, there are multiple subunits or molecules that could be the targets for an individual drug. Each molecular target is recorded in the database with relevant molecular entries, such as molecular protein name and gene name, EC number, subcellular location, involved reactions, and modulators (Fig. 4b). The drugs that target mitochondria can affect multiple factors to induce mitochondrial toxicity [29]. Therefore, drug targets related to mitochondria and related metabolism or signaling pathways are cross-referenced and linked by toxin-target associations based on multiple experimental data from the literature and databases.

Common screening assays for mitochondrial toxicity include evaluation of cytotoxicity of the compounds incubated with cultured cells in glucose or galactose conditioned media (Glu/Gal assay) [30], and measurements of mitochondrial oxygen consumption [31, 32], mitochondrial membrane potential [33, 34], reactive oxygen species [31, 32], mitochondrial permeability transition [35], and cellular ATP content [21, 36, 37]. Many of these mitochondrial functional assays have been applied to high-throughput screening for mitochondrial toxicity [38]. Mitochondrial membrane potential ($\Delta\Psi m$) is determined by a balance between the proton gradient and other ion gradients across the inner mitochondrial membrane maintained by proton pumps and electrogenic ion channels or transporters. Since mitochondrial membrane potential is a part of the proton-motive force that generates ATP, $\Delta\Psi m$ is an important indicator of mitochondrial function and cell health. Lipophilic cationic compounds (such as TMRM, TMRE, and JC1) accumulating in the mitochondrial membrane matrix space are often used to monitor $\Delta\Psi m$ in high-throughput screening assays. Oxygen consumption profile is another important indicator of mitochondrial function. Microplate-based assays using oxygen-sensitive fluorescent probes provide a means for high-throughput evaluation of oxygen consumption rate (OCR), reserve capacity, and extracellular acidification rate (ECAR). Drugs or compounds that significantly change the mitochondrial membrane potential and OCR profiles are included in the MitoTox database and are labeled as positive or negative entries in the assay results based on the experimental criteria. The experimental systems and conditions, including the dosage, exposure time, substrates, or growth conditions, are also recorded along with their corresponding references (Fig. 5).

Discussion

In addition to pathological functional changes in the mitochondria, mitochondrial dysfunction resulting from drug treatment or chemical exposure is an important concern in the area of biomedical and ecological health. Mitochondrial health is essential to the survival and fitness of the organisms, and detailed investigations of mitochondrial functions are imperative. To further our understanding of the pharmacological and toxicological mechanisms influencing mitochondrial toxicity, it is necessary to have a database capable of integrating information about mitochondrial toxicity and recording data acquired by the measurements of mitochondrial functions, particularly those functions that have been perturbed by extrinsic and intrinsic factors. Many assays have been developed and implemented to investigate the functional and morphological changes in mitochondria. Respiration rate and ATP content are important indicators of mitochondrial health. Mitochondrial stress can be assessed

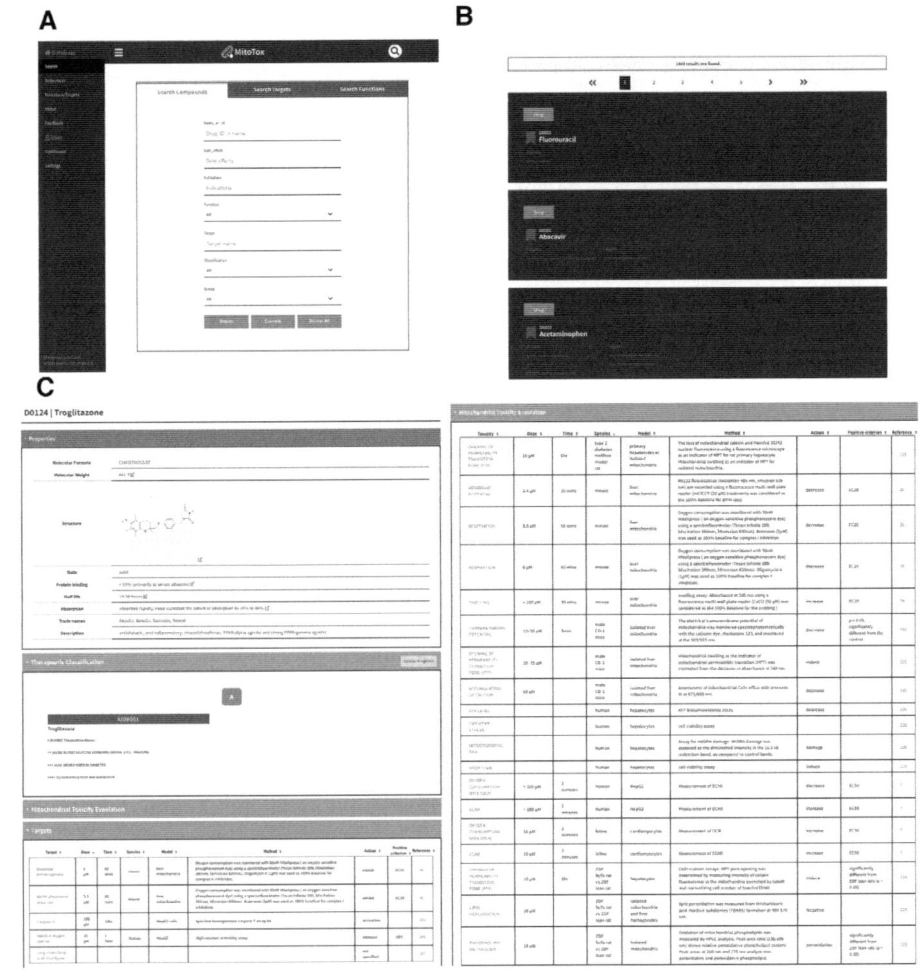

Fig. 5 Screenshot of the MitoTox database. **a** A screenshot of the Mitochondria Toxicity Database (MitoTox) showing several search and display tools of MitoTox. **b** A screenshot of the search results. **c** Example of the data fields or data types found in MitoTox describing the mitochondrial toxin "Troglitazone."

based on the parameters of ROS production, mitochondrial dynamics, permeability transition, and quality control. In particular, oxidative stress is an important issue in mitochondrial toxicity. ROS are the side-products of mitochondrial electron transfer and are regulators of cellular signaling during growth, differentiation, and metabolism in low physiological doses; however, excessive levels of ROS cause oxidative damage and cell death. Nevertheless, many chemical compounds and environmental pollutants are involved in the generation of reactive species and oxidative stress, thus influencing mitochondrial and cellular functions [32, 39, 40].

An electronic database that provides a searchable index and keywords is constructed and presented in this paper to facilitate drug safety and toxicity prevention. By collecting such data, we could examine the association between physiological responses, such as membrane potential changes and ATP production, and known drug toxicity and clinical side effects. These data should produce a repository of useful information for future mitochondrial toxicity studies, build an algorithm for mitochondrial toxicity prediction, and shed light on candidate drug selection during drug discovery.

Development of high-throughput technologies increased availability and accessibility of mitochondrial toxicity-related data. Big data containing mitochondria-related information obtained from high-throughput or high-content systems provide opportunities for statistical analysis and machine learning to classify the mechanism of action and discover or predict potential mitochondrial toxicants and therapeutics. Statistical methods, such as Fisher's exact test, can be used to define associations between the mechanisms of mitochondrial toxicity and side effects. In addition to enhanced insight into mitochondrial toxicity, the recent rapid development of machine learning technologies can be used to analyze and predict side effects [41, 42] or identify toxicophores that are predictive of mitochondrial toxicity [43].

Our purpose in building MitoTox is to study how drug-induced mitochondrial toxicity causes side effects in clinical research. The MitoTox database bridges the gap between the two sides (mitochondrial toxicity vs. side effects) and could help predict potential mitochondrial toxicity for future drug discovery. The combined clinical information, including clinical indications, side effects, frequency of side effects, and therapeutic indications with molecular toxicity mechanisms, enables us to learn and prevent potential side effects.

Conclusion

MitoTox database is intended to provide novel strategies for the preclinical screening of mitochondrial toxicity, which will be of immediate benefit to drug safety and global health. Furthermore, the planned investigations of the mitochondrial toxicity subtypes are expected to elucidate how the morphological features and functional indicators of mitochondria may serve as a pathological feature of disease onset and disease progression. MitoTox will help advance our understanding of mitochondrial-associated toxicity and provide new possibilities for the early diagnosis of mitochondrial diseases, as well as innovative therapeutic strategies against these diseases. Last but not least, understanding the mitochondrial toxicity of drugs could help in the development of drugs that target mitochondria to treat common pathologies [44].

Abbreviations

ATP: Adenosine triphosphate; MPTP: 1-Methyl-4-phenyl-1, 2, 3, 6-tetrahydropyridine; $\Delta\Psi m$: Mitochondrial membrane potential; TCA cycle: Tricarboxylic acid cycle; IMM: Inner mitochondrial membrane; OMM: Outer mitochondrial membrane; IMS: Intermembrane space; OCR: Oxygen consumption rate; ECAR: Extracellular acidification rate.

Acknowledgements

Not applicable.

About this supplement

This article has been published as part of BMC Bioinformatics Volume 22 Supplement 10 2021: Selected articles from the 19th Asia Pacific Bioinformatics Conference (APBC 2021): bioinformatics. The full contents of the supplement are available at https://bmcbioinformatics.biomedcentral.com/articles/supplements/volume-22-supplement-10.

Authors' contributions

AW and YL designed and implemented the database. YL, AW, KL, and CH performed data acquisition, analysis and interpretation. YL developed the database and designed the web site. AW and KL conceived the idea and wrote the manuscript. All authors have read and approved the final manuscript.

Funding

All work and publication costs were funded by the Ministry of Science and Technology in Taiwan (MOST-108-2636-B- 002 -001 and MOST-109-2636-B- 002 -001 Grants to AW). The funders had no role in: the design of the study and collection, analysis or interpretation of data or in writing the manuscript.

Availability of data and materials
MitoTox is an open access database at http://www.mitotox.org.

Declarations

Ethics approval and consent to participate
Not applicable.

Consent to publish
Not applicable.

Competing interests
The authors declare that they have no competing interests.

Author details
[1]Graduate Institute of Biomedical Electronics and Bioinformatics, National Taiwan University, Taipei, Taiwan. [2]Department of Electrical Engineering, National Taiwan University, Taipei, Taiwan.

Received: 13 June 2021 Accepted: 6 July 2021
Published online: 15 July 2021

References
1. Mottis A, Herzig S, Auwerx J. Mitocellular communication: shaping health and disease. Science. 2019;366:827–32. https://doi.org/10.1126/science.aax3768.
2. Osellame LD, Blacker TS, Duchen MR. Cellular and molecular mechanisms of mitochondrial function. Best Pract Res Clin Endocrinol Metab. 2012;26:711–23. https://doi.org/10.1016/j.beem.2012.05.003.
3. Robinson BH. Lactic acidemia and mitochondrial disease. Mol Genet Metab. 2006;89:3–13. https://doi.org/10.1016/j.ymgme.2006.05.015.
4. Dimauro S. A history of mitochondrial diseases. J Inherit Metab Dis. 2011;34:261–76. https://doi.org/10.1007/s10545-010-9082-x.
5. Chan DC. Mitochondrial dynamics and its involvement in disease. Annu Rev Pathol. 2020;15:235–59. https://doi.org/10.1146/annurev-pathmechdis-012419-032711.
6. Nadanaciva S, Will Y. Current concepts in drug-induced mitochondrial toxicity. Curr Protoc Toxicol. 2009. https://doi.org/10.1002/0471140856.tx0215s40.
7. Chan K, Truong D, Shangari N, O'Brien PJ. Drug-induced mitochondrial toxicity. Expert Opin Drug Metab Toxicol. 2005;1:655–69. https://doi.org/10.1517/17425255.1.4.655.
8. Wallace KB. Mitochondrial toxicity. Toxicology. 2017;391:1. https://doi.org/10.1016/j.tox.2017.08.005.
9. Wallace KB, Starkov AA. Mitochondrial targets of drug toxicity. Annu Rev Pharmacol Toxicol. 2000;40:353–88. https://doi.org/10.1146/annurev.pharmtox.40.1.353.
10. Dykens JA, Will Y. The significance of mitochondrial toxicity testing in drug development. Drug Discov Today. 2007;12:777–85. https://doi.org/10.1016/j.drudis.2007.07.013.
11. Vuda M, Kamath A. Drug induced mitochondrial dysfunction: Mechanisms and adverse clinical consequences. Mitochondrion. 2016;31:63–74. https://doi.org/10.1016/j.mito.2016.10.005.
12. Will Y, Dykens J. Mitochondrial toxicity assessment in industry—a decade of technology development and insight. Expert Opin Drug Metab Toxicol. 2014;10:1061–7. https://doi.org/10.1517/17425255.2014.939628.
13. Schirris TJJ, Renkema GH, Ritschel T, Voermans NC, Bilos A, van Engelen BGM, et al. Statin-induced myopathy is associated with mitochondrial complex III inhibition. Cell Metab. 2015;22:399–407. https://doi.org/10.1016/j.cmet.2015.08.002.
14. Siasos G, Tsigkou V, Kosmopoulos M, Theodosiadis D, Simantiris S, Tagkou NM, et al. Mitochondria and cardiovascular diseases-from pathophysiology to treatment. Ann Transl Med. 2018;6:256. https://doi.org/10.21037/atm.2018.06.21.
15. Pereira GC, Silva AM, Diogo CV, Carvalho FS, Monteiro P, Oliveira PJ. Drug-induced cardiac mitochondrial toxicity and protection: from doxorubicin to carvedilol. Curr Pharm Des. 2011;17:2113–29. https://doi.org/10.2174/138161211796904812.
16. Meyer JN, Leung MCK, Rooney JP, Sendoel A, Hengartner MO, Kisby GE, et al. Mitochondria as a target of environmental toxicants. Toxicol Sci. 2013;134:1–17. https://doi.org/10.1093/toxsci/kft102.
17. Zolkipli-Cunningham Z, Falk MJ. Clinical effects of chemical exposures on mitochondrial function. Toxicology. 2017;391:90–9. https://doi.org/10.1016/j.tox.2017.07.009.
18. Nelms MD, Mellor CL, Cronin MTD, Madden JC, Enoch SJ. Development of an in silico profiler for mitochondrial toxicity. Chem Res Toxicol. 2015;28:1891–902. https://doi.org/10.1021/acs.chemrestox.5b00275.
19. Attene-Ramos MS, Huang R, Michael S, Witt KL, Richard A, Tice RR, et al. Profiling of the Tox21 chemical collection for mitochondrial function to identify compounds that acutely decrease mitochondrial membrane potential. Environ Health Perspect. 2015;123:49–56. https://doi.org/10.1289/ehp.1408642.
20. Attene-Ramos MS, Huang R, Sakamuru S, Witt KL, Beeson GC, Shou L, et al. Systematic study of mitochondrial toxicity of environmental chemicals using quantitative high throughput screening. Chem Res Toxicol. 2013;26:1323–32. https://doi.org/10.1021/tx4001754.
21. Xia M, Huang R, Shi Q, Boyd WA, Zhao J, Sun N, et al. Comprehensive analyses and prioritization of Tox21 10K chemicals affecting mitochondrial function by in-depth mechanistic studies. Environ Health Perspect. 2018;126: 077010. https://doi.org/10.1289/EHP2589.

22. Judson R, Richard A, Dix D, Houck K, Elloumi F, Martin M, et al. ACToR–Aggregated computational toxicology resource. Toxicol Appl Pharmacol. 2008;233:7–13. https://doi.org/10.1016/j.taap.2007.12.037.
23. Lim E, Pon A, Djoumbou Y, Knox C, Shrivastava S, Guo AC, et al. T3DB: a comprehensively annotated database of common toxins and their targets. Nucleic Acids Res. 2010. https://doi.org/10.1093/nar/gkp934.
24. Davis AP, Murphy CG, Saraceni-Richards CA, Rosenstein MC, Wiegers TC, Mattingly CJ. Comparative toxicogenomics database: a knowledgebase and discovery tool for chemical-gene-disease networks. Nucleic Acids Res. 2009. https://doi.org/10.1093/nar/gkn580.
25. Potera C. Potential mitochondrial toxicants: Tox21 screen identifies structures of interest. Environ Health Perspect. 2015;123:A23. https://doi.org/10.1289/ehp.123-A23.
26. Wishart DS, Feunang YD, Guo AC, Lo EJ, Marcu A, Grant JR, et al. DrugBank 5.0: a major update to the DrugBank database for 2018. Nucleic Acids Res. 2018;46:D1074–82. https://doi.org/10.1093/nar/gkx1037.
27. Krämer A, Green J, Pollard J, Tugendreich S. Causal analysis approaches in ingenuity pathway analysis. Bioinformatics. 2014;30:523–30. https://doi.org/10.1093/bioinformatics/btt703.
28. Dykens JA, Marroquin LD, Will Y. Strategies to reduce late-stage drug attrition due to mitochondrial toxicity. Expert Rev Mol Diagn. 2007;7:161–75. https://doi.org/10.1586/14737159.7.2.161.
29. Wallace KB. Multiple targets for drug-induced mitochondrial toxicity. Curr Med Chem. 2015;22:2488–92. https://doi.org/10.2174/0929867322666150514095424.
30. Swiss R, Will Y. Assessment of mitochondrial toxicity in HepG2 cells cultured in high-glucose- or galactose-containing media. Curr Protoc Toxicol. 2011. https://doi.org/10.1002/0471140856.tx0220s49.
31. Dikalov SI, Harrison DG. Methods for detection of mitochondrial and cellular reactive oxygen species. Antioxid Redox Signal. 2014;20:372–82. https://doi.org/10.1089/ars.2012.4886.
32. Sieprath T, Corne TDJ, Willems PHGM, Koopman WJH, De Vos WH. Integrated high-content quantification of intracellular ROS levels and mitochondrial morphofunction. Adv Anat Embryol Cell Biol. 2016;219:149–77. https://doi.org/10.1007/978-3-319-28549-8_6.
33. Sakamuru S, Li X, Attene-Ramos MS, Huang R, Lu J, Shou L, et al. Application of a homogenous membrane potential assay to assess mitochondrial function. Physiol Genomics. 2012;44:495–503. https://doi.org/10.1152/physiolgenomics.00161.2011.
34. Scaduto RC, Grotyohann LW. Measurement of mitochondrial membrane potential using fluorescent rhodamine derivatives. Biophys J. 1999;76(1 Pt 1):469–77. https://doi.org/10.1016/S0006-3495(99)77214-0.
35. Šileikytė J, Forte M. Shutting down the pore: The search for small molecule inhibitors of the mitochondrial permeability transition. Biochim Biophys Acta. 2016;1857:1197–202. https://doi.org/10.1016/j.bbabio.2016.02.016.
36. Eakins J, Bauch C, Woodhouse H, Park B, Bevan S, Dilworth C, et al. A combined in vitro approach to improve the prediction of mitochondrial toxicants. Toxicol In Vitro. 2016;34:161–70. https://doi.org/10.1016/j.tiv.2016.03.016.
37. Brand MD, Nicholls DG. Assessing mitochondrial dysfunction in cells. Biochem J. 2011;435:297–312. https://doi.org/10.1042/BJ20110162.
38. Wills LP. The use of high-throughput screening techniques to evaluate mitochondrial toxicity. Toxicology. 2017;391:34–41. https://doi.org/10.1016/j.tox.2017.07.020.
39. Al-Gubory KH. Environmental pollutants and lifestyle factors induce oxidative stress and poor prenatal development. Reprod Biomed Online. 2014;29:17–31. https://doi.org/10.1016/j.rbmo.2014.03.002.
40. Leni Z, Künzi L, Geiser M. Air pollution causing oxidative stress. Curr Opin Toxicol. 2020;20–21:1–8. https://doi.org/10.1016/j.cotox.2020.02.006.
41. Mayr A, Klambauer G, Unterthiner T, Hochreiter S. DeepTox: toxicity prediction using deep learning. Front Environ Sci. 2016. https://doi.org/10.3389/fenvs.2015.00080.
42. Hirohara M, Saito Y, Koda Y, Sato K, Sakakibara Y. Convolutional neural network based on SMILES representation of compounds for detecting chemical motif. BMC Bioinf. 2018;19(Suppl 19):526. https://doi.org/10.1186/s12859-018-2523-5.
43. Wills LP, Beeson GC, Trager RE, Lindsey CC, Beeson CC, Peterson YK, et al. High-throughput respirometric assay identifies predictive toxicophore of mitochondrial injury. Toxicol Appl Pharmacol. 2013;272:490–502. https://doi.org/10.1016/j.taap.2013.06.014.
44. Murphy MP, Hartley RC. Mitochondria as a therapeutic target for common pathologies. Nat Rev Drug Discov. 2018;17:865–86. https://doi.org/10.1038/nrd.2018.174.

Publisher's Note

Shu *et al. BMC Bioinformatics* (2021) 22:371
https://doi.org/10.1186/s12859-021-04287-1

BMC Bioinformatics

METHODOLOGY

Open Access

MSCFS: inferring circRNA functional similarity based on multiple data sources

Liang Shu[1], Cheng Zhou[1], Xinxu Yuan[2], Jingpu Zhang[3] and Lei Deng[1*]

From The 19th Asia Pacific Bioinformatics Conference (APBC 2021) Tainan, Taiwan. 3-5 February 2021

*Correspondence: leideng@csu.edu.cn
[1] School of Computer Science and Engineering, Central South University, Lushangnan Road, Changsha, China
Full list of author information is available at the end of the article

Abstract

Background: More and more evidence shows that circRNA plays an important role in various biological processes and human health. Therefore, inferring the circRNA's potential functions and obtaining circRNA functional similarity has become more and more significant. However, there is no effective approach to explore the functional similarity of circRNAs.

Methods: In this paper, we propose a new approach, called MSCFS, to calculate the functional similarity of circRNA by integrating multiple data sources. We combine circRNA-disease association, circRNA-gene-Gene Ontology association, and circRNA sequence information to explore the functional similarity of circRNA. Firstly, we employ different learning representation methods from three data sources to establish three circRNA functional similarity networks. Then we integrate the three networks to obtain the final circRNA functional similarity.

Results: We utilize circRNA–miRNA association similarity and circRNA co-expression similarity to evaluate the performance of MSCFS. The results show a positive correlation with miRNA association ($R = 0.213$) and circRNA co-expression similarity ($R = 0.8991$). Finally, we construct a circRNA functional similarity network and perform case analysis. The result shows our method can be applied to infer new potential functions of circRNA and other associations.

Conclusions: MSCFS combines multiple data sources related to circRNA functions. Correlation analysis and case analyses prove that MSCFS is a useful method to explore circRNA functional similarity.

Keywords: CircRNA functional similarity, Multiple data sources, Multiple representations

Background

Circular RNAs, a class of endogenous non-coding RNAs, are characterized by their covalently closed-loop structures without a $5'$ cap or a $3'$ Poly A tail [1]. Sanger et al. [2] first found CircRNAs in 1976. However, the circRNAs were thought to be splicing artifact; and were continuously considered as "junk" RNAs for about two decades [3]. More and more researches have corroborated that circRNAs play an essential role in many cell activities, affecting arteriosclerosis and participating in mRNA expression variable splicing regulation [4–9]. In recent years, circRNAs have been identified as biomarkers and therapeutic targets for various acute diseases. CircRNA is associated with a variety of chronic diseases, such as lung cancer, Alzheimer's disease, diabetes, cardiovascular disease and other [10, 11]. The emerging experimental results have corroborated that circRNA molecules have abundant miRNA binding sites and act as miRNA sponges in cells to releasing the inhibitory effect of miRNA on their target genes and improving the expression level of target genes [12–14]. Identifying the targets of circRNAs helps to understand the functions of circRNAs. Several efforts have been developed to identify circRNA targets [15–17]. For example, Lin et al. [17] designed Analysis of common targets (ACT) to facilitate the identification of potential circRNA targets.

Functional similarity can be defined as an association, such as co-expression similarity, co-Gene Ontology (GO) term similarity, co-similar disease similarity, and co-literature similarity [18]. Analogous to the methods of studying the functional similarity of microRNA, Wang et al. [19] obtained the functional similarity of miRNA through the DAG of disease and microRNA-disease association. The Gene Ontology (GO) project provides the most comprehensive resource currently available for computable knowledge regarding the functions of genes and gene products. Gene Ontology provides the logical structure of the biological functions ('terms') and their relationships to one another, manifested as a directed acyclic graph [20]. Yang et al. [21] obtained the functional similarity of microRNA by calculating GO semantic similarity and miRNA-GO association. There are many methods to acquire the semantic similarity of GO, such as the measures proposed by Resnik et al. [22], Jiang et al. [23], Lin et al. [24], Wang et al. [25], and Wu et al. [26]. Obtaining the functional similarity of RNA can also be obtained through sequence information. Sequence similarity can be calculated by methods such as K-mer [27] or LSTM [28].

However, there is no valid method to calculate the functional similarity of circRNA, and a single circRNA data source can't effectively explore the circRNA functional similarity. In this paper, we propose a novel method called MSCFS by integrating multiple biological data sources to calculate the functional similarity between circRNAs. Firstly, we obtain the circRNA functional similarity matrix by using the DAG graph and association information of the disease. Secondly, we construct the corpus through circRNA-gene-GO associations and GO annotations and employ word2vec to obtain the circRNA functional similarity matrix. Thirdly, we adopt chaos game representation to get circRNA functional similarity by circRNA sequence information. Finally, the circRNA functional similarity is obtained by integrating the three networks. The results show that MSCFS is efficacious and accurate, and it can infer the potential functions of circRNA. The flowchart of our proposed model is shown in Fig. 1.

Methods

Dataset

We downloaded the MeSH descriptor from the National Library of Medicine (http://www.nlm.nih.gov/) [29]. MeSH descriptors are divided into 16 categories: category A is anatomical terms, category B is organisms, category C is diseases, category D

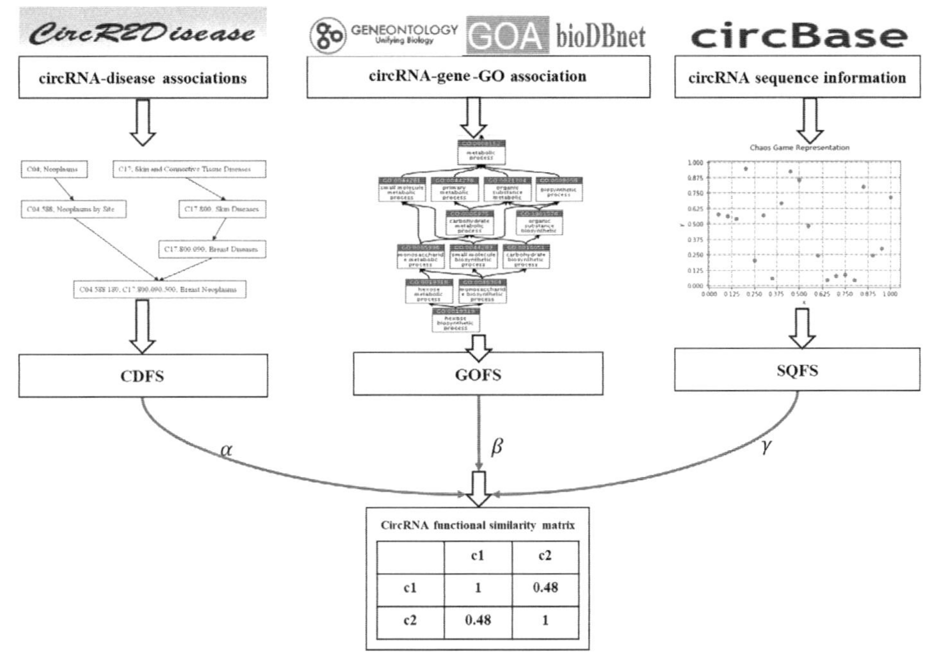

Fig. 1 Overview of MSCFS demonstrating the basic ideas of measuring circRNAs functional similarity

is drugs and chemicals, etc. Then, we obtained the relationship of various diseases based on DAG diseases from the MeSH descriptor of category C.

Many benchmark databases contain circRNA-disease association data, such as circR2Disease [30], circRNADisease [31], circFunBase [32], and Circ2Disease [33], which contain experimentally verified associations between circRNAs and diseases. We utilize circR2Disease as the benchmark data set. Circ2Disease is a database that can manually manage human circRNA supported by experiments and provide the association between circRNA and human diseases. We obtained 418 confirmed circRNA-disease associations consisting of 365 circRNA and 71 diseases after removing the circRNAs in which the gene symbol could not be found.

We downloaded the Gene Ontology (GO) in OWL format from the Gene Ontology Consortium (GOC) [34] and GO annotations in the Gene Ontology Annotation (GOA) Database [35]. We used the OWL API version 4.2.6 to process the GO in OWL format.

We extracted 321 genes associated with circRNA and the circRNA sequence information from the circBase [36]. We obtained 7321 GO-gene associations from multiple versions of the database bioDBnet [37].

Overview of MSCFS

In this article, we combine the three data sources of circRNA to calculate the functional similarity of circRNA. Specifically, we obtain three circRNA functional similarity matrices from the circRNA-disease association, circRNA-gene-GO association, and circRNA sequence information. Finally, we integrate three networks to obtain the final circRNAs functional similarity, and the formula is as follows:

$$\begin{cases} \mathrm{FS} = \alpha * \mathrm{CDFS} + \beta * \mathrm{GOFS} + \gamma * \mathrm{SQFS} \\ \alpha + \beta + \gamma = 1 \end{cases} \tag{1}$$

where CDFS, GOFS, SQFS are circRNA functional similarity matrices obtained through circRNA-disease association, circRNA-gene-GO association, and circRNA sequence information, respectively. α, β, and γ are the weighting coefficients of the three networks severally.

Functional similarity based on circRNA-disease association

Genes with similar functions are known to be associated with similar diseases. A structure of a directed acyclic graph (DAG) can represent the relationship between different diseases. Therefore, we can calculate the functional similarity of circRNA through circRNA-disease association. The process is shown in Fig. 2.

In the MeSH database, the relationship between diseases is described in the form of a directed acyclic graph (*DAG*), where nodes represent diseases and edges represent relationships between diseases. Given a disease *D*, we have defined a DAG graph $DAG_D = (D, T_a, E_a)$ based on the other diseases it is associated with and related edges, where T_a is the set of ancestor nodes containing itself, and E_a is the set of corresponding edges connecting these diseases. If disease *d* is in the *DAG*, its contribution to disease *A* can be calculated as follows:

$$\begin{cases} D_D(D) = 1 \\ D_D(d) = \max\left\{\Delta * D_A\left(d'\right) \mid d' \in children\ of\ d\right\}\ if\ d \neq A \end{cases} \tag{2}$$

where Δ is the semantic contribution factor of disease d and its child nodes. In DAG, the semantic value of disease D itself is defined as 1. Therefore, through the following formula, we calculate the semantic value *DV*(*D*) of disease *D*:

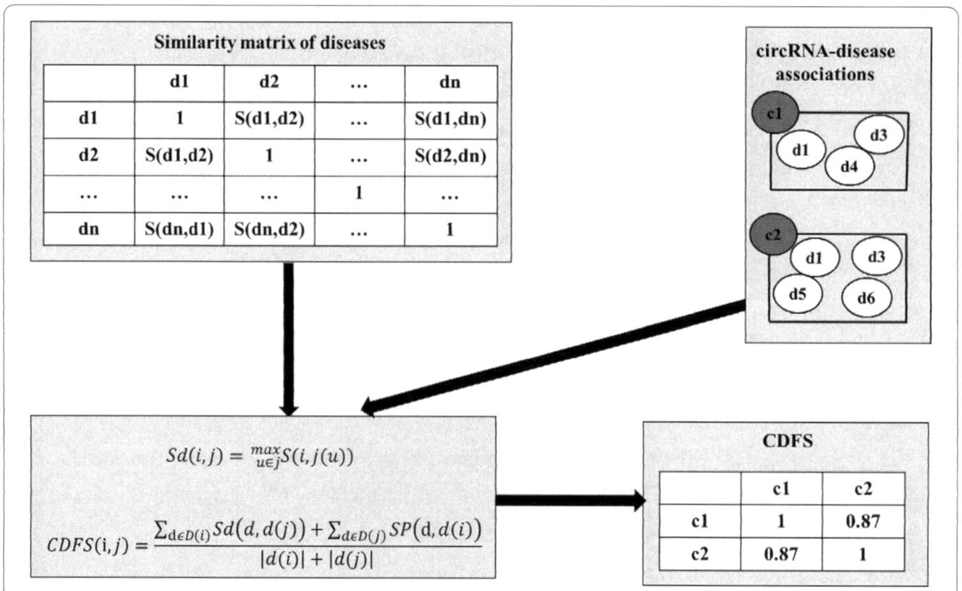

Fig. 2 Calculating circRNA functional similarity based on circRNA-disease association

$$\mathrm{DV(D)} = \sum_{d \in T_A} D_A(d). \tag{3}$$

Here, we assume that the more *DAG* shared parts of the two diseases, the higher the semantic similarity, so according to the position of the two diseases in the *DAG* graph and the semantic relationship with the ancestral diseases, the formula for calculating the semantic similarity of the two diseases *M* and *N* is as follows:

$$S(M,N) = \frac{\sum_{d \in T_M \cap T_N} (D_M(d) + D_N(d))}{DV(M) + DV(N)}, \tag{4}$$

where $D_M(d)$ is the semantic value of disease d related to disease *M*, and $D_N(d)$ is the semantic value of disease d related to disease *N*.

$$Sd(d, DT) = \max_{u \in DT} S(d, DT(u)), \tag{5}$$

where *DT* is a group of diseases, *u* is any disease in *DT*. After obtaining the semantic similarity of disease combinations, we use circRNA-disease correlation to obtain the functional similarity of circRNA, CDFS.

$$CDFS(i,j) = \frac{\sum_{d \in D(i)} Sd(d, d(j)) + \sum_{d \in D(j)} Sd(d, d(i))}{|D(i)| + |D(j)|}, \tag{6}$$

where *CDFS*(*i*, *j*) is the similarity between the *i*th circRNA and the *j*th circRNA, *D*(*i*) is the *i*th circRNA associated disease set.

Functional similarity based on circRNA-gene-GO association

Onto2Vec is a measure that combines formal ontology axioms and annotation axioms in ontology metadata to generate a vector representation of biological entities in the ontology [38]. Gene Ontology contains the representation of the essence of the knowledge system in the field of biology. Ontologies are usually composed of a set of categories (or terms or concepts) with relationships between them. In order to explore the functional similarity of circRNA, we use the circRNA-gene-GO association. We add circRNA as new entities and apply the *has−function* relationship to connect them with their functions to generate a corpus. Then we use Onto2Vec to generate a vector representation for each class (using a corpus to only be based on axioms), and further, generate a joint representation of circRNA and classes (using a corpus-based axiom and circRNA and its annotations).

In the end, we constructed 230,699 corpus, with 50,409 categories, using the Skip-gram model in word2vec. Word2Vec is a set of neural network-based tools that can generate vector representations of words from a large corpus. There are two models: the continuous bag of word (CBOW), which uses a context to predict a target word, and the Skip-gram model that tries to maximize the classification of a word based on another word from the same sentence. Figure 3 shows the flow of this section.

The Skip-gram model is chosen because the Skip-gram model generates higher quality rare word representations in the corpus. The Skip-gram model learns more detailed word vectors and has a large number of low-frequency words in the corpus

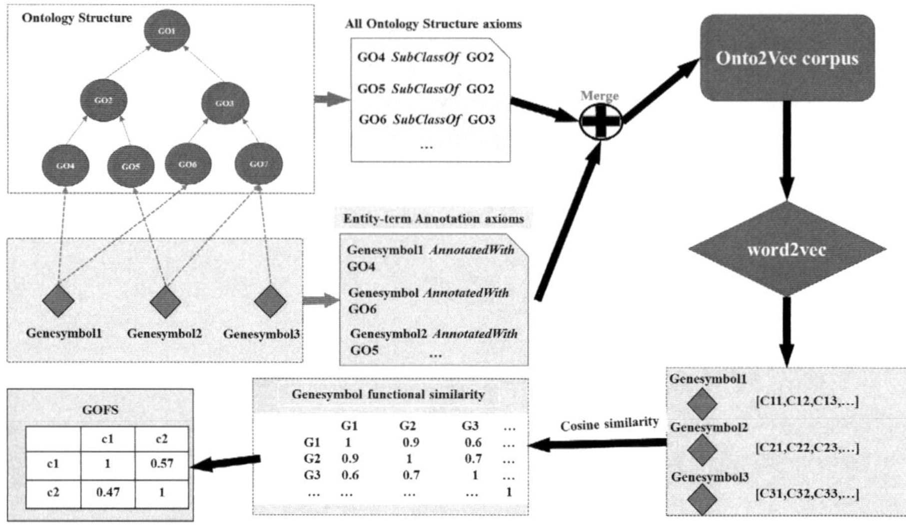

Fig. 3 Using word2vec to get circRNA functional similarity based on circRNA-gene-GO association and GO annotations

Table 1 Parameters of Word2Vec model

Parameter	Definition	Default value
sg	Choice of training algorithm (sg = 1:skip gram; sg = 0:CBOW)	1
min_count	Words with frequency lower than this value will be ignored	1
size	Dimension of the obtained vectors	200
window	Maximum distance between the current and the predicted word	10
iter	Number of iterations	5
negative	Whether negative sampling will be used and how many 'noise words' would be drawn	4

to produce high-quality representations of all biological entities occurring in our large corpus, including uncommon ones. Given a set of training word sequences $w_1, w_2 \ldots, w_N$, Skip-gram the goal is to maximize the following average logarithmic likelihood values:

$$\frac{1}{N}\sum_{t=1}^{N}\sum_{-s\leq i\leq s, j\neq 0}\log p(\omega_{t+j} \mid \omega_t), \tag{7}$$

where s means the size of the training context, N means the size of the set of the training words, and w_i is the ith training word in the sequence. In our research, the parameters of word2vec used are shown in Table 1.

Through the training, we get the similarity of the genes, and then through the circRNA-gene relationship GOFS, the calculation formula is as follows:

$$\mathrm{GOFS}(i,j) = \frac{\sum_{g\in G(i)} S(g, g(j)) + \sum_{g\in G(j)} S(g, g(i))}{|g(i)| + |g(j)|}, \tag{8}$$

where $g(i)$ is the set of genes associated with the ith circRNA, and $g(j)$ is the set of genes associated with the jth circRNA.

Functional similarity based on circRNA sequence

Different from the K-mer [27], PSSM method [39], chaos game representation [40] combines position information and nonlinear relationship to obtain vector representation of sequences. Finally, the Pearson correlation is used to quantify their correlation. The advantage of the algorithm is that the original information of the sequence is completely restored in the coordinate system, and the information will not be lost in the mapping. Secondly, the position information will be retained as a mapping. Figure 4 shows the workflow of this section.

The position of each nucleotide in the plane:

$$P_i = 0.5 * (P_{i-1} + S_i) \quad i = 1 \ldots N, \tag{9}$$

where P_0 is any given starting point ($P_0 = (0.5, 0.5)$), N represents the length of the sequence. S_i represents the ith nucleotide in the sequence, which corresponds to the fixed vertex coordinates of $A = (0,0), C = (1,0), G = (1,1)$ and $U = (0,1)$ respectively.

In this way, the CGR graph is transformed into a N_g grid $\left(N_g = 2^s \times 2^s, \mathrm{s} = 3\right)$ digital matrix, which is called the frequency matrix of CGR graph (FCGR). And grid can be represented as follows:

$$grid_i = (X_i, Y_i, Z_i) \tag{10}$$

We use the x-axis, y-axis direction and their digital features to construct the feature vector of the sequence, the calculation formula is as follows: the abscissa point.x

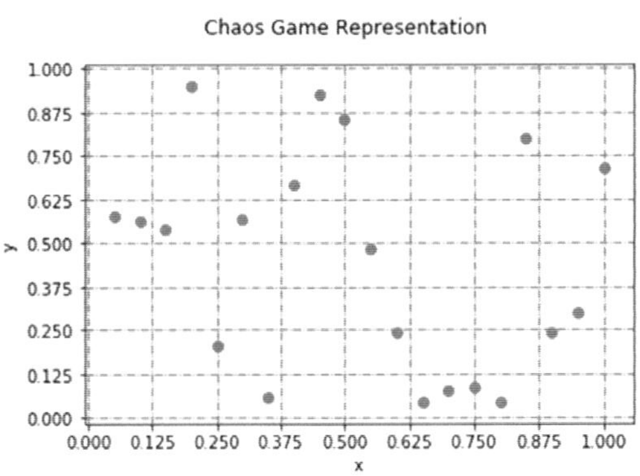

$$X_i = \sum point.x\,, if\ points\ in\ grid_i$$

$$Y_i = \sum point.y\,, if\ points\ in\ grid_i$$

$$Num_i = number\ of\ points\ in\ grid_i$$

$$Z_i = \frac{Num_i - \frac{\sum_{m=1}^{N_g} Num_m}{N_g}}{\sqrt{\frac{1}{N_g}\sum_{k=1}^{N_g}\left(Num_k - \frac{\sum_{n=1}^{N_g} Num_n}{N_g}\right)^2}}$$

$$grid_i = (X_i, Y_i, Z_i)$$

$$vector(c(i)) = (grid_1, grid_2, \ldots, grid_{N_g})$$

Cosine similarity

SQFS

	c1	c2
c1	1	0.37
c2	0.37	1

Fig. 4 Utilizing Chaos game representation to obtain circRNA functional similarity though circRNA sequence information

and ordinate point.y in each grid are accumulated respectively to quantify position information.

$$\begin{cases} X_i = \sum point.x & if\ points\ in\ grid_i \\ Y_i = \sum point.y & if\ points\ in\ grid_i \end{cases}. \tag{11}$$

Then, we obtain the z-scores of each grid Z_i to quantify potential features.

$$\begin{cases} Z_i = \dfrac{Num_i - \frac{\sum_{m=1}^{Ng} Num_m}{Ng}}{\sqrt{\frac{1}{Ng}\sum_{k=1}^{Ng}\left(Num_k - \frac{\sum_{n=1}^{Ng} Num_n}{Ng}\right)^2}} \\ Num_i = number\ of\ points\ in\ grid_i \end{cases} \tag{12}$$

Finally, each grid can be represented as three attributes, and we fused the attributes to construct the vectors *vector*(*c*(*i*)) to define the sequence functional similarity of circRNAs *SQFS*(*c*(*i*), *c*(*j*)) by Pearson correlation coefficient. Where *c*(*i*) represents the *A* *i*th cricRNA.

$$\begin{cases} SQFS(c(i), c(j)) = \frac{vector(c(i)) \cdot vector(c(j))}{\|vector(c(i))\| \|vector(c(j))\|} \\ vevtor(c(i)) = \left(grid_1, grid_2, \ldots, grid_{N_g}\right) \end{cases}, \tag{13}$$

where *vector*(*c*(*i*)) means the sequence feature vector of the *i*th circRNA, $vector(c(i)) \cdot vector(c(i))$ is the dot product of *vector*(*c*(*i*)) and *vector*(*c*(*i*)).

Results

Parameters selection

circRNA can be used as the sponge of microRNA to play a role in biological processes [12]. We downloaded circRNA–miRNA association data from the starbase [41]. There are 267 types of circRNAs that match the 365 types of circRNA we calculated. We obtained 267 pairs of circRNA–miRNA association similarities through the Jaccard similarity method, which were compared with the functional similarities we calculated. Denote *CMS* as the circRNA–miRNA similarity matrix, and its entry *CMS*(*i*, *j*) can be obtained by the following formula:

$$CMS(i,j) = \frac{|CM_i \cap CM_j|}{|CM_i \cup CM_j|}, \tag{14}$$

where CM_i is the set of microRNAs associated with the *i*th circRNA, and CM_j is the set of microRNAs associated with the *j*th circRNA.

We set the parameter step size to 0.1. Because circRNA has little data associated with diseases, we set α the value range from 0 to 0.2, and the value range for β and γ from 0 to 1.We used the grid search method to obtain the optimal parameters through 30 sets of experiments and selected two groups for display. The results are shown in Fig. 5. The experimental results with parameters of 0.1, 0.4, 0.5 and 0.1, 0.3, 0.6 are $(R = 0.205, P = 8.2e^{-3})$, $(R = 0.213, P = 4.6e^{-4})$. The results prove that our circRNA functional similarity is related to microRNA similarity. We selected the optimal value of

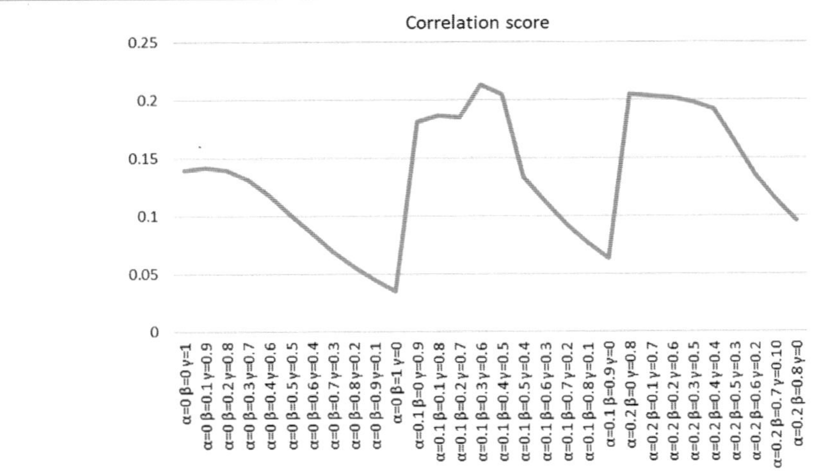

Fig. 5 Parameters selection

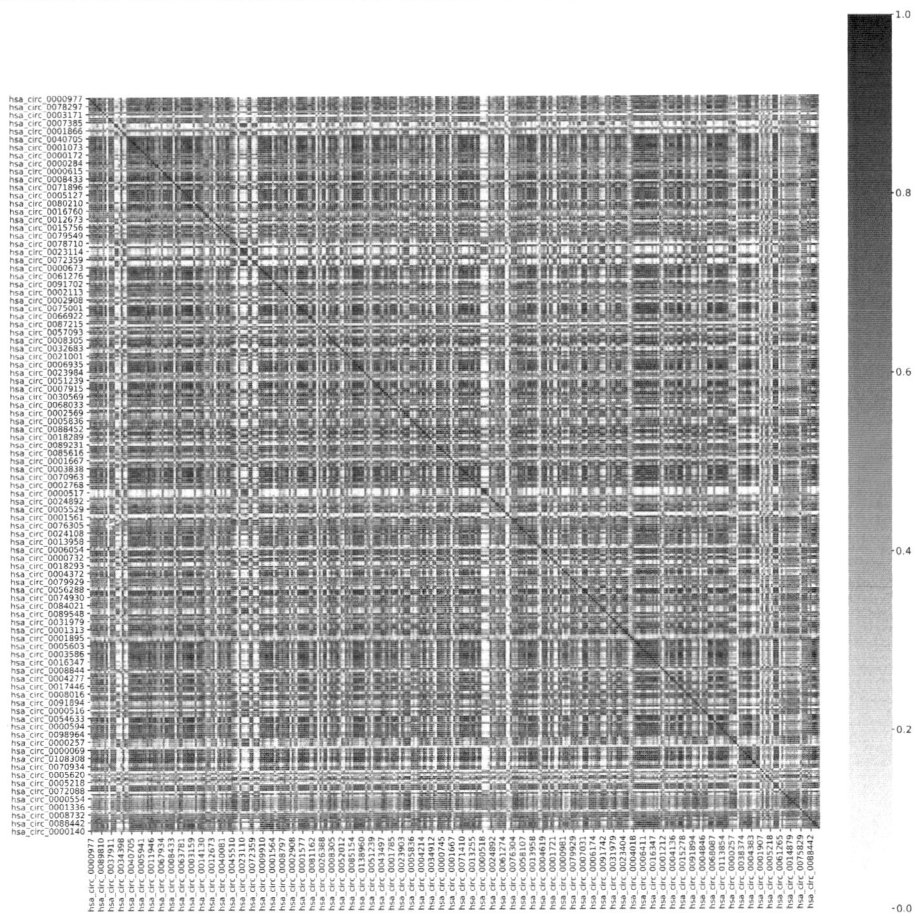

Fig. 6 CircRNA functional similarity heatmap shows the functional similarity of 365 kinds of circRNAs

the parameter α, β, γ are 0.1, 0.3, 0.6. Figure 6 shows the heat map of circRNA functional similarity calculated with parameters of 0.1, 0.3 and 0.6.

CircRNA functional similarity is correlated with expression similarity

CircRNAs with semblable functions squint towards participate in semblable biological processes and interact with semblable cellular components. To verify circRNAs with similar functions may squint towards having similar expression profiles, we seeked the relationship between circRNA functional similarity calculated by MSCFS and expression similarity. In this study, we used the absolute Pearson's correlation coefficient (PCC) to measure circRNA expression similarity.

We finally obtain circRNA expression profiling data from Peng et al.'s work [42], which consists of expression profiles of 2932 circRNAs. Then, we calculated the conducted a comprehensive analysis of circRNA expression in papillary thyroid carcinoma PCC score as the co-expression similarity of each pair of circRNA expression profiles and obtained the co-expression similarity of 8049 pairs of circRNAs, and then performed the correlation analysis of circRNA expression similarity and circRNA functional similarity. As a result, the functional similarity of circRNA confirmed positive correlation with circRNA co-expression similarity ($R = 0.076, P = 9.05e^{-4}$, Pearson correlation). We grouped 8049 pairs of circRNAs into different groups according to functional similarity in steps of 0.1 and calculated the average expression similarity and functional similarity of each group. Clearly, the functional similarity of circRNA is positively correlated with the expression similarity ($R = 0.8991, P = 9.73e^{-4}$, Fig. 7). Results inform that circRNA functional similarity obtained by our method is correlated with circRNA expression similarity, which is well known to be associated with circRNA functional similarity.

A circRNA functional similarity network

Figure 8 shows the distribution of circRNA functional similarity scores. We have structured a partial graph of the circRNA network with a threshold of 0.7 (Fig. 9). Some circRNAs are less associated with other circRNAs, while some circRNAs are more associated with other circRNAs. We can do more research on those circRNAs that are more related and explore the potential functions of circRNAs related to them.

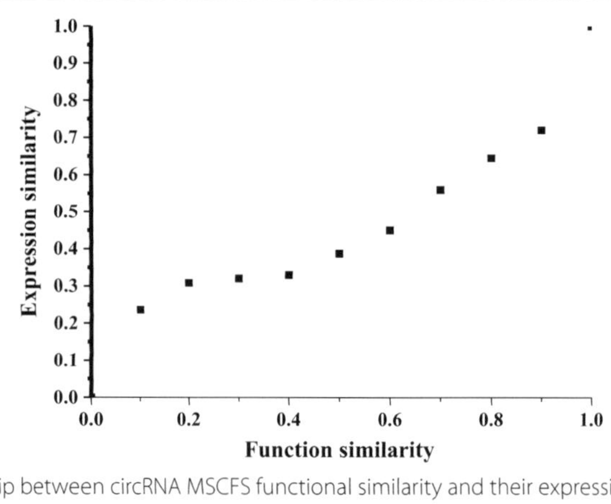

Fig. 7 The relationship between circRNA MSCFS functional similarity and their expression similarity

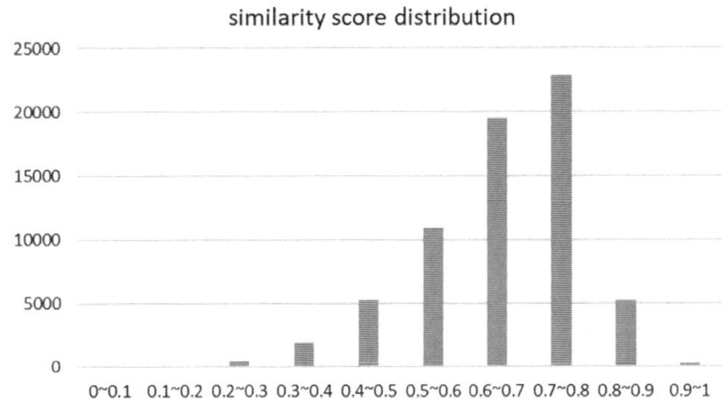

Fig. 8 The circRNA funtional similarity score distribution

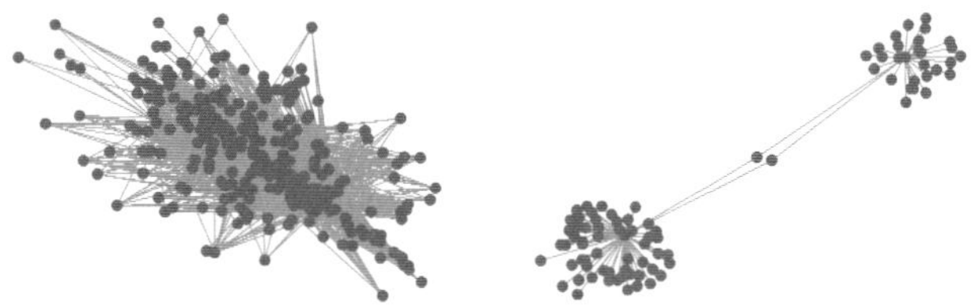

Fig. 9 A circRNA functional network based on the functional similarity of MSCFS. Each node and edge connecting any two nodes (circRNAs) indicates that the functional similarity of the two circRNAs is equal to or greater than the similarity cutoff value

Case study

To verify our results, we executed case analysis on the circRNA function annotation in the CircFunBase database. CircFunBase is a web-accessible database that aims to provide a high-quality functional circRNA resource, including experimentally validated and computationally predicted functions [32].

Amongst the 365 kinds of circRNAs, hsa_circ_0000140 has the highest correlation score with hsa_circ_0001946 in other 364 circRNAs. In the CircFunBase database, the functional annotations of these two circRNAs are related to gastric cancer. Then we select circRNA pairs with high similarity scores for analysis. For example, the functional similarity score of hsa_circ_0043278 and hsa_circ_0006220 is 0.82, and it can be found in the CircFunBase database that both are related to hypertension [differential expression (hypertensive patients and healthy controls)]. The functional similarity score of hsa_circ_0005927 and hsa_circ_0138960 is 0.83, both of which are related to gastric cancer. Through case analysis, we can know the practicality and accuracy of the MSCFS method.

Discussion and conclusion

CircRNAs have peculiar biological structures and have proven to play essential roles in biological processes and human health. Inferring the functional similarity of circRNAs can help analyze the function of circRNAs and predict the association of circRNA-disease. However, due to the lack of functional annotations for circRNAs in public databases, it is not straightforward to calculate the functional similarity of circRNAs using existing single data sources.

This paper proposes a new algorithm, MSCFS, to calculate the functional similarity of circRNA by integrating multiple circRNA associated biological data. The results showed that the circRNAs associated with the same miRNA have a high similarity score. The circRNA co-expression similarity was positively correlated with our calculated results. We also found that circRNAs with a high similarity score were also similar in the function of the disease. By calculating the functional similarity of circRNAs, we can explore more potential functions and associations of circRNAs.

We have integrated multiple biological data sources related to circRNAs, and the results will be somewhat biased due to the quality of some data sources. In the future, we will integrate more and more reliable data to further improve the accuracy of circRNA functional similarity calculations.

Abbreviations
circRNAs: Circular RNAs; GO: Gene Ontology; RBPs: RNA binding proteins; DAG: Directed acyclic graphs; PSSM: Position specific scoring matrix; CMS: CircRNA–miRNA similarity matrix; PCC: Pearson's correlation coefficient.

Acknowledgements
The authors are very grateful to the anonymous reviewers for their constructive comments which have helped significantly in revising this work. We would like to thank the Experimental Center of School of Computer Science and Engineering of Central South University, for providing computing resources.

About this supplement
This article has been published as part of BMC Bioinformatics Volume 22 Supplement 10 2021: Selected articles from the 19th Asia Pacific Bioinformatics Conference (APBC 2021): bioinformatics. The full contents of the supplement are available athttps://bmcbioinformatics.biomedcentral.com/articles/supplements/volume-22-supplement-10.

Authors' contributions
LS, CZ, XY, LD and JP designed the study and conducted experiments. LS, CZ, LD and JP performed statistical analyses. LS, XY and JP drafted the manuscript. JP and LD prepared the experimental materials and benchmarks. All authors have read and approved the final manuscript.

Funding
This work was supported by National Natural Science Foundation of China under Grant No. 61972422. The funding body has not played any roles in the design of the study and collection, analysis and interpretation of data in writing the manuscript.

Availability of data and materials
The Python source codes and the datasets in this work are freely available in the GitHub (https://github.com/CJNabla/MultiSourcCFS).

Declarations

Ethics approval and consent to participate
Not applicable.

Consent for publication
Not applicable.

Competing interests
The authors declare that they have no competing interests.

Author details
[1]School of Computer Science and Engineering, Central South University, Lushangnan Road, Changsha, China. [2]Department of Chemical and Life Science Engineering, Virginia Commonwealth University, Richmond, VA 23284, USA. [3]School of Computer and Data Science, Henan University of Urban Construction, Longxiang Road, Pingdingshan 467000, China.

Received: 20 June 2021 Accepted: 6 July 2021
Published online: 16 July 2021

References

1. Meng S, Zhou H, Feng Z, Xu Z, Tang Y, Li P, Wu M. CircRNA: functions and properties of a novel potential biomarker for cancer. Mol Cancer. 2017;16(1):1–8.
2. Sanger HL, Klotz G, Riesner D, Gross HJ, Kleinschmidt AK. Viroids are single-stranded covalently closed circular RNA molecules existing as highly base-paired rod-like structures. Proc Natl Acad Sci. 1976;73(11):3852–6.
3. Zhang Z, Yang T, Xiao J. Circular RNAs: promising biomarkers for human diseases. EBioMedicine. 2018;34:267–74.
4. Du WW, Fang L, Yang W, Wu N, Awan FM, Yang Z, Yang BB. Induction of tumor apoptosis through a circular RNA enhancing Foxo3 activity. Cell Death Differ. 2017;24(2):357–70.
5. Armakola M, Higgins MJ, Figley MD, Barmada SJ, Scarborough EA, Diaz Z, Fang X, Shorter J, Krogan NJ, Finkbeiner S, et al. Inhibition of RNA lariat debranching enzyme suppresses TDP-43 toxicity in ALS disease models. Nat Genet. 2012;44(12):1302.
6. Li Z, Huang C, Bao C, Chen L, Lin M, Wang X, Zhong G, Yu B, Hu W, Dai L, et al. Exon-intron circular RNAs regulate transcription in the nucleus. Nat Struct Mol Biol. 2015;22(3):256.
7. Zhang Y, Zhang X-O, Chen T, Xiang J-F, Yin Q-F, Xing Y-H, Zhu S, Yang L, Chen L-L. Circular intronic long noncoding RNAs. Mol Cell. 2013;51(6):792–806.
8. Xu H, Guo S, Li W, Yu P. The circular RNA Cdr1as, via miR-7 and its targets, regulates insulin transcription and secretion in islet cells. Sci Rep. 2015;5(1):1–12.
9. Li F, Zhang L, Li W, Deng J, Zheng J, An M, Lu J, Zhou Y. Circular RNA ITCH has inhibitory effect on ESCC by suppressing the WNT/β-catenin pathway. Oncotarget. 2015;6(8):6001.
10. Lukiw W. Circular RNA (circRNA) in Alzheimer's disease (AD). Front Genet. 2013;4:307.
11. Greene J, Baird A-M, Brady L, Lim M, Gray SG, McDermott R, Finn SP. Circular RNAs: biogenesis, function and role in human diseases. Front Mol Biosci. 2017;4:38.
12. Hansen TB, Jensen TI, Clausen BH, Bramsen JB, Finsen B, Damgaard CK, Kjems J. Natural RNA circles function as efficient microRNA sponges. Nature. 2013;495(7441):384–8.
13. Thomas LF, Sætrom P. Circular RNAs are depleted of polymorphisms at microRNA binding sites. Bioinformatics. 2014;30(16):2243–6.
14. Kulcheski FR, Christoff AP, Margis R. Circular RNAs are miRNA sponges and can be used as a new class of biomarker. J Biotechnol. 2016;238:42–51.
15. Dudekula DB, Panda AC, Grammatikakis I, De S, Abdelmohsen K, Gorospe M. CircInteractome: a web tool for exploring circular RNAs and their interacting proteins and microRNAs. RNA Biol. 2016;13(1):34–42.
16. Dori M, Bicciato S. Integration of bioinformatic predictions and experimental data to identify circRNA–miRNA associations. Genes. 2019;10(9):642.
17. Lin Y-C, Lee Y-C, Chang K-L, Hsiao K-Y. Analysis of common targets for circular RNAs. BMC Bioinform. 2019;20(1):372.
18. Li J, Zhang S, Wan Y, Zhao Y, Shi J, Zhou Y, Cui Q. MISIM v2.0: a web server for inferring microRNA functional similarity based on microRNA-disease associations. Nucleic Acids Res. 2019;47(W1):536–41.
19. Wang D, Wang J, Lu M, Song F, Cui Q. Inferring the human microRNA functional similarity and functional network based on microRNA-associated diseases. Bioinformatics. 2010;26(13):1644–50.
20. Ashburner M, Ball CA, Blake JA, Botstein D, Butler H, Cherry JM, Davis AP, Dolinski K, Dwight SS, Eppig JT, et al. Gene ontology: tool for the unification of biology. Nat Genet. 2000;25(1):25–9.
21. Yang Y, Fu X, Qu W, Xiao Y, Shen H-B. MiRGOFS: a GO-based functional similarity measurement for miRNAs, with applications to the prediction of miRNA subcellular localization and miRNA-disease association. Bioinformatics. 2018;34(20):3547–56.
22. Resnik P. Semantic similarity in a taxonomy: an information-based measure and its application to problems of ambiguity in natural language. J Artif Intell Res. 1999;11:95–130.
23. Jiang JJ, Conrath DW. Semantic similarity based on corpus statistics and lexical taxonomy. arXiv preprint arXiv:cmp-lg/9709008. 1997.
24. Lin D, et al. An information-theoretic definition of similarity. Icml. 1998;98:296–304.
25. Wang JZ, Du Z, Payattakool R, Yu PS, Chen C-F. A new method to measure the semantic similarity of GO terms. Bioinformatics. 2007;23(10):1274–81.
26. Wu H, Su Z, Mao F, Olman V, Xu Y. Prediction of functional modules based on comparative genome analysis and gene ontology application. Nucleic Acids Res. 2005;33(9):2822–37.
27. Fletez-Brant C, Lee D, McCallion AS, Beer MA. kmer-SVM: a web server for identifying predictive regulatory sequence features in genomic data sets. Nucleic Acids Res. 2013;41(W1):544–56.
28. Hochreiter S, Schmidhuber J. Long short-term memory. Neural Comput. 1997;9(8):1735–80.
29. Lord PW, Stevens RD, Brass A, Goble CA. Investigating semantic similarity measures across the gene ontology: the relationship between sequence and annotation. Bioinformatics. 2003;19(10):1275–83.
30. Fan C, Lei X, Fang Z, Jiang Q, Wu F-X. CircR2Disease: a manually curated database for experimentally supported circular RNAs associated with various diseases. Database. 2018;2018:bay044.
31. Zhao Z, Wang K, Wu F, Wang W, Zhang K, Hu H, Liu Y, Jiang T. circRNA disease: a manually curated database of experimentally supported circRNA-disease associations. Cell Death Dis. 2018;9(5):1–2.
32. Meng X, Hu D, Zhang P, Chen Q, Chen M. CircFunBase: a database for functional circular RNAs. Database. 2019;2019:baz003.
33. Yao D, Zhang L, Zheng M, Sun X, Lu Y, Liu P. Circ2Disease: a manually curated database of experimentally validated circRNAs in human disease. Sci Rep. 2018;8(1):1–6.
34. Consortium GO. Gene ontology consortium: going forward. Nucleic Acids Res. 2015;43(D1):1049–56.
35. Camon E, Magrane M, Barrell D, Lee V, Dimmer E, Maslen J, Binns D, Harte N, Lopez R, Apweiler R. The gene ontology annotation (GOA) database: sharing knowledge in uniprot with gene ontology. Nucleic Acids Res. 2004;32(suppl–1):262–6.
36. Glažar P, Papavasileiou P, Rajewsky N. circBase: a database for circular RNAs. RNA. 2014;20(11):1666–70.

37. Mudunuri U, Che A, Yi M, Stephens RM. bioDBnet: the biological database network. Bioinformatics. 2009;25(4):555–6.
38. Smaili FZ, Gao X, Hoehndorf R. Onto2vec: joint vector-based representation of biological entities and their ontology-based annotations. Bioinformatics. 2018;34(13):52–60.
39. Kelley LA, MacCallum RM, Sternberg MJ. Enhanced genome annotation using structural profiles in the program 3D-PSSM. J Mole Biol. 2000;299(2):501–22.
40. Jeffrey HJ. Chaos game representation of gene structure. Nucleic Acids Res. 1990;18(8):2163–70.
41. Li J-H, Liu S, Zhou H, Qu L-H, Yang J-H. starBase v2.0: decoding miRNA–ceRNA, miRNA–ncRNA and protein–RNA interaction networks from large-scale CLIP-Seq data. Nucleic acids Res. 2014;42(D1):92–7.
42. Peng N, Shi L, Zhang Q, Hu Y, Wang N, Ye H. Microarray profiling of circular RNAs in human papillary thyroid carcinoma. PLoS One. 2017;12(3):0170287.

Publisher's Note

Springer Nature remains neutral with regard to jurisdictional claims in published maps and institutional affiliations.

Zhang and Zhang *BMC Bioinformatics* (2021) 22:378
https://doi.org/10.1186/s12859-021-04284-4

BMC Bioinformatics

METHODOLOGY Open Access

METAMVGL: a multi-view graph-based metagenomic contig binning algorithm by integrating assembly and paired-end graphs

Zhenmiao Zhang and Lu Zhang*

From The 19th Asia Pacific Bioinformatics Conference (APBC 2021) Tainan, Taiwan. 3-5 February 2021

*Correspondence:
ericluzhang@hkbu.edu.hk
Department of Computer Science, Hong Kong Baptist University, Hong Kong, SAR, China

Abstract

Background: Due to the complexity of microbial communities, de novo assembly on next generation sequencing data is commonly unable to produce complete microbial genomes. Metagenome assembly binning becomes an essential step that could group the fragmented contigs into clusters to represent microbial genomes based on contigs' nucleotide compositions and read depths. These features work well on the long contigs, but are not stable for the short ones. Contigs can be linked by sequence overlap (assembly graph) or by the paired-end reads aligned to them (PE graph), where the linked contigs have high chance to be derived from the same clusters.

Results: We developed METAMVGL, a multi-view graph-based metagenomic contig binning algorithm by integrating both assembly and PE graphs. It could strikingly rescue the short contigs and correct the binning errors from dead ends. METAMVGL learns the two graphs' weights automatically and predicts the contig labels in a uniform multi-view label propagation framework. In experiments, we observed METAMVGL made use of significantly more high-confidence edges from the combined graph and linked dead ends to the main graph. It also outperformed many state-of-the-art contig binning algorithms, including MaxBin2, MetaBAT2, MyCC, CONCOCT, SolidBin and GraphBin on the metagenomic sequencing data from simulation, two mock communities and *Sharon* infant fecal samples.

Conclusions: Our findings demonstrate METAMVGL outstandingly improves the short contig binning and outperforms the other existing contig binning tools on the metagenomic sequencing data from simulation, mock communities and infant fecal samples.

Keywords: Contig binning, Assembly graph, Paired-end graph, Dead end, Multi-view label propagation

Background

During long-term genetic evolution, animals, including humans, have formed complex ecosystems of symbiotic relationships with diverse microbes. The gut microbiome is a community with the highest microbial density in the human body, including thousands of microbial species mixed in varying proportions and constituting a dynamic system. Most gut microbes are difficult to be isolated and cultured in vitro. Metagenomic sequencing is designed to directly sequence a mixture of microbes and explore microbial compositions and abundances by data post-processing.

Due to the paucity of high-quality microbial reference genomes, current pipelines commonly target single genes or species using species-specific markers [1, 2]. But novel microbes would be lost by the alignment-based approaches. Metagenome assembly is a promising strategy to explore the novel species by concatenating the short-reads into contigs. But these contigs could be fragmented and can be only regarded as pieces of the target genomes. Contig binning algorithms provide a supplement to genome assembly that group the contigs into clusters to represent the complete microbial genomes. This strategy has been widely adopted to explore the novel microbes from the human gut metagenomic sequencing data [3–10].

Many state-of-the-art contig binning algorithms have been developed by considering contig nucleotide compositions (tetranucleotide frequencies (TNF), *k*-mer frequencies) and read depths. MaxBin2 [11] uses Expectation–Maximization algorithm to maximize the probability of a contig belonging to the local cluster centers using TNF and read depth. These two types of information are also used in MetaBAT2 [12] to calculate the contig similarities. MetaBAT2 constructs a graph using contigs as vertices and their similarities as the edges' weights, which is further partitioned into subgraphs by applying a modified label propagation algorithm. CONCOCT [13] applies Gaussian mixture models for contig clustering based on *k*-mer frequencies and read depths across multiple samples. Besides considering TNF and read depths, MyCC [14] aggregates the contigs with complementary marker genes by affinity propagation; SolidBin [15] develops a spectral clustering algorithm using taxonomy alignments as must-links between contigs; BMC3C [16] applies codon usage in an ensemble clustering algorithm. All these methods are helpless in labeling short contigs (< 1 kb), because the limited number of nucleotides might lead to unstable TNF distributions and read depths. We observed a majority of the contigs (89.55%, Additional file 1: metaSPAdes assembly of *Sharon* dataset) in the assembly graph were shorter than 1 kb, which would be dropped by most of the existing binning algorithms.

To rescue those short contigs, Mallawaarachchi et al. developed GraphBin [17] to label the short contigs and correct the potential binning errors by employing label propagation on the assembly graph. In principle, the assembly graph should include *s* disconnected subgraphs, each representing one species. In practice, the subgraphs could be linked by the repeat sequences and some contigs are isolated from the main graph (the largest graph component) due to sequencing errors, imbalanced reads coverage, named dead ends. The performance of label propagation heavily relies on the number of edges and label density in the graph. The labels of short contigs would be significantly affected by the dead ends in two ways: (1) contigs are failed to be labeled if the dead end contains no label before propagation (Fig. 1 dead end 1); (2) erroneous

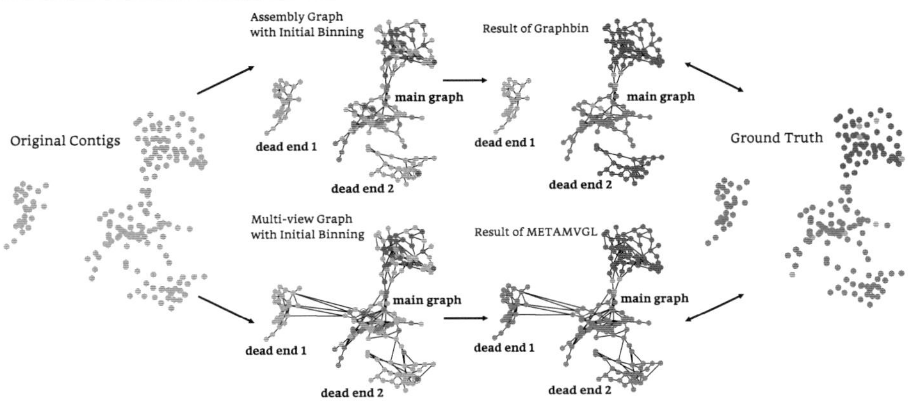

Fig. 1 Visualization of the running process of METAMVGL compared with GraphBin in the simulated data. METAMVGL connected dead end 1 and 2 to the main graph by paired-end reads, also enhanced its connectivity. We observed (1) GraphBin failed to correct the two blue labels in the central of the graph, because it could not remove them before propagation due to lack of connectivity; (2) GraphBin mislabeled all the contigs in the dead end 2, caused by a small number of wrongly labeled contigs in the dead end; (3) METAMVGL labeled all the contigs in the dead end 1 but GraphBin did not

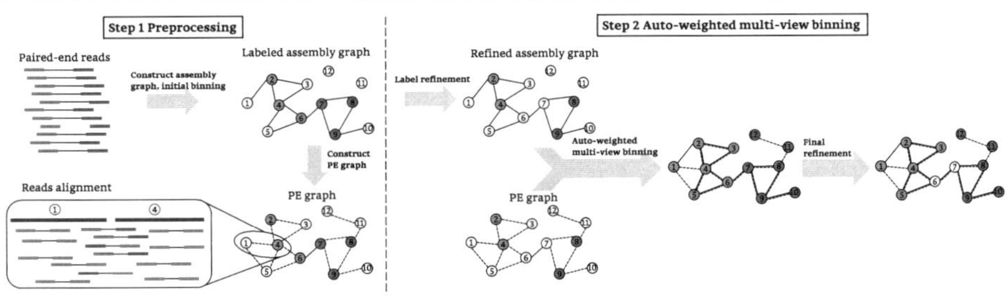

Fig. 2 Workflow of METAMVGL. In step 1, METAMVGL constructs the assembly graph and PE graph by aligning paired-end reads to the contigs. The contigs are initially labeled by the existing binning tools (vertices in orange and blue). In step 2, the ambiguous labels are removed if their neighbors are labeled as belonging to the other binning groups in the assembly graph. METAMVGL applies the auto-weighted multi-view graph-based algorithm to optimize the weights of the two graphs and predict binning groups for the unlabeled contigs. Finally, it performs the second round ambiguous labels removal on the combined graph

labeling happens if only a small proportion of nodes are labeled in the dead end (Fig. 1 dead end 2).

Here we present METAMVGL (Fig. 2), a multi-view graph-based metagenomic contig binning algorithm to address the above-mentioned issues. METAMVGL not only considers the contig sequence overlaps from the assembly graph but also involves the paired-end graph (PE graph), representing the shared paired-end reads between two contigs. The two graphs are integrated together by auto-weighting, where the weights together with the predicted contig labels are updated in a uniform framework [18] (*Methods*). Figure 1 gives a proof-of-concept example on the simulated data, where the paired-end reads connect the two dead ends (dead end 1 and 2) to the main graph. Our experiments indicate METAMVGL substantially improves the binning performance of the state-of-the-art algorithms, including MaxBin2, MetaBAT2, MyCC, CONCOCT, SolidBin and GraphBin in the simulated, mock and *Sharon* datasets (Figs. 3, 4, Additional files 3, 4, 5,

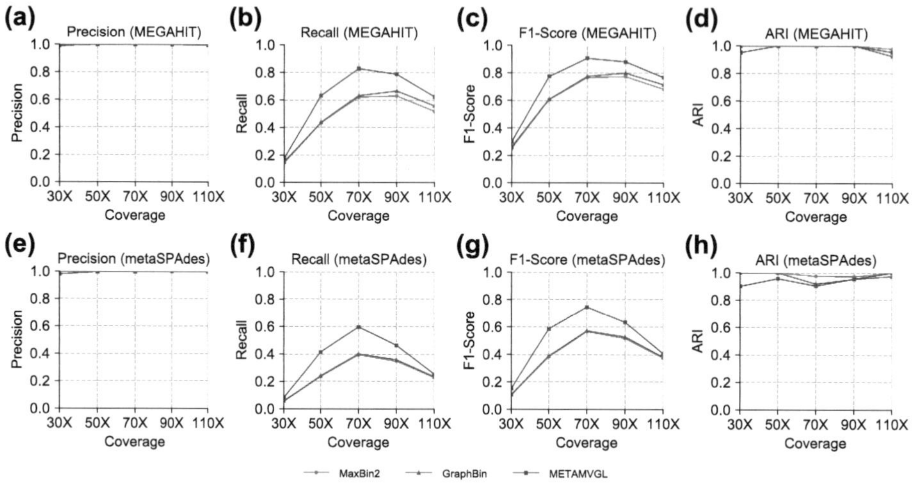

Fig. 3 The performance of MaxBin2, GraphBin and METAMVGL on the simulated datasets. **a–d** Results based on the assembly by MEGAHIT, and **e–h** results based on the assembly by metaSPAdes. The initial binning tool is MaxBin2

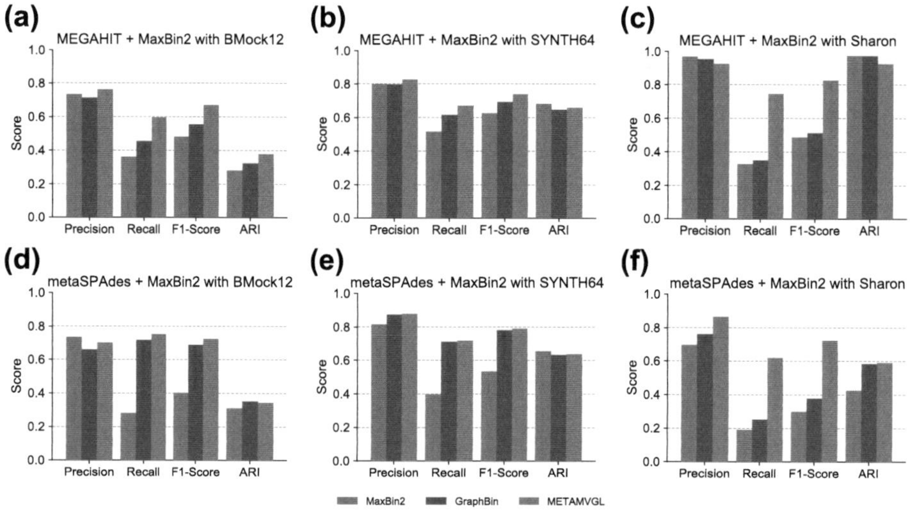

Fig. 4 The performance of MaxBin2, GraphBin and METAMVGL on the *BMock12*, *SYNTH64* and *Sharon* datasets: **a**, **d** for *BMock12* dataset; **b**, **e** for *SYNTH64* dataset; **c**, **f** for *Sharon* dataset. MEGAHIT and metaSPAdes are used to generate the assembly graphs. The initial binning tool is MaxBin2

6: Figures). Comparing with assembly graph, the combined graph adds up to 8942.37% vertices and 15,114.06% edges to the main graph (Additional file 2: The assembly graph from MEGAHIT for *Sharon* dataset).

Methods

Figure 2 demonstrates the workflow of METAMVGL, which consists of two steps. In step 1, METAMVGL constructs the assembly graph and PE graph with contig labels generated by the existing binning tools. In step 2, we remove the ambiguous labels of contigs if their neighbors are labeled as belonging to the other binning groups. The two graphs are integrated based on the automatic weights and the unlabeled contigs will be further predicted by label propagation on this combined graph. Finally, METAMVGL removes the ambiguous labels and generates the binning results.

Step 1: Preprocessing

Construct assembly graph

We define the assembly graph as $\mathcal{G}_1(\mathcal{V}, \mathcal{E}_1)$, where the vertex $v_i \in \mathcal{V}$ represents the contig i, and an edge $e_{i,j} \in \mathcal{E}_1$ exists if v_i and v_j are connected in the assembly graph and with $k-1$ mer (continuous nucleotide of length $k-1$) overlap. In principle, the assembly graph should include s unconnected subgraphs, each representing one species and we can easily recognize contig binning groups. In practice, the subgraphs could be linked due to the inter-species repeat sequences and complicated by the sequencing errors and imbalanced genomic coverage. Commonly the assembly graph includes a main graph and several dead ends. Figure 2 illustrates an assembly graph with two dead ends (vertices 11 and 12). METAMVGL uses the assembly graph from metaSPAdes [19] or MEGAHIT [20]. The original assembly graph of metaSPAdes is a unitig-based graph, where each vertex represents a unitig. The contigs are sets of unitigs after resolving short repeats. Hence we convert the unitig-based graph to contig-based graph by adding the edge $e_{i,j}$, if at least two unitigs connect to each other and belong to v_i and v_j, respectively. MEGAHIT would not provide the assembly graph directly, so METAMVGL uses contig2fastg module in megahit_toolkit to generate the graph in fastg format.

Construct PE graph

In order to deal with the dead ends in assembly graph, METAMVGL constructs the PE graph by aligning paired-end reads to the contigs by BWA-MEM[21]. For every two contigs v_i, v_j, we maintain a read name set $\mathcal{RS}_{i,j}$ to keep the names of read pairs, where the forward and reverse reads are aligned to the two contigs, respectively. The library insert size $\mathcal{IS}$ is calculated based on the uniquely aligned paired-end reads in the same contigs. To alleviate the influence of chimeric reads, METAMVGL links v_i and v_j if at least half of the reads in $\mathcal{RS}_{i,j}$ come from the two stretches with length $\mathcal{IS}$ in v_i and v_j, respectively [22].

We denote the PE graph as $\mathcal{G}_2(\mathcal{V}, \mathcal{E}_2)$, where $\mathcal{V}$ represents contigs, and $\mathcal{E}_2$ the edges linked by the paired-end reads (PE links). According to our observation, PE graph is complementary to the assembly graph to some extent, because the edges in assembly graph ($\mathcal{E}_1$) only capture the overlaps between contigs, while the PE graph edges ($\mathcal{E}_2$) link the contigs with gaps. Figure 2 illustrates how dead ends of assembly graph can be linked to the main graph using PE links.

Initial binning

The contigs' initial labels are generated by any existing binning tools. In experiments, we evaluated the performance of METAMVGL with the initial labels from MaxBin2, MetaBAT2, MyCC, CONCOCT and SolidBin in SolidBin-SFS mode. We used the default parameters for these algorithms except the MetaBAT2, where the minimum contig length was set to 1.5*kb* to label more contigs.

Step 2: Auto-weighted multi-view binning

METAMVGL applies a multi-view label propagation algorithm [18] to learn the weights of assembly and PE graphs automatically and predict the unlabeled contigs in a uniformed framework. We remove the ambiguous labels for two times before and after label propagation.

Remove ambiguous labels

The initial contig labels could be incorrect especially for the ones from the inter-species repeat sequences and their influence would be amplified in label propagation. METAMVGL computes the distance between two vertices as the length of shortest path between them. Let $\mathcal{CLV}(v)$ be the set of labels from vertex v's closest labeled neighbors in graph $\mathcal{G}$ and v's label is ambiguous if $\mathcal{CLV}(v)$ contains a label that is different from v [17]. Let $\mathcal{VA}(\mathcal{G})$ denote the set including all the vertices with ambiguous labels in graph $\mathcal{G}$, and we remove the labels in $\mathcal{VA}(\mathcal{G})$. In Fig. 2, the closest labeled vertices of v_6 are $\{v_4, v_7\}$. Because v_6 and v_7 have different labels, the v_6's label is marked ambiguous. Algorithm 1 shows the procedure to remove ambiguous labels. As shown in Fig. 2, we applied Algorithm 1 on the assembly and combined graphs before and after label propagation, respectively. We only use assembly graph to mark ambiguous labels in the preprocessing step for keeping more labels before propagation.

Algo. 1: Remove Ambiguous Labels

```
Input: Labels of Contigs L = {l_1, l_2, ..., l_n}, Graph G
Output: Refined Labels: L_out
VA(G) ← ∅
for i ← 1, 2, ..., n do
    ambiguous ← false
    if l_i ≠ null then
        foreach l ∈ CLV(v_i) do
            if l ≠ l_i then
                ambiguous ← true
            end
        end
    end
    if ambiguous then
        VA(G).insert(i)
    end
end
foreach i ∈ VA(G) do
    l_i ← null
end
L_out ← {l_1, l_2, ..., l_n}
```

Auto-weighted multi-view binning algorithm

Assume l contigs are initially labeled with s groups, denoted as $Y_l = [y_1, y_2, \ldots, y_l]^T \in \mathbb{R}^{l\times s}$, where $y_{ij} \in \{0, 1\}$, and $y_{ij} = 1$ indicates the contig v_i is labeled from group j. We define a indicator matrix $F = [F_l; F_u] \in \mathbb{R}^{n\times s}$, where $F_l = Y_l$ and $F_u = [f_{l+1}, f_{l+2}, \ldots, f_n]^T$ are labels to be inferred. Let $D_i, W_i \in \mathbb{R}^{n\times n}$ denote the degree and adjacent matrices of $\mathcal{G}_i$ ($i \in \{1, 2\}$), respectively. The normalized Laplacian matrix of $\mathcal{G}_i$ has the formulation

$\mathcal{L}_i = D_i^{-1/2}(D_i - W_i)D_i^{-1/2}$. According to [18], the inference of F_u by label propagation can be modeled as the following optimization problem:

$$\arg\min_{F_u} \sum_{i=1}^{2} \sqrt{\mathcal{TR}(F^T \mathcal{L}_i F)},\ s.t.\ F_l = Y_l, \tag{1}$$

where $\mathcal{TR}(\cdot)$ computes the trace of a matrix. The optimization problem is converted to

$$\arg\min_{F_u} \mathcal{TR}(F^T \mathcal{L} F),\ s.t.\ F_l = Y_l, \tag{2}$$

where $\mathcal{L} = \sum_{i=1}^{2} \alpha_i \mathcal{L}_i$. α_i is the weight of $\mathcal{G}_i$, with initial values of 1/2. We partition $\mathcal{L}$ from $(l+1)$th row and column into four blocks as $[\mathcal{L}_{ll}, \mathcal{L}_{lu}; \mathcal{L}_{ul}, \mathcal{L}_{uu}]$. F_u and α_i can be updated alternatively until convergence by the following equations [18]:

$$F_u = \mathcal{L}_{uu}^{-1} \mathcal{L}_{ul} Y_l,\ F = [Y_l; F_u], \tag{3}$$

$$\alpha_i = \frac{1}{2\sqrt{\mathcal{TR}(F^T \mathcal{L}_i F)}},\ i \in \{1, 2\}. \tag{4}$$

Equation 3 can be considered as performing label propagation in the combined graph with iteratively updated weight α_i, hence α_i implies the confidence of each graph. After obtaining F_u, we infer the labels of all the contigs by $l_i = \arg\max_j F_{ij}, i \in \{1, 2, \ldots, n\}, j \in \{1, 2, \ldots, s\}$. Algorithm 2 shows the procedure of auto-weighted multi-view binning, and Fig. 2 is an illustration of this algorithm.

Algo. 2: Auto-weighted multi-view binning

```
Input: Initial Labels of Contigs L, Graphs {G_1, G_2}
Output: Final Labels L_out
L ← Remove Ambiguous Labels (L, G_1) (Algo. 1)
Calculate Y_l from L
for i ← 1, 2 do
    α_i ← 1/2
    Calculate D_i and W_i for G_i
    L_i ← D_i^{-1/2} (D_i − W_i) D_i^{-1/2}
end
repeat
    L ← Σ_{i=1}^{2} α_i L_i
    F_u ← L_uu^{-1} L_ul Y_l,  F ← [Y_l; F_u] (Eq. 3)
    for i ← 1, 2 do
        α_i ← 1/(2 sqrt(TR(F^T L_i F))) (Eq. 4)
    end
until Convergence;
for i ← 1, 2, ..., n do
    l_i ← arg max_j F_ij
end
G ← α_1 G_1 + α_2 G2
L_out ← {l_1, l_2, ..., l_n}
L_out ← Remove Ambiguous Labels (L_out, G) (Algo. 1)
```

Datasets

Simulated datasets

We simulated metagenomic sequencing data for a mixture of three strains with low, medium and high abundances. The components are:

- Acinetobacter baumannii: 0.90%,
- Streptococcus agalactiae: 9.01%,
- Streptococcus mutans: 90.09%.

We downloaded the complete reference genomes of the three strains from the NCBI Nucleotide Database (Taxonomy ID: 400667, 208435, 210007). CAMISIM [23] generated short-reads for the three strains with corresponding abundances. Five simulated datasets were generated with read depths as 30x (SIM_30x), 50x (SIM_50x), 70x (SIM_70x), 90x (SIM_90x) and 110x (SIM_110x).

Mock datasets

We evaluated the performance of METAMVGL on the metagenomic sequencing from two mock communities:

- *BMock12* refers to the metagenomic sequencing for a mock community with 12 bacterial strains sequenced by Illumina HiSeq 2500 [24] (NCBI acc. no. SRX4901583). It contains 426.8 million 150 bp short-reads with a total size of 64.4G bases.
- *SYNTH64* is a metagenomic sequencing dataset for a synthetic community with 64 diverse bacterial and archaea species [25] (NCBI acc. no. SRX200676), sequenced by Illumina HiSeq 2000 with read length 101*bp* and total size 11.1G bases.

Real dataset

Sharon dataset [26] (NCBI acc. no. SRX144807) contains the metagenomic sequencing data of infant fecal samples from 18 time points, sequenced by Illumina HiSeq 2000 with a total of 274.4 million 100 bp short-reads. We combined all the 18 datasets for co-assembly and referred them as *Sharon*.

Evaluation criteria

We annotated the potential species the contigs came from as ground truth to compare METAMVGL with the other tools. For the simulated and mock datasets, we aligned the contigs to the available reference genomes and selected the ones with unique alignments. For the *Sharon* dataset, we used Kraken2 [27] to annotate the contigs according to *k*-mer similarities with the species from the build-in database.

Assume there are s ground truth species, and the binning result have k groups. To evaluate the binning result, we define the assessment matrix $[n_{i,j}]^{(k+1)\times(s+1)}$, where $n_{i,j}$ represents the number of contigs in ith binning group that are annotated jth ground truth species. The $(k+1)$th row denotes unbinned contigs. The $(s+1)$th column indicates contigs without ground truth annotations. We applied (1) Precision, (2) Recall,

(3) F1-Score and (4) Adjusted Rand Index (ARI) to evaluate the performance of binning algorithms. Let $N = \sum_{i=1}^{k} \sum_{j=1}^{s} n_{i,j}$ be the number of contigs; the four metrics are calculated as follows:

1. $\text{Precision} = \frac{1}{N} \sum_{i=1}^{k} \max_{j \le s}(n_{i,j})$,
2. $\text{Recall} = \frac{1}{N + \sum_{j=1}^{s} n_{k+1,j}} \sum_{j=1}^{s} \max_{i \le k}(n_{i,j})$,
3. $\text{F1-Score} = \frac{2 \times Precision \times Recall}{Precision + Recall}$,
4. $\text{ARI} = \frac{\sum_{i=1}^{k} \sum_{j=1}^{s} \binom{n_{i,j}}{2} - t}{\frac{1}{2}\left(\sum_{i=1}^{k} \binom{\sum_{j=1}^{s} n_{i,j}}{2} + \sum_{j=1}^{s} \binom{\sum_{i=1}^{k} n_{i,j}}{2} \right) - t}$,

where $t = \frac{1}{\binom{N}{2}} \sum_{i=1}^{k} \binom{\sum_{j=1}^{s} n_{i,j}}{2} \sum_{j=1}^{s} \binom{\sum_{i=1}^{k} n_{i,j}}{2}$.

Results

METAMVGL was compared to six binning tools, MaxBin2, MetaBAT2, MyCC, CONCOCT, SolidBin in SolidBin-SFS mode and GraphBin. We analyzed their binning results on the five simulated datasets with various read depths, two mock communities (*BMock12* and *SYNTH64*) and a real metagenomic sequencing dataset (*Sharon* dataset).

Evaluation on the simulated datasets

Figure 3 shows the binning results of the simulated datasets. The contigs and assembly graph were generated by MEGAHIT (Fig. 3a–d) and metaSPAdes (Fig. 3e–h). MaxBin2 was applied as the initial binning tool for GraphBin and METAMVGL.

All the three binning algorithms (MaxBin2, GraphBin and METAMVGL) yielded extremely high precision and ARI (Fig. 3a, d, e, h), due to the low complexity of the simulated datasets. Because of considering assembly and PE graphs jointly, METAMVGL labeled more contigs than GraphBin and MaxBin2 across various sequence depths, as shown in Fig. 3b, f. We also found both Recall and F1-Score were improved as read depth became higher until SIM_70x (Fig. 3b, c, f, g). This observation was analog to the results from CAMISIM [23], suggesting too high read depth would introduce assembly noise even it might help in detecting the low-abundance microbes.

Evaluation on the mock communities

We illustrate the binning results for two mock communities with initial binning tool of MaxBin2 in Fig. 4a, b, d, e. In general, the graph-based methods (METAMVGL and GraphBin) were better than MaxBin2, but their performance would be influenced by the assembly graph. We observed the recalls could be significantly improved using the assembly graph from metaSPAdes (Fig. 4d, e), but the elevation became unobvious by the one from MEGAHIT (Fig. 4a, b). This was probably because metaSPAdes could generate more accurate and complete assembly graph than MEGAHIT. In the mock communities, METAMVGL was just slightly better than GraphBin, suggesting the PE graph was largely overlapped with the assembly graph (Additional file 2: *BMock12* and *SYNTH64* datasets). This observation only occurred if a perfect assembly graph was generated due to a low microbial complexity in the community. The results for the other initial binning tools (MetaBAT2, MyCC, CONCOCT and SolidBin) could be found in Additional files 3, 4, 5, 6: Figures, which were akin to the observations from MaxBin2.

Evaluation on *Sharon* datasets

Figure 4c, f describe the binning results of MaxBin2, GraphBin and METAMVGL in *Sharon* dataset. METAMVGL substantially improved the recalls comparing with GraphBin (2.12 and 2.46 times on the assembly graph from MEGAHIT and metaSPAdes, respectively) and MaxBin2 (2.29 and 3.24 times on the assembly graph from MEGAHIT and metaSPAdes, respectively). METAMVGL showed the highest precision on the assembly by metaSPAdes (Fig. 4f). All the three binning algorithms were comparable in ARI.

The outstanding recalls from METAMVGL validate the capability of PE graph to connect the dead ends to the main graph when the assembly graph is incomplete in the complex microbial community. We observed that MEGAHIT produced very fragmented assembly graph in the *Sharon* dataset, in which the main graph only had 59 vertices with 64 edges, while a total of 15,660 vertices existed in the whole graph (Additional file 2: MEGAHIT assembly of *Sharon* dataset). The fragmented assembly graph was also mentioned as a limitation of GraphBin [17]. With PE links, METAMVGL yielded 5335 vertices and 9737 edges in the main graph (Additional file 2: MEGAHIT assembly of *Sharon* dataset), rescuing a large number of unlabeled contigs from the dead ends (Fig. 4c). Although the assembly graph was more complete (23.69% vertices in the main graph) from metaSPAdes, the PE graph still added 28.97% edges to the main graph (Additional file 2: metaSPAdes assembly of *Sharon* dataset) and improved the recall substantially.

Discussion

De novo assembly together with contig binning offer a practical way to explore the novel microbes from metagenomic sequencing. But the current binning algorithms work stably merely on long contigs; the shorter ones are commonly neglected in the subsequent analysis. We observed a large proportion of contigs were shorter than 1 kb, which resulted in low completeness of the binning groups. A recent study [17] suggests the short contigs could be rescued from the assembly graph by considering their connections with the labeled ones. Assembly graph is accurate, but its connectivity relies heavily on the complexity of microbial community. Extremely low or high read depth, sequencing errors and imbalanced coverage could generate considerable dead ends, which would introduce both missing labels and labeling errors (Fig. 1).

In experiments, we observed a slightly lower ARI of METAMVGL comparing with MaxBin2 and GraphBin when read depth was low (Fig. 3d, h). It might because METAMVGL could only retrieve very few and low-confidence PE links. First, the label propagation would perform poorly on the contigs with low read depth. Comparing with GraphBin, METAMVGL included more edges in the graph, but these edges were sparse and cannot guarantee the good performance of label propagation, e.g. the unlabeled contigs might only have one neighbor. Second, the label refinement after label propagation could remove a majority of erroneous labels generated by METAMVGL based on our experience. Due to the paucity of edges, this step also performed inefficiently. Third, the quality of the contigs assembled from the sequencing data with low read depth was poor, making difficulties in aligning paired-end reads correctly.

In this paper, we developed METAMVGL, a multi-view graph-based contig binning algorithm to integrate both assembly and PE graphs to label short contigs and correct initial labeling errors. PE graph could link the dead ends to the main graph and increase the graph connectivity. METAMVGL automatically weights the two graphs and performs label propagation to label the short contigs. In experiments, we observed METAMVGL could substantially improve the recalls without loosing any precision comparing with the existing contig binning tools, especially for the metagenomic data from the complex microbial community (Fig. 4c, f). We also evaluated METAMVGL: 1. on the assembly graphs from metaSPAdes and MEGAHIT; and 2. using the initial binning labels from different tools. All these results support METAMVGL outperform GraphBin in different experimental configurations. On average, METAMVGL could finish the contig binning in 3.38 min and requires 2.81 Gb RAM to store the two graphs and perform label propagation. It requires a little bit more computational resources than GraphBin due to the analysis of more complex and complete graph (Additional file 7). Sometimes, we found the combined graph was still incomplete after incorporating both assembly and PE graphs and there still required to consider other information to reveal contig long-range connectivity from various long-fragment sequencing (PacBio and Oxford Nanopore sequencing) or linked-read sequencing (Tell-seq and stLFR sequencing) technologies.

Conclusion

Metagenomic sequencing has been proved as an efficient technology to explore and recognize the novel microbes in the environmental and human fecal samples. Due to the scarcity of the reference genomes, the genomes of novel species could be obtained by de novo assembly. Because only fragmented contigs can be assembled from the mainstream short-read sequencing technologies, the interests rise quickly in developing efficient contig binning algorithms. But most of the available algorithms can only handle long contigs based on their sequence contexts and read depths. In this study, we developed METAMVGL, a multi-view graph-based contig binning algorithm to integrate both assembly and PE graphs to label short contigs and correct initial labeling errors. METAMVGL could weight the two graphs automatically and connect the dead ends to the main graph efficiently. Our experiments proved it can significantly improve the recalls without loosing any precision comparing with the existing contig binning tools on the metagenomic sequencing data from simulation, mock communities and infant fecal samples. We believe METAMVGL would attract more interests of the fast-growing metagenomic research field and pave the way to future understanding the microbial genome dark matter.

Abbreviations
METAMVGL: Metagenomic contig binning using multi-view graph learning; PE graph: Paired-end graph; PE link: Paired-end link; TNF: Tetranucleotide frequencies; ARI: Adjusted rand index; RAM: Random-access memory.

Supplementary Information

The online version contains supplementary material available at https://doi.org/10.1186/s12859-021-04284-4.

Additional file 1. The statistics of assemblies from MEGAHIT andmetaSPAdes for the simulated, *BMock12*, *SYNTH64* and *Sharon* datasets.

Additional file 2. The statistics of the largest component (main graph) inthe assembly graph, PE graph, and combined graph for the simulated,*BMock12*, *SYNTH64* and *Sharon* datasets.

Additional file 3. The performance of MetaBAT2, GraphBin andMETAMVGL on the BMock12, SYNTH64 and Sharon datasets: (a) and(d) for *BMock12* dataset; (b) and (e) for *SYNTH64* dataset; (c) and (f)for *Sharon* dataset. MEGAHIT and metaSPAdes are used to generate theassembly graphs. The initial binning tool is **MetaBAT2**.

Additional file 4. The performance of MyCC, GraphBin and METAMVGLon the *BMock12*, *SYNTH64* and *Sharon* datasets: (a) and (d) for*BMock12* dataset; (b) and (e) for *SYNTH64* dataset; (c) and (f) for*Sharon* dataset. MEGAHIT and metaSPAdes are used to generate theassembly graphs. The initial binning tool is **MyCC**.

Additional file 5. The performance of CONCOCT, GraphBin andMETAMVGL on the *BMock12*, *SYNTH64* and *Sharon* datasets: (a) and(d) for *BMock12* dataset; (b) and (e) for *SYNTH64* dataset; (c) and (f)for *Sharon* dataset. MEGAHIT and metaSPAdes are used to generate theassembly graphs. The initial binning tool is **CONCOCT**.

Additional file 6. The performance of SolidBin, GraphBin and METAMVGLon the *BMock12*, *SYNTH64* and *Sharon* datasets: (a) and (d) for*BMock12* dataset; (b) and (e) for *SYNTH64* dataset; (c) and (f) for*Sharon* dataset. MEGAHIT and metaSPAdes are used to generate theassembly graphs. The initial binning tool is **SolidBin**.

Additional file 7. Running time and memory usage of GraphBin andMETAMVGL.

Authors' contribution
LZ conceived the study. ZMZ implemented METAMVGL. ZMZ and LZ analyzed the results. ZMZ and LZ wrote the paper. Both authors have read and approved the final manuscript.

Funding
Publication costs are funded by Research Grant Council Early Career Scheme (HKBU 22201419). This work is supported by Research Grant Council Early Career Scheme (HKBU 22201419), IRCMS HKBU (No. IRCMS/19-20/D02), Guangdong Basic and Applied Basic Research Foundation (No. 2019A1515011046). The funders did not play any role in the design of the study, the collection, analysis, and interpretation of data, or in writing of the manuscript.

Availability of data and materials
The source code of METAMVGL is publicly available at https://github.com/ZhangZhenmiao/METAMVGL. The Illumina short-reads of ***BMock12***, ***SYNTH64*** and ***Sharon*** datasets are available in NCBI Sequence Read Archive (SRA); the accession numbers are SRX4901583, SRX200676 and SRX144807, respectively.

About this supplement
This article has been published as part of BMC Bioinformatics Volume 22 Supplement 10 2021: Selected articles from the 19th Asia Pacific Bioinformatics Conference (APBC 2021): bioinformatics. The full contents of the supplement are available at https://bmcbioinformatics.biomedcentral.com/articles/supplements/volume-22-supplement-10.

Declarations

Ethics approval and consent to participate
Not applicable.

Consent for publication
Not applicable.

Competing interests
The authors have no conflicts of interest.

Received: 13 June 2021 Accepted: 6 July 2021
Published online: 22 July 2021

References

1. Li J, Jia H, Cai X, Zhong H, Feng Q, Sunagawa S, Arumugam M, Kultima JR, Prifti E, Nielsen T, et al. An integrated catalog of reference genes in the human gut microbiome. Nat Biotechnol. 2014;32(8):834–41.
2. Truong DT, Franzosa EA, Tickle TL, Scholz M, Weingart G, Pasolli E, Tett A, Huttenhower C, Segata N. Metaphlan2 for enhanced metagenomic taxonomic profiling. Nat Methods. 2015;12(10):902–3.
3. Almeida A, Mitchell AL, Boland M, Forster SC, Gloor GB, Tarkowska A, Lawley TD, Finn RD. A new genomic blueprint of the human gut microbiota. Nature. 2019;568(7753):499–504.
4. Almeida A, Nayfach S, Boland M, Strozzi F, Beracochea M, Shi ZJ, Pollard KS, Sakharova E, Parks DH, Hugenholtz P, et al. A unified catalog of 204,938 reference genomes from the human gut microbiome. Nat Biotechnol. 2020; 1–10.
5. Poyet M, Groussin M, Gibbons S, Avila-Pacheco J, Jiang X, Kearney S, Perrotta A, Berdy B, Zhao S, Lieberman T, et al. A library of human gut bacterial isolates paired with longitudinal multiomics data enables mechanistic microbiome research. Nat Med. 2019;25(9):1442–52.
6. Nayfach S, Shi ZJ, Seshadri R, Pollard KS, Kyrpides NC. New insights from uncultivated genomes of the global human gut microbiome. Nature. 2019;568(7753):505–10.
7. Zou Y, Xue W, Luo G, Deng Z, Qin P, Guo R, Sun H, Xia Y, Liang S, Dai Y, et al. 1520 reference genomes from cultivated human gut bacteria enable functional microbiome analyses. Nat Biotechnol. 2019;37(2):179–85.
8. Forster SC, Kumar N, Anonye BO, Almeida A, Viciani E, Stares MD, Dunn M, Mkandawire TT, Zhu A, Shao Y, et al. A human gut bacterial genome and culture collection for improved metagenomic analyses. Nat Biotechnol. 2019;37(2):186–92.
9. Consortium HMJRS, et al. A catalog of reference genomes from the human microbiome. Science. 2010;328(5981):994–9.
10. Pasolli E, Asnicar F, Manara S, Zolfo M, Karcher N, Armanini F, Beghini F, Manghi P, Tett A, Ghensi P, et al. Extensive unexplored human microbiome diversity revealed by over 150,000 genomes from metagenomes spanning age, geography, and lifestyle. Cell. 2019;176(3):649–62.
11. Wu Y-W, Simmons BA, Singer SW. Maxbin 2.0: an automated binning algorithm to recover genomes from multiple metagenomic datasets. Bioinformatics. 2016;32(4):605–7.
12. Kang DD, Li F, Kirton E, Thomas A, Egan R, An H, Wang Z. Metabat 2: an adaptive binning algorithm for robust and efficient genome reconstruction from metagenome assemblies. PeerJ. 2019;7:7359.
13. Alneberg J, Bjarnason BS, De Bruijn I, Schirmer M, Quick J, Ijaz UZ, Lahti L, Loman NJ, Andersson AF, Quince C. Binning metagenomic contigs by coverage and composition. Nat Methods. 2014;11(11):1144–6.
14. Lin H-H, Liao Y-C. Accurate binning of metagenomic contigs via automated clustering sequences using information of genomic signatures and marker genes. Sci Rep. 2016;6:24175.
15. Wang Z, Wang Z, Lu YY, Sun F, Zhu S. Solidbin: improving metagenome binning with semi-supervised normalized cut. Bioinformatics. 2019;35(21):4229–38.
16. Yu G, Jiang Y, Wang J, Zhang H, Luo H. Bmc3c: binning metagenomic contigs using codon usage, sequence composition and read coverage. Bioinformatics. 2018;34(24):4172–9.
17. Mallawaarachchi V, Wickramarachchi A, Lin Y. Graphbin: refined binning of metagenomic contigs using assembly graphs. Bioinformatics. 2020;36(11):3307–13.
18. Nie F, Li J, Li X, et al. Parameter-free auto-weighted multiple graph learning: a framework for multiview clustering and semi-supervised classification. In: IJCAI, 2016; p. 1881–7.
19. Nurk S, Meleshko D, Korobeynikov A, Pevzner PA. metaspades: a new versatile metagenomic assembler. Genome Res. 2017;27(5):824–34.
20. Li D, Liu C-M, Luo R, Sadakane K, Lam T-W. Megahit: an ultra-fast single-node solution for large and complex metagenomics assembly via succinct de Bruijn graph. Bioinformatics. 2015;31(10):1674–6.
21. Li H. Aligning sequence reads, clone sequences and assembly contigs with bwa-mem. arXiv preprint arXiv:1303.3997 (2013).
22. Bishara A, Moss EL, Kolmogorov M, Parada AE, Weng Z, Sidow A, Dekas AE, Batzoglou S, Bhatt AS. High-quality genome sequences of uncultured microbes by assembly of read clouds. Nat Biotechnol. 2018;36(11):1067–75.
23. Fritz A, Hofmann P, Majda S, Dahms E, Dröge J, Fiedler J, Lesker TR, Belmann P, DeMaere MZ, Darling AE, et al. Camisim: simulating metagenomes and microbial communities. Microbiome. 2019;7(1):1–12.
24. Sevim V, Lee J, Egan R, Clum A, Hundley H, Lee J, Everroad RC, Detweiler AM, Bebout BM, Pett-Ridge J, et al. Shotgun metagenome data of a defined mock community using oxford nanopore, pacbio and illumina technologies. Sci data. 2019;6(1):1–9.
25. Shakya M, Quince C, Campbell JH, Yang ZK, Schadt CW, Podar M. Comparative metagenomic and rrna microbial diversity characterization using archaeal and bacterial synthetic communities. Environ Microbiol. 2013;15(6):1882–99.
26. Sharon I, Morowitz MJ, Thomas BC, Costello EK, Relman DA, Banfield JF. Time series community genomics analysis reveals rapid shifts in bacterial species, strains, and phage during infant gut colonization. Genome Res. 2013;23(1):111–20.
27. Wood DE, Lu J, Langmead B. Improved metagenomic analysis with kraken 2. Genome Biol. 2019;20(1):257.

Publisher's Note

Yang *et al. BMC Bioinformatics* 2021, **22**(10):387
https://doi.org/10.1186/s12859-021-04292-4

BMC Bioinformatics

RESEARCH

Open Access

Mining a stroke knowledge graph from literature

Xi Yang[1,2,3], Chengkun Wu[2], Goran Nenadic[3*], Wei Wang[1] and Kai Lu[1]

From The 19th Asia Pacific Bioinformatics Conference (APBC 2021) Tainan, Taiwan. 3-5 February 2021

*Correspondence:
g.nenadic@manchester.ac.uk
[3] Department of Computer Science, University of Manchester, Manchester M13 9PL, UK
Full list of author information is available at the end of the article

Abstract

Background: Stroke has an acute onset and a high mortality rate, making it one of the most fatal diseases worldwide. Its underlying biology and treatments have been widely studied both in the "Western" biomedicine and the Traditional Chinese Medicine (TCM). However, these two approaches are often studied and reported in insolation, both in the literature and associated databases.

Results: To aid research in finding effective prevention methods and treatments, we integrated knowledge from the literature and a number of databases (e.g. CID, TCMID, ETCM). We employed a suite of biomedical text mining (i.e. named-entity) approaches to identify mentions of genes, diseases, drugs, chemicals, symptoms, Chinese herbs and patent medicines, etc. in a large set of stroke papers from both biomedical and TCM domains. Then, using a combination of a rule-based approach with a pre-trained BioBERT model, we extracted and classified links and relationships among stroke-related entities as expressed in the literature. We construct StrokeKG, a knowledge graph includes almost 46 k nodes of nine types, and 157 k links of 30 types, connecting diseases, genes, symptoms, drugs, pathways, herbs, chemical, ingredients and patent medicine.

Conclusions: Our Stroke-KG can provide practical and reliable stroke-related knowledge to help with stroke-related research like exploring new directions for stroke research and ideas for drug repurposing and discovery. We make StrokeKG freely available at http://114.115.208.144:7474/browser/ (Please click "Connect" directly) and the source structured data for stroke at https://github.com/yangxi1016/Stroke

Keywords: Stroke, Knowledge graph, Biomedical text mining, Traditional Chinese Medicine

Background

Stroke, also known as cerebrovascular accident (CVA), is a group of diseases with three major types (hemorrhagic stroke, ischemic stroke, and TIA transient ischemic attack) and with cerebral infarction being the most common phenotype [1]. In the past decades, stroke treatment and prevention have seen significant advances in particular in declining stroke mortality [2]. Western therapeutic such as drug injection and endovascular therapy [3], as well as traditional Chinese treatment such as herbal medicine and acupuncture [4], have made tremendous efforts for preventing stroke and recovery after stroke. However, stroke is still one of the most critical fatal diseases worldwide (the second leading cause of death) [5] because of acute onset, with the enormous economic burden of recovery for those who survive. So, there is a need to investigate potential pathogenic genes, risk factors further, and aura symptoms of stroke to find efficient preventative and therapeutic approaches.

There are some existing structured knowledge sources focused on stroke [6–8]. Still, a large amount of stroke-related information is available in scientific articles. For example, a recent search for '*stroke*' in PubMed resulted in over 327 K papers. In this study, we aim to develop a stroke-related knowledge base by combining information extracted from these scientific papers and existing knowledge bases. The large volume of texts requires automated and computational methods to extract useful information from these unstructured data to build structured databases.

Knowledge graphs (KGs) [9] are widely known as knowledge domain visualization or knowledge domain mapping graphs in the library and information industry [10]. They are often represented as a series of different graphs with the relationships between development processes and the structure of knowledge. Visualization technology is used to describe, analyze, construct, and display knowledge and inter-relationships [11]. Such representation methods can promote the understanding of relations between biomedical entities, which is vital for scientific researchers to refine their research scope and improve personalized medicine. It is also possible to discover new knowledge (e.g., new drugs [12] and effective prevention/treatment methods [13, 14]). However, it is a laborious and time-consuming process to construct a KG manually. Therefore, automated approaches to assist an automated/semi-automated construction of knowledge graphs in specific domains have been used [15, 16].

In this paper we introduce a stroke-related knowledge graph (StrokeKG) by combining information extracted from these scientific papers and existing knowledge bases. In addition to biomedical entities, we also add entities from Traditional Chinese medicine (TCM) [17], which pays close attention to the medical characteristics of the entire system of the human body, which makes it a promising candidate for the treatment of stroke [18]. We use a suite of tools to extract genes, diseases, drugs, symptoms, Chinese herbal medicine, and other entities and link them using relationship extraction methods. As a result, StrokeKG includes 46,983 nodes of 9 types, and 157,302 relationships of 30 types, connecting diseases, genes, symptoms, drugs, pathways, Chinese Patent Medicines (CPMs), Herbs, Chemical, ingredients. Besides, we marked 265 CPM entities and 404 CPM-Disease relationships through verification and manual annotation of existing databases to provide practical and accurate stroke-related knowledge. The graph can be used to facilitate our understanding of this complex disease, for example, by exploring precursor symptoms and sequelae of stroke, therapeutic drugs, and the pathway for treating related diseases.

Related work

In the field of biomedicine, knowledge bases (KBs) such as Gene ontology [19], disease ontology [20], reference terms for national drug archives [21], and basic models of anatomy [22] have been prominent examples of efforts to provide structured knowledge systematically. Some of these KBs, e.g. OpenKG [23], BenevolentAI [24], and KnowLife [14] have made significant contributions to the development of the biomedical field, including recent drug repositioning for COVID-19 [25], SemaTyP [26], and protein-drug target KG [15] have been used. Despite many efforts to provide more structured data, vast amounts of relevant knowledge are still hidden in the biomedical literature [27]. There are three main limitations to previous work on KB construction [9]. First, most biomedical KBs are manually constructed and curated, which defer them from keeping up with the pace of novel discoveries. Second, potentially useful text sources such as health portals, online communities, or other sources of information are often ignored. Finally, most previous works focused on one molecular level or chemical genomics, such as protein–protein interactions [28], gene-drug relationships [29], or just highly specific topics such as drug effects.

Natural language processing tools are indispensable to extract useful information from biomedical literature [30]. We need to start with the named entity recognition process and then relationship extraction. Biomedical Named Entity Recognition (NER) [31] aims to identify specific biomedical concepts in the text. NER consists of two steps: (1) classifying specific substrings obtained from the text to determine whether it is the name of a specific type of entity; (2) selecting a standard name or a unique identifier for one kind of entity [32]. There are already many NER tools available for different types of biomedical entities, such as genes/proteins [33], diseases [34, 35], species [36], mutations [37], chemicals [38, 39] and biological pathways [40]. Still, many essential concept types such as RNAs, phenotypes, Chinese Patent Medicines (CPMs), and herbal medicines do not have corresponding NER tools.

The task of Relation Extraction (RE) is has been in the focus of research in recent years. Due to the inherent complexity of the biomedical text, most relation extraction systems work at the sentence-based level. Common relationships include protein–protein interactions [28], drug–drug interactions [41], gene regulatory events [42], associations between mutations and diseases [43]. Early relationships used a co-occurrence approach [44], while pattern-based systems [45] rely on a set of manually or automatically collected patterns to extract relations and classify relation types between entities. Rule-based methods [46, 47] use a set of processes or some heuristic algorithms to manually define or build a set of rules based on domain experts and automatically generated from the training data. It adds multiple constraints to scope specific relationships: for example, BioNLP'09 [48] focused on nine common molecular events. More recently, with the improvement of the accuracy and expanded availability of curated corpora, deep learning models are widely used in the field of natural language processing. Convolutional Neural Network (CNN) [49], Recurrent Neural Network (RNN) [50], Long Short Term Memory Network (LSTM) [41], Capsule Network, CapsNet [51], Graph Neural Networks [52, 53], and BERT [54] are prevalent models employed in relation extraction, making great contribution to biomedical text mining.

For the field of traditional Chinese medicine, Manually organized TCM database, TCMID [55], TCM-MESH [56], and *Chinese medicine network pharmacology* ETCM [57, 58], TCMSP brings convenience to the research of Chinese medicine. However, to the best of our knowledge, there is no text mining tool specifically for Traditional Chinese medicine, and there is also the non-disclosure or incomplete knowledge in stroke-related knowledge [6–8]. Therefore, in this research, we will enrich the application of text mining in the construction of Chinese medicine knowledge, and based on this and the-start-of-art, construct a stroke-related knowledge graph.

Methods

In this work, we designed a computational workflow to mine the stroke-related and TCM-related literature for the identification of biomedical entities and the relations between them. We split stroke-related abstracts into 463,225 sentences, the analysis pipeline tags the mentions of the following entities: drugs, chemicals, genes, pathways, and diseases, as well as traditional Chinese treatments like Herbs, Chinese Patent Medicines (CPMs), and ingredients. To increase the data set of Chinese medicine on stroke-related disease, we then split TCM-related abstracts into sentences for extract disease, CPM and herbs. We then use several approaches to relations between entities. After verifying and cleaning of the results, we use NEO4J to construct StrokeKG.

The steps of our workflow are explained below (Fig. 1).

Data source

A search for "Stroke AND treatment OR gene OR Herbs OR TCM" in PubMed resulted in 45,080 stroke-related and "Traditional Chinese Medicine" 72,410 TCM related abstracts, which we used as a dataset to extract information from. In addition, manually created databases and annotated corpora, drug-disease relation database: CDR [59], CTD [60], gene-disease relation corpus: EU-ADR [42], and TCMID ETCM [57],

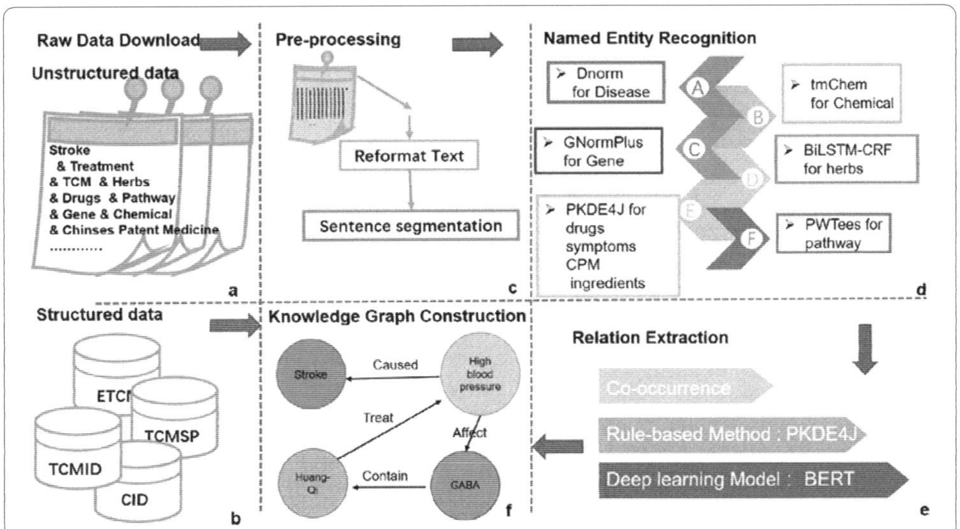

Fig. 1 The workflow for constructing StrokeKG. **a** Unstructured data search, **b** structured data download, **c** pre-processing, **d** Named Entity Recognition, **e** Relation Extraction, **f** Knowledge Graph construction

TCMSP [58] are also the main source of our knowledge graph data. Table 1 details the data source of our research.

Pre-processing

We re-formatted the PubMed abstracts into the PubTator [63] format to match the data for NER tools and then split sentences by NLTK [64].

Named entity recognition

We extract mentions of nine named-entity types (diseases, drugs, genes, symptoms, pathways, Chinese Patent Medicines (CPMs), Herbs, Chemicals, Ingredients). We use state-of-the-art NER methods, including DNorm [34] to extract and normalize disease words, tmChem [38] as a chemical named entity identifier, GNormPlus [33] to handle both gene mentions and identifier detection, and pathways through PWTEES [40].

We used a pre-trained BiLSTM-CRF [65] model with the Plant-disease corpus [62] to build a NER classifier to identify Herbs. The lack of annotated corpora poses a considerable challenge to using deep learning methods to build other NERs needed for our study. We have therefore developed dictionary- and rule-based methods for other entity types. A rule-based method PKDE4J [46] was used to modify the Stanford CoreNLP pipeline to extract entities based on drug dictionaries. Which we have collected Symptoms and ingredients are recognized by collecting terms from download the CPM database [55] and the ingredient database[56], construct a symptoms dictionary, and which are then inserted the dictionary into the PKDE4J model applied a dictionary-based method for NER.

Table 1 Data resource of our StrokeKG

Data source	Types	Number of entities	Number of relations	Number of documents
PubMed	English abstracts	–	–	52,175 abstracts
CTD [60]	Western medicine databases	Chemical (19, stroke-related)	19 (stroke-related)	19 (stroke-related) 470(total)
CDR [59]	Western medicine corpus	Chemical (14, stroke-related) Chemical (1279, total) Disease (1188, total)	15 (stroke-related) 3116 [42] (total)	10 (stroke-related) 1500 (total)
TCMID	TCM databases	CPM (681)	–(therapeutic)681	–
ETCM [57]	TCM databases	Gene(10), Herbs (498), CPM (3419)	–	–
TCMSP [58]	TCM databases	Gene (63), Herbs (114)	–	–
DDI corpus [61]	Western medicine corpus	–	Drug-drug:14,281	
EU-ADR [42] corpus	Western medicine corpus	–	Gene–disease:355	100 Medline abstracts
Plant-disease corpus [62]	TCM databases	Disease:100 Plant:1102	Plant-disease: Therapeutic (Treatment):708 Induce(Cause):486	180 abstracts

Stroke MESHID:D020521, cerebrovascular ischemia/Ischemic stroke/ brain ischemia D007511, cerebral ischemia D002545, transient ischemic attack(TIA): D002546, haemorrhagic stroke/subarachnoid haemorrhage MESH:D013345, cerebrovascular accident(CVA) D002544, Subarachnoid hemorrhageMESH:D01334 D002543

Table 2 Keyword classification rules and corpus for drugs/TCM/Herbs/chemicals and diseases

Entity pairs	Relation types	Associated Keywords	Corpus
CPM-Disease/Symptoms Herbs-Disease/Symptoms Drug-Disease/Symptoms Chemical-Disease/Symptoms	Treatment	Therapy, treating, cure, remedy Inhibit…	Plant-disease Corpus
	Cause	Induce, cause, side effect, influence, dynamic…	
	other	No special key word…	
CPM-Herbs CPM-Chemicals Herbs-Chemicals CPM-Drugs Herbs-Drugs	advise	Avoid, should not be	Drug–Drug Interaction Corpus
	int	Interaction, and, between	
	effect	Enhance, against, demonstrated	
	mechanism	Metabolize, decrease, increase	
	negative	no	
	Entity-Origin	Include, contain,	
Gene-Disease	Positive association (PA)	Effect, induce, target	EU-ADR
	Negative association (NA)	Indifference, no	

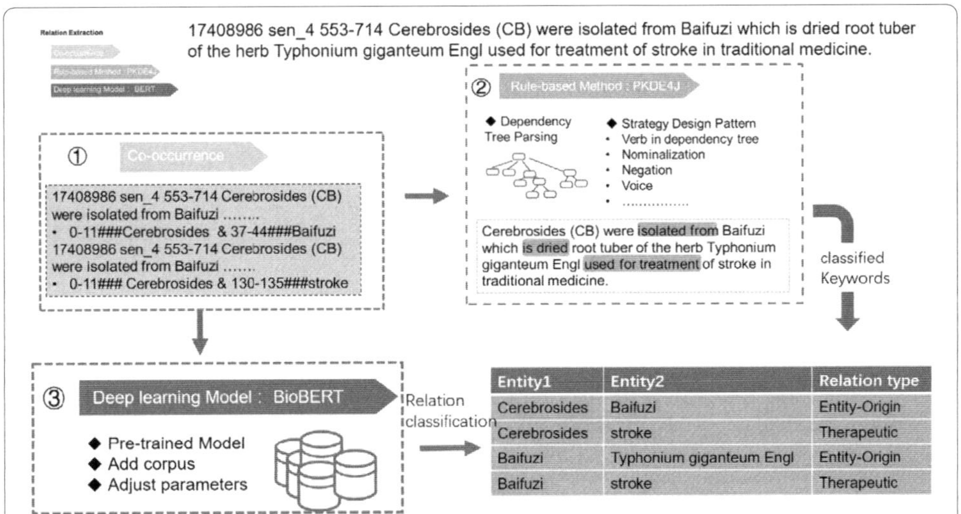

Fig. 2 Pipeline and details about relation extraction

To eliminate the occurrence of an entity by accident, we determine the threshold based on the number of occurrences of the entity. When the number of occurrences of the entity is smaller than 3, We will manually determine whether the entity is related to stroke.

Relation extraction

We focus on eleven relationship types as specified in Table 2. These have been taken from the existing databases and from existing corpora (see below).

The relation extraction process is shown in Fig. 2. We first use a simple co-occurrence method. When two entities appear in the same sentence, we consider that there is a particular relationship between them. Secondly, a rule-based method has been used to extract 'evidence' for the relationship between two entities. Finally, we developed a machine-learning model to further classify relation types according to existing databases or corpora.

- Co-occurrence extraction

  We use NLTK [64] to segment each sentence and match the position of each entity in the sentence according to the entity positions determined by multiple NER model (see Fig. 2①).

- Rule-based approach
  We used PKDE4J [46] to create a dependency tree containing syntactic and grammatical structures. We rely on standard features and structures in sentences that may represent relationships and extract the keywords that may express the relationship between two entities identified via co-occurrence (Fig. 2②). We then designed a set of matching rules to classify these keywords to elven relationship types (e.g., positive association; therapeutic; induce; etc.) between specific pairs of entities (e.g., Gene-Disease; Herb-Chemical) as specified in the existing biomedical databases (e.g., TCMID [55], CTD database [60]) (as shown in Table 1).

- Extracting relation by Bio-BERT

We chose Bio-Bert [54] as a pre-trained model, which shares potential latent features with our data as it was re-trained on biomedical corpora. According to the parameter configuration of BioBERT, we use the gold standard data sets [42, 60, 61] as the training sets and our **Co-occurrence** results as the test set and select the result of the 20th epoch as the final result of our relation extraction. The corpora for relationship extraction we use can be seen in Table 2.

The co-occurrence method proves that two entities appear in the same sentence, indicating that there is a possible relationship between the entities. Rule-based methods can classify entity relationships well if keywords are extracted. When the keywords cannot be extracted, we use BioBERT's classification results, which can classify all relationships, but it much depends on the richness of the corpus and the accuracy of the model.

Because entity pairs may appear in different sentences, the classification results may differ. To find all relations between the pair of two entities, we calculated the confidence for the pair related by a particular relation, overall the sentences in which that pair of entities co-occur. We select only those relationships with confidence more considerably than the threshold to eliminate the noisy relationships that happen by accident. Afterward, we analyze the final relationship results for the entity.

Manual annotations for TCM corpus

Entity annotation

To verify the effectiveness of our Chinese herbal medicine related entity mining tool. This work mainly focuses on the annotation of herbs and Chinese Patent medicine in 450 TCM-related abstracts. We regard mentions mined by the tool as pre-annotation of entities. Therefore, according to the vocabulary provided by TCMID [55] and ETCM [57], we only need to modify the incorrect annotations and add annotations to the undetected entities, instead of annotating entities from scratch.

The definition of the target entities we are concerned with is as follows:

Chinese Patent medicine: including clinical prescription, TCM formulas and CPM.

Relation annotation

In relation annotation task, we only considered two relations between entities. For each relationship, we classified the type of relationship based on the two annotation guidelines. Once two target entities appear in the same sentence, we label the relationship between them.

Chinese patent medicine-disease this indicates the drug will treat the disease or induce the disease. According to Plant-disease corpus, the relationships are divided into 3 categories: treatment, cause and others.

Evaluation of text-mined results

The evaluation of NER and RE was to compare the extracted results with the existing databases or manually annotated corpus.

For TCM-related NER tools, we compare whether the results we extracted overlap with the existing database. Secondly, for CPM entities, we will compare the results by dictionary-based tool and the results we manually annotated.

For relation extraction results, also check the overlap of the entity pair we extracted with the existing database, and then calculate the correct rate (CR) of relationship classification in the overlap section.

$$\text{Correct rate} = \frac{\text{correct classification relationships}}{\text{Overlapped relationships}}$$

Knowledge graph construction

The construction of a knowledge graph is a compelling visual representation of entities and relationships. These are embedded in the knowledge graph to carry information about entities and relationships and are widely used in learning tasks to accelerate the completion and recommendation of the knowledge graph. By mapping the stroke-related entities from our results and existing data source (TCMID [55], CDR [66], CTD [60], TCMSP [58] and ETCM [57]) in a common ID space, we can combine these triplets into one single dataset to construct a comprehensive stroke-related repurposing knowledge graph.

Results

Results statistics

The results mainly include the entities and relations we mined. The statistical results and specific results of drugs, chemicals, symptoms, pathways, etc. are shown in Table 3 and https://github.com/yangxi1016/Stroke.

Relation extraction results statistics are in Table 4.

Table 3 Models and results of named entity recognition

Entity type	Entity_type/(> 3)	Entity_mentions
Disease	3250/2733	220,144
Symptoms	728/485	350,833
Chemical	7062/2246	166,701
Drugs	2156/119	102,072
Chinese Patent medicine	242/manual(265)	914
Herbs	2402/2138	7386
Ingredients	270/170	29,301
Gene	5953/3978(	180,280
Pathway	18,842	105,337

Table 4 Number of relation results by using co-occurrence and PKDE4J

	Entity1	Entity2	Relation pairs (unique pairs)	Co-occurrence	PKDE4J Keywords
1	CPM	Herbs	190	478	120
2	CPM	Ingredients	29	5	2
3	CPM	Drugs	19	30	9
4	CPM	Chemical	220	498	326
5	CPM	Symptoms	138	600	371
6	CPM	Disease	404/704(CPM-Dis++)	1031/7771	637/5439
7	CPM	Pathway	94	22	11
8	CPM	Gene	194/982(CPM-Gene++)	118	53
9	Herbs	Ingredients	515	1185	311
10	Herbs	Drugs	1545	2873	1034
11	Herbs	Chemical	2382	3926	894
12	Herbs	Symptoms	1222	4927	1974
13	Herbs	Disease	2012/12115((Herb-DIS++)	7872	2847
14	Herbs	Pathway	1303	2221	841
14	Herbs	Gene	1686	17,829	12,938
15	Ingredients	Drugs	6775	9485	1337
16	Ingredients	Chemical	3718	14,623	1965
17	Ingredients	Symptoms	1339	5513	2064
18	Ingredients	Disease	341	5224	2008
19	Ingredients	Pathway	1932	2652	870
20	Drugs	Chemical	225	844	564
21	Drugs	Symptoms	9212	27,593	12,888
22	Drugs	Disease	14,756	59,924	41,095
23	Drugs	Pathway	20,053	23,848	7836
	Drugs	Gene	22,873		
24	Chemical	Symptoms	4427	6752	2335
25	Chemical	Disease	8835	21,858	10,062
26	Chemical	Pathway	33,614	58,426	17,584
27	Symptoms	Disease	26,756	101,842	24,683
28	Symptoms	Pathway	23,536	53,599	37,897
29	Disease	GENE	10,084	84,003	16,668

Table 5 F1-score of the rule-based CPM NER tool

Models	Recall	Precision	F1-score
CPM NER	86.04	94.21	90.06

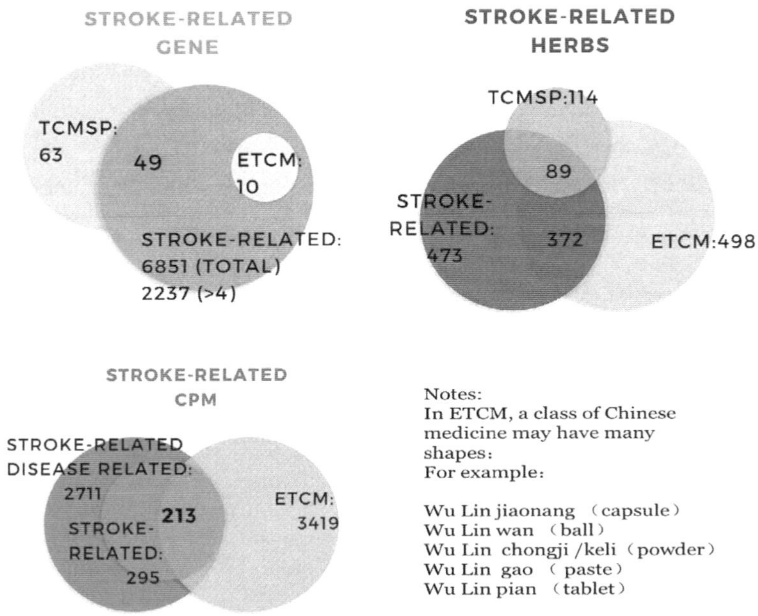

Fig. 3 The comparison of NER results with ETCM [57] and TCMSP [58]

Evaluation

Evaluation for NER

Compared with our manually labeled CPM results, the recall, precision and F1-score of the rule-based CPM NER are shown in Table 5.

The reason for the low recall mainly because of the lack of abbreviation (CY-Tang: Chungsim-Yeunja-Tang) and the different spelling of TCM caused by different pronunciation. (For example, Hwangryun-Hae-Dok-tang and Huanglian-Jie-Du-Tang).

Compare with existing database

To assess how validity the literature-derived knowledge represented data, we compared the results to two Chinese Medicine Pharmacology Knowledge Base: ETCM and TCMSP to those obtained in StrokeKG. Including stroke-related CPM, herbs, and genes. Figure 3 shows the result of comparison with ETCM [57] and TCMSP [58].

Compared with the existing database, our name recognition results partially overlap with the existing database, which indicates that our entity recognition results are effective. More importantly, we have unearthed many stroke-related entities that do not exist in the database. Which provides a new direction for future research.

Evaluation for RE

Compared with our manually labeled CPM-Disease relations, the recall, precision and F1-score of the CPM-Disease RE are shown in Table 6.

On some relationship pairs, the model cannot judge whether it is a Treatment or a Cause, and is classified as Other, which is most of the reason for the error.

As shown in Fig. 4a and Table 7, our mining results include 190 pairs of CPM-Herbs, 4 pairs of CPM-Ingredients, and 515 pairs of Herbs-Ingredients, compared with the existing TCMID (only CPM components and Herbs component table) database, there are 275 pairs of relationships that overlap and the correct rate of the relationship classification results is 91.42%. Secondly, our mining results include 404 pairs of CPM-Disease (with 704 CPM for stroke-related disease) compared with TCMID (only comparing whether herbal medicine has a therapeutic effect on the disease). The rate is 84.37%. The correct rates of the relation between genes-diseases and drugs-diseases are 90.47% and 88.86%, respectively.

To determine if classification of overlaps can and made Table 8.

By detailed analysis, we found our relation extraction method can accurately extract two entities in the same sentence, but there will be errors in the classification of the relations. The main reason is the inability to identify keywords in relation extraction.

Table 6 F1-score of the CPM-Disease RE tool

Models	Recall	Precision	F1-score
PKDE4J + BioBERT	88.92%	80.06%	84.26%

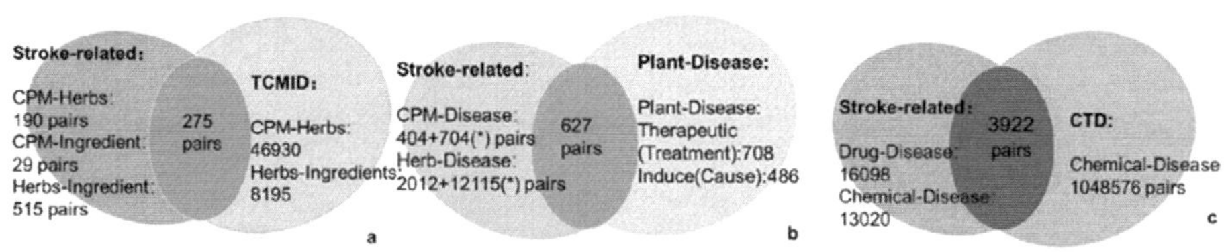

Fig. 4 The part of our mining results that is a duplicate (comparable) with the existing database. **a** The relation between CPM-Herbs, CPM-Ingredients, and Herbs-Ingredients (Comparative database: TCMID). **b** The relation between CPM-Disease, Ingredients-Disease, and Herbs—Disease (Comparative database: Plant-disease). **c** The relation between Drugs-Disease and Chemical-Disease(Comparative database: CTD: Chemical-Disease)

Table 7 The correct rate of our RE results

Relation	Comparative database	number of relation overlapped	Number of correct classification	Correct rate
CPM-Herbs, CPM-Ingredients, and Herbs-Ingredients	TCMID, TCM-Mesh	275	269	97.81%
CPM-Disease, Ingredients-Disease, and Herbs–Disease	Plant-Disease	687	621	90.39%
Gene-Disease	CTD Gene-Disease	378	342	90.47%
Drugs-Disease and Chemical-Disease	CTD + CID: Chemical-Disease	3922	3485	88.86%

Table 8 Comparison of the text mining results of Drug-disease relation and existing databases (CID + CTD)

Relation in CID + CTD database	Relation in our results			
	Treatment	Cause	Other	Total
Treatment/therapeutic	1193	3	318	1514
Cause/induce	0	65	6	71
Other	82	28	2227	2337
Total	1275	96	2551	3922

At the same time, the other purpose of the construction of our knowledge graph is to extract knowledge that may be useful but not included in the existing data set in the vast ocean of data. For this, we compare the size of the data set with the existing biomedical common knowledge base and proposed new possible clinical medical research directions.

StrokeKG

StrokeKG (http://114.115.208.144:7474/browser/) contains a total of 46,983 entities belonging to K=9 entity-types. The type-wise distribution of the entities. StrokeKG contains a total of 157,302 triplets belonging to R=30 edge-types with 659,838 properties. A part of the results, as shown in Fig. 5a, using entities as graphs nodes, and the entities contain entity ID, entity name, and standard classification (MESH). As shown in Fig. 5b, the PMID number of the article where two entities co-exist is used as the edge of the graph. In particular, the edge also contains the keyword (RelationKeyword) extracted by PKDE4J and the relationship classification result (RelationType) based on the BERT model.

To enhance the effectiveness of our knowledge graph, we also annotated the reliable 32,031 nodes of 9 types and 4,800 relationships of 16 types with evidence from the entirely correct part of the evaluation results and the information in the existing database.

Discussion

Detail results of NER and RE

Stroke-related disease/symptoms

In total, we mined 4210 kinds of diseases (401,644 entity mentions) in downloaded documents. (Results detail shows in Additional file 1) According to our results, the expression of stroke in related literature includes synonyms(e.g., Apoplexy (105), Brain ischemia (605), Cerebral ischemia (3183), Cerebrovascular Accident (227), Hemorrhage, Transient Ischemic Attack), abbreviations (e.g., CVA, TIA (722)), lexical changes, and word order changes. The generation of a stroke may be related to other diseases, such as atrial fibrillation (MESH:D001281,2732) diabetes (MESH:D006973, 2571), heart disease(MESH:D006333, 2590) etc., or it may have some sequelae after a stroke, such as acute gastrointestinal bleeding, hypertension, cerebral heart syndrome, pulmonary infection, and acute Pulmonary edema, etc. There are 728 types of symptoms (350,833 mentions), (Results detail shows in Additional file 2) Among them, aging (4103)

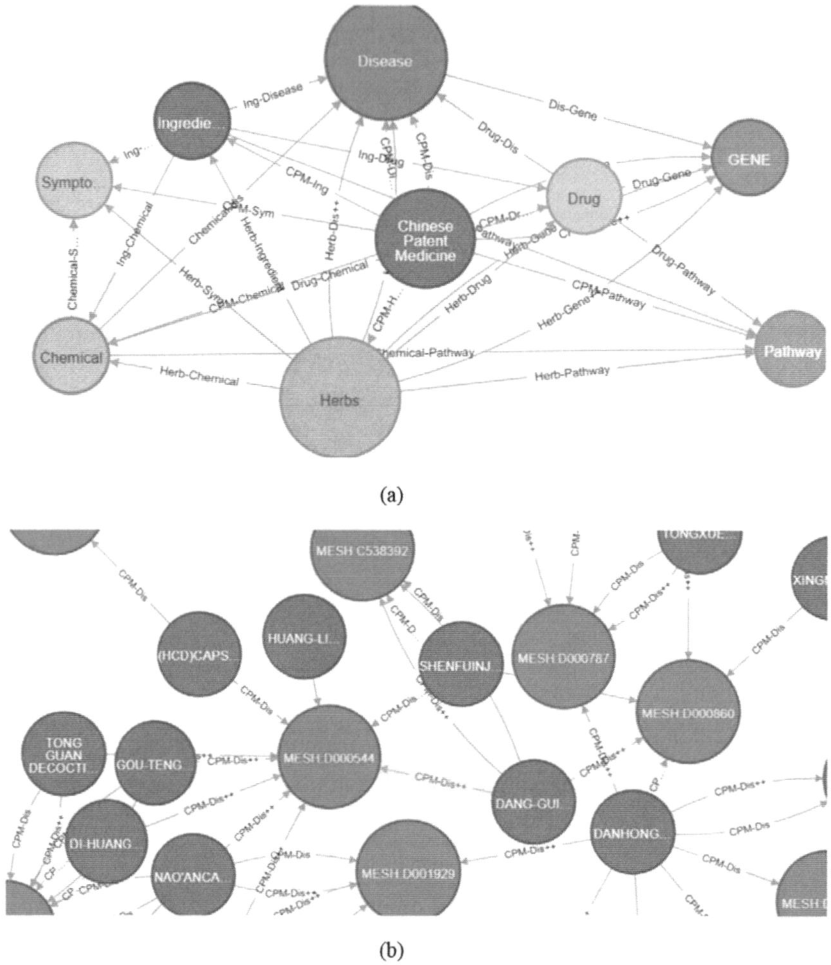

DANG-GUI MESH:D002544
@@ PMID: 15074085 #En1: Angelica #En2: acute cerebral infarction *KeyWords: treating RelationType:Therapeutic @Sen: OBJECTIVE: To study the therapeutic effect of Angelica Injection (AI) in treating acute cerebral infarction (ACI)./

(c)

Fig. 5 Screenshots of our knowledge graph. **a** Overall schema, **b** An example of details about the entity. **c** The relation between Danggui and acute cerebral infarction

mentioned in the 4041 abstract, which is also consistent with the fact that 64% of strokes occur in people aged 55 to 75 years. Virus (731) infection is a possible factor for sudden stroke. At the same time depression (1067), anxiety (301), and other unfavorable psychological conditions are also common complications that we need for stroke patients.

Stroke-related gene and relation between stroke-genes and stroke-related disease-genes

Gene mutations are related to the incidence of stroke. By relation extraction in disease-genes, we found 5953 types of genes (included 180,280 mentions). We linked 1238 diseases and 1574 genes, created 10,094 relationships. The results show that small changes in 588 genes can affect the risk of stroke and nearly 1000 genes affect stroke-related diseases. Specifically, changes in ACE (Angiotensin Converting Enzyme) (803), Collagen

Binding Protein (437), or MTHFr (326) affect the risk of stroke. Secondly, VEGF (558) can be used as a drug target for the treatment of stroke patients. At the same time, the regulation of the brain protein of UCHL1, Hypoxia-inducible factor 1alpha (HIF-1α, 239) may be crucial for how nerve cells repair themselves after a stroke.

Western medicine for treating stroke and stroke-related disease

tmChem system successfully identified 11,129 types of Chemical entities (201,234 mentions) from the abstracts we downloaded. Among them, Ticlopidine, Nimodipine, Triphenyltetrazolium chloride has been mentioned many times and are ingredients contained in various medicines for the treatment of stroke and related diseases. It can be seen from the number of mentions that the aspirin (1475) is main chemical for relieving/ alleviating the risk of stroke, and angiotensin (435) causes vasoconstriction and increased blood pressure, which ultimately leads to stroke. Secondly, oxygen (2290), iron (1524) calcium (1918), glucose (1193) cholesterol (1177), nucleotide (1144), the index of these main compounds on the impact of human stroke and related diseases is the most concerned by the medical community.

According to the drug list provided by Drugbank, we have normalized and classified 2156 kinds of drugs for entities. In addition to the individual elements of statistics in the chemical, The drugs with the greatest impact on stroke are aspirin (DB00945,1475), warfarin (DB00682,1034), clopidogrel (DB00758,666).

TCM for treating stroke and stroke-related disease

We have identified 294 Chinese patent medicines that have played a role in the prevention and treatment of stroke and related diseases. From our mining results, GUALOUGUIZHI DECOCTION (10), KUDIEZIINJECTION(10), DANHONGINJECTION(20), and BUYANGHUANWU DECOCTION(36) are potent medicine in treating stroke. We also extracted 420 species of Herbs (11,671 mentions). DAN-SHEN (58), Chuan-Xiong (50), Dang-Gui (23), Huang-Lian (21), and Bai-Fu-Zi (19) are in various Chinese patent medicines or prescriptions for the treatment of stroke and related diseases. In ingredients extraction, except the ingredients like glucose (1947) cholesterol (1394) glutamate (767) dopamine (478), the unique ingredients in Chinese herbal medicine such as Hyperin (265) and catechol (207) are important for treating stoke-related diseases.

Pathways

In our results, a total of 105,337 pathways mentions were identified. In the subsequent relation extraction process, we use the results for analyzing what kind of molecular pair does the chemical in the medicine or the herbal ingredient play in the disease and identify what the key genes and pathways involved in stroke-related diseases are.

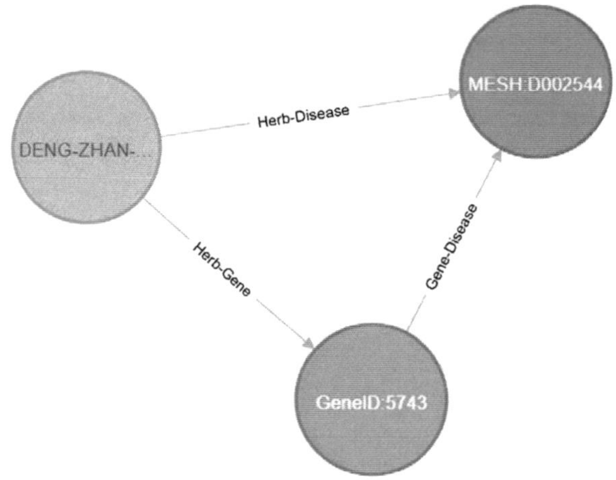

Fig. 6 An example of expressed as a link between any pair of biological entities

For example, the ERK1/2 activity generated by cytokines and free radicals or other inflammatory factors after stroke may worsen ischemic damage, whereas the ERK1/2 activity produced by exogenous growth factors, estrogen, and preconditioning favors neuroprotection.

Discover possible existing CPM to treat stroke

StrokeKG Construction can discover possible existing drugs/CPM/herbs to treat stroke-related diseases to reduce the risk of stroke. Such a task can be expressed as a direct link prediction between the drug and disease entity, or indirectly expressed as a link between any pair of biological entities involved in a particular pathway. For example, *31348992 Intersection analysis between DZXXI's putative targets with ischemic stroke-associated genes identified two important targets (PTGS1, PTGS2)* (Fig. 6).

Conclusions

In this study, we analyzed stroke-related literature with natural language processing, including named entity recognition and relation extraction. We showed that the-state-of-the-art text mining tools could efficiently extract the critical information hidden behind the unstructured data in the biomedical domain.

Through the knowledge base and knowledge graph, we have a clearer understanding of stroke-related diseases, symptoms, gene mutations that cause stroke, and the vital role of Chinese and Western medicine in preventing and treating stroke. We constructed StrokeKG, representing the relation among stroke-related entities successfully.

In future research, we will optimize the relationship mining model in the field of biomedicine, apply the model to all aspects of various diseases, and establish a larger and more comprehensive map of medical knowledge.

Abbreviations
NER: Named Entity Recognition; RE: Relation Extraction; TCM: Traditional Chinese Medicine.

Supplementary Information

The online version contains supplementary material available at https://doi.org/10.1186/s12859-021-04292-4.

Additional file 1. The list of stroke-related diseases.

Additional file 2. The list of stroke-related symptoms.

Acknowledgements
Not applicable.

About this supplement
This article has been published as part of BMC Bioinformatics Volume 22 Supplement 10 2021: Selected articles from the 19th Asia Pacific Bioinformatics Conference (APBC 2021): bioinformatics. The full contents of the supplement are available at https://bmcbioinformatics.biomedcentral.com/articles/supplements/volume-22-supplement-10.

Authors' contributions
GN set up the general idea of this project. XY and WW developed the codes and drafted the manuscript, drafted the discussion and revised the whole manuscript together with CW and KL. All authors have read and approved the manuscript.

Funding
Computing resources are supported by National High-level Personnel for Defense Technology Program (2017-JCJQ-ZQ-013), and NSF 61902405. ***Publication costs are funded by*** the National Key R&D project by Ministry of Science and Technology of China (2018YFB1003203), the open fund from the State Key Laboratory of High Performance Computing (No. 201901-11), and National Science Foundation of China (U1811462). The funder CW and KL took part in the formulation and development of methodology, and provided financial support for this study.

Availability of data and materials
The code files are available at: https://github.com/yangxi1016/Stroke/.

Declarations

Ethics approval and consent to participate
Not applicable.

Consent for publication
Not applicable.

Competing interests
The authors declare that they have no competing interests.

Author details
[1]College of Computer, National University of Defence Technology, Changsha 410073, China. [2]State Key Laboratory of High-Performance Computing, National University of Defence Technology, Changsha 410073, China. [3]Department of Computer Science, University of Manchester, Manchester M13 9PL, UK.

Received: 13 June 2021 Accepted: 6 July 2021
Published online: 29 July 2021

References

1. Lau AY, Wong EH, Leung TW, Mok VC, Wong KS. Intravenous alteplase for Chinese patients with stroke and borderline eligibility. J Clin Neurosci. 2012;19(10):1383–6.
2. Brainin M, Feigin VL, Norrving B, Martins SCO, Hankey GJ, Hachinski V. Global prevention of stroke and dementia: the WSO declaration. Lancet Neurol. 2020;19(6):487–8.
3. Peisker T, Koznar B, Stetkarova I, Widimsky P. Acute stroke therapy: a review. Trends Cardiovasc Med. 2017;27(1):59–66.
4. Ton G, Liao H-Y, Chiang J-H, Chen Y-H, Lee Y-C. Chinese herbal medicine and acupuncture reduced the risk of stroke after Bell's Palsy: a population-based retrospective cohort study. J Altern Complement Med. 2019;25(9):946–56.
5. Katan M, Luft A. Global burden of stroke. In: Seminars in neurology: 2018. Thieme Medical Publishers, 208–211.
6. International Stroke Database [Internet]. 2006 Nov 1 [updated 2015 Jan 1; cited 2020 Dec 8]. Available from: http://www.nmr.mgh.harvard.edu/stroke/index.html
7. OSR Acute Stroke Databases [Internet]. 2015 Oct 4 [updated 2021 May 1; cited 2021 May 6]. Available from: https://www.ices.on.ca/Research/Research-programs/Cardiovascular/Ontario-Stroke-Registry/OSR-Acute-Stroke-Databases
8. SSNAP [Internet]. 2020 Jan 1 [updated 2021 May 4; cited 2021 May 6]. Available from: https://www.strokeaudit.org/

9. Wang Q, Mao Z, Wang B, Guo L. Knowledge graph embedding: a survey of approaches and applications. IEEE Trans Knowl Data Eng. 2017;29(12):2724–43.
10. Ji S, Pan S, Cambria E, Marttinen P, Yu PS. A survey on knowledge graphs: representation, acquisition and applications. arXiv preprint arXiv:2002.00388 2020.
11. Wang X, He X, Cao Y, Liu M, Chua T-S. Kgat: knowledge graph attention network for recommendation. In: Proceedings of the 25th ACM SIGKDD international conference on knowledge discovery & data mining: 2019. 950–958.
12. Bean DM, Wu H, Iqbal E, Dzahini O, Ibrahim ZM, Broadbent M, Stewart R, Dobson RJ. Knowledge graph prediction of unknown adverse drug reactions and validation in electronic health records. Sci Rep. 2017;7(1):1–11.
13. Gyrard A, Gaur M, Shekarpour S, Thirunarayan K, Sheth A. Personalized health knowledge graph. 2018.
14. Ernst P, Siu A, Weikum G. Knowlife: a versatile approach for constructing a large knowledge graph for biomedical sciences. BMC Bioinform. 2015;16(1):157.
15. Mohamed SK, Nováček V, Nounu A. Discovering protein drug targets using knowledge graph embeddings. Bioinformatics. 2020;36(2):603–10.
16. Yuan J, Jin Z, Guo H, Jin H, Zhang X, Smith T, Luo J. Constructing biomedical domain-specific knowledge graph with minimum supervision. Knowl Inf Syst. 2020;62(1):317–36.
17. Teschke R, Zhang L, Long H, Schwarzenboeck A, Schmidt-Taenzer W, Genthner A, Wolff A, Frenzel C, Schulze J, Eickhoff A. Traditional Chinese Medicine and herbal hepatotoxicity: a tabular compilation of reported cases. Ann Hepatol. 2015;14(1):7–19.
18. Wu S, Wu B, Liu M, Chen Z, Wang W, Anderson CS, Sandercock P, Wang Y, Huang Y, Cui L. Stroke in China: advances and challenges in epidemiology, prevention, and management. Lancet Neurol. 2019;18(4):394–405.
19. Consortium GO. Gene ontology consortium: going forward. Nucleic Acids Res. 2015;43(D1):D1049–56.
20. Schriml LM, Arze C, Nadendla S, Chang Y-WW, Mazaitis M, Felix V, Feng G, Kibbe WA. Disease ontology: a backbone for disease semantic integration. Nucleic Acids Res. 2012;40(D1):D940–6.
21. Petry NM, Peirce JM, Stitzer ML, Blaine J, Roll JM, Cohen A, Obert J, Killeen T, Saladin ME, Cowell M. Effect of prize-based incentives on outcomes in stimulant abusers in outpatient psychosocial treatment programs: a national drug abuse treatment clinical trials network study. Arch Gen Psychiatry. 2005;62(10):1148–56.
22. Gregory JK, Lachman N, Camp CL, Chen LP, Pawlina W. Restructuring a basic science course for core competencies: an example from anatomy teaching. Med Teach. 2009;31(9):855–61.
23. Yuanzhuo W, Yantao J, Zeya Z. OpenKG-knowledge computing engine in the era of network big data. Commun Chin Comput Fed. 2014;10(11):30–5.
24. Fauqueur J, Thillaisundara A, Togia T. Constructing large scale biomedical knowledge bases from scratch with rapid annotation of interpretable patterns. arXiv preprint arXiv:1907.01417 2019.
25. Stebbing J, Phelan A, Griffin I, Tucker C, Oechsle O, Smith D, Richardson P. COVID-19: combining antiviral and anti-inflammatory treatments. Lancet Infect Dis. 2020.
26. Sang S, Yang Z, Wang L, Liu X, Lin H, Wang J. SemaTyP: a knowledge graph based literature mining method for drug discovery. BMC Bioinform. 2018;19(1):193.
27. Cohen AM, Hersh WR. A survey of current work in biomedical text mining. Brief Bioinform. 2005;6(1):57–71.
28. Roux KJ, Kim DI, Burke B, May DG. BioID: a screen for protein–protein interactions. Curr Protoc Prot Sci. 2018;91(1):19.23.11–5.
29. Lee K, Kim B, Choi Y, Kim S, Shin W, Lee S, Park S, Kim S, Tan AC, Kang J. Deep learning of mutation-gene-drug relations from the literature. BMC Bioinform. 2018;19(1):21.
30. Zhang Y, Lin H, Yang Z, Wang J, Zhang S, Sun Y, Yang L. A hybrid model based on neural networks for biomedical relation extraction. J Biomed Inform. 2018;81:83–92.
31. Yadav V, Bethard S. A survey on recent advances in named entity recognition from deep learning models. arXiv preprint arXiv:1910.11470 2019.
32. Krauthammer M, Nenadic G. Term identification in the biomedical literature. J Biomed Inform. 37(6):512–26.
33. Wei C-H, Kao H-Y, Lu Z. GNormPlus: an integrative approach for tagging genes, gene families, and protein domains. BioMed Res Int. 2015;2015.
34. Leaman R, Islamaj Doğan R, Lu Z. DNorm: disease name normalization with pairwise learning to rank. Bioinformatics. 2013;29(22):2909–17.
35. Dang TH, Le H-Q, Nguyen TM, Vu ST. D3NER: biomedical named entity recognition using CRF-biLSTM improved with fine-tuned embeddings of various linguistic information. Bioinformatics. 2018;34(20):3539–46.
36. Gerner M, Nenadic G, Bergman CM. LINNAEUS: a species name identification system for biomedical literature. BMC Bioinform. 2010;11(1):85.
37. Perera D, Poulos RC, Shah A, Beck D, Pimanda JE, Wong JW. Differential DNA repair underlies mutation hotspots at active promoters in cancer genomes. Nature. 2016;532(7598):259–63.
38. Leaman R, Wei C-H, Lu Z. tmChem: a high performance approach for chemical named entity recognition and normalization. J Cheminform. 2015;7(S1):S3.
39. Rocktäschel T, Weidlich M, Leser U. ChemSpot: a hybrid system for chemical named entity recognition. Bioinformatics. 2012;28(12):1633–40.
40. Wu C, Schwartz J-M, Brabant G, Peng S-L, Nenadic G. Constructing a molecular interaction network for thyroid cancer via large-scale text mining of gene and pathway events. BMC Syst Biol. 2015;9(6):S5.
41. Wang W, Yang X, Yang C, Guo X, Zhang X, Wu C. Dependency-based long short term memory network for drug–drug interaction extraction. BMC Bioinform. 2017;18(16):578.
42. Van Mulligen EM, Fourrier-Reglat A, Gurwitz D, Molokhia M, Nieto A, Trifiro G, Kors JA, Furlong LI. The EU-ADR corpus: annotated drugs, diseases, targets, and their relationships. J Biomed Inform. 2012;45(5):879–84.
43. Trifirò G, Patadia V, Schuemie MJ, Coloma PM, Gini R, Herings R, Hippisley-Cox J, Mazzaglia G, Giaquinto C, Scotti L. EU-ADR healthcare database network vs. spontaneous reporting system database: preliminary comparison of signal detection. Stud Health Technol Inform. 2011;166:25–30.
44. Junge A, Jensen LJ. CoCoScore: context-aware co-occurrence scoring for text mining applications using distant supervision. Bioinformatics. 2020;36(1):264–71.

45. Sarhan I, El-Sonbaty Y, El-Nasr MA. Semi-supervised pattern based algorithm for arabic relation extraction. In: 2016 IEEE 28th international conference on tools with artificial intelligence (ICTAI): 2016. IEEE: 177–183.
46. Song M, Kim WC, Lee D, Heo GE, Kang KY. PKDE4J: Entity and relation extraction for public knowledge discovery. J Biomed Inform. 2015;57:320–32.
47. Ravikumar K, Rastegar-Mojarad M, Liu H. BELMiner: adapting a rule-based relation extraction system to extract biological expression language statements from bio-medical literature evidence sentences. Database. 2017;2017.
48. Kim J-D, Ohta T, Pyysalo S, Kano Y. 2009. Overview of bionlp'09 shared task on event extraction. In: Proceedings of natural language processing in biomedicine (BioNLP) NAACL 2009 workshop. Citeseer.
49. Liu C, Sun W, Chao W, Che W. Convolution neural network for relation extraction. In: International conference on advanced data mining and applications: 2013. Springer: 231–242.
50. Zhang D, Wang D: Relation classification via recurrent neural network. arXiv preprint arXiv:1508.01006 2015.
51. Xi E, Bing S, Jin Y: Capsule network performance on complex data. arXiv preprint arXiv:1712.03480 2017.
52. Sun M, Zhao S, Gilvary C, Elemento O, Zhou J, Wang F. Graph convolutional networks for computational drug development and discovery. Brief Inform. 2019.
53. Zitnik M, Agrawal M, Leskovec J. Modeling polypharmacy side effects with graph convolutional networks. Bioinformatics. 2018;34(13):i457–66.
54. Yu X, Hu W, Lu S, Sun X, Yuan Z. BioBERT Based Named Entity Recognition in Electronic Medical Record. In: 2019 10th international conference on information technology in medicine and education (ITME): 2019. IEEE: 49–52.
55. Huang L, Xie D, Yu Y, Liu H, Shi Y, Shi T, Wen C. TCMID 2.0: a comprehensive resource for TCM. Nucleic Acids Res. 2017;46(D1):D1117–20.
56. Zhang R-z, Yu S-j. Bai H, Ning K: TCM-Mesh: the database and analytical system for network pharmacology analysis for TCM preparations. Sci Rep. 2017;7(1):2821.
57. Xu H-Y, Zhang Y-Q, Liu Z-M, Chen T, Lv C-Y, Tang S-H, Zhang X-B, Zhang W, Li Z-Y, Zhou R-R. ETCM: an encyclopaedia of traditional Chinese medicine. Nucleic Acids Res. 2019;47(D1):D976–82.
58. Ru J, Li P, Wang J, Zhou W, Li B, Huang C, Li P, Guo Z, Tao W, Yang Y. TCMSP: a database of systems pharmacology for drug discovery from herbal medicines. J Cheminform. 2014;6(1):13.
59. Taboureau O, Nielsen SK, Audouze K, Weinhold N, Edsgärd D, Roque FS, Kouskoumvekaki I, Bora A, Curpan R, Jensen TS. ChemProt: a disease chemical biology database. Nucleic Acids Res. 2010;39(suppl_1):D367–72.
60. Mattingly C, Rosenstein M, Colby G, Forrest J Jr, Boyer J. The Comparative Toxicogenomics Database (CTD): a resource for comparative toxicological studies. J Exp Zool Part A Comp Exp Biol. 2006;305(9):689–92.
61. Herrero-Zazo M, Segura-Bedmar I, Martínez P, Declerck T. The DDI corpus: an annotated corpus with pharmacological substances and drug–drug interactions. J Biomed Inform. 2013;46(5):914–20.
62. Kim B, Choi W, Lee H. A corpus of plant–disease relations in the biomedical domain. PLoS ONE 2019;14(8).
63. Wei C-H, Kao H-Y, Lu Z. PubTator: a web-based text mining tool for assisting biocuration. Nucleic Acids Res. 2013;41(W1):W518–22.
64. Loper E, Bird S. NLTK: the natural language toolkit. arXiv preprint arXiv:cs/0205028 2002.
65. Luo L, Yang Z, Yang P, Zhang Y, Wang L, Lin H, Wang J. An attention-based BiLSTM-CRF approach to document-level chemical named entity recognition. Bioinformatics. 2018;34(8):1381–8.
66. Li J, Sun Y, Johnson RJ, Sciaky D, Wei C-H, Leaman R, Davis AP, Mattingly CJ, Wiegers TC, Lu Z. BioCreative V CDR task corpus: a resource for chemical disease relation extraction. Database. 2016;2016.

Publisher's Note

Springer Nature remains neutral with regard to jurisdictional claims in published maps and institutional affiliations.

Liu *et al. BMC Bioinformatics* (2021) 22:394
https://doi.org/10.1186/s12859-021-04290-6

BMC Bioinformatics

RESEARCH Open Access

Maximum parsimony reconciliation in the DTLOR model

Jingyi Liu[1], Ross Mawhorter[3], Nuo Liu[1,2], Santi Santichaivekin[1], Eliot Bush[2] and Ran Libeskind-Hadas[1*]

From The 19th Asia Pacific Bioinformatics Conference (APBC 2021) Tainan, Taiwan. 3-5 February 2021

*Correspondence:
hadas@cs.hmc.edu
[1] Department of Computer Science, Harvey Mudd College, Claremont, CA, USA
Full list of author information is available at the end of the article

Abstract

Background: Analyses of microbial evolution often use reconciliation methods. However, the standard duplication-transfer-loss (DTL) model does not account for the fact that species trees are often not fully sampled and thus, from the perspective of reconciliation, a gene family may enter the species tree from the outside. Moreover, within the genome, genes are often rearranged, causing them to move to new syntenic regions.

Results: We extend the DTL model to account for two events that commonly arise in the evolution of microbes: *origin* of a gene from outside the sampled species tree and *rearrangement* of gene syntenic regions. We describe an efficient algorithm for maximum parsimony reconciliation in this new DTLOR model and then show how it can be extended to account for non-binary gene trees to handle uncertainty in gene tree topologies. Finally, we describe preliminary experimental results from the integration of our algorithm into the existing xenoGI tool for reconstructing the histories of genomic islands in closely related bacteria.

Conclusions: Reconciliation in the DTLOR model can offer new insights into the evolution of microbes that is not currently possible under the DTL model.

Keywords: Microbial evolution, Phylogenetic trees, Maximum parsimony reconciliation

Background

Microbes occupy a vast range of ecological niches [1]. Understanding how particular species have come to occupy their niches requires us to reconstruct how their genomes have evolved over time.

In a clade of closely related microbes with a known gene and species tree, inferring the genetic history can be done through a process called *reconciliation*. This process maps the gene tree to the species tree, and in doing so implies genetic events that explain the discordance between the two trees. The DTL model considers duplication, horizontal gene transfer, and loss events whereas some models consider a subset of these events (e.g., only duplication and loss) or different types of events (e.g,. incomplete lineage sorting).

While the DTL model is applicable to evolution in microbes, it only allows horizontal transfer between species that are part of the species tree. In the analysis of the evolution of microbes in particular, it is quite common that the species tree is not fully sampled. Thus, from the perspective of performing a reconciliation analysis, a gene family may effectively enter the given species tree via transfer from the outside [2, 3].

In this paper we describe the DTLOR model that addresses this issue by extending the DTL model to allow some or all of the evolution of a gene family to occur outside of the given species tree and for transfers events to occur from the outside. To facilitate the recognition of such entry events, the model also keeps track of the *syntenic region* of each gene as it evolves in the species tree. Two genes are said to be in the same syntenic region if they share a substantial fraction of core genes in a relatively large window around them and, second, they share a certain amount of similarity among all genes in a smaller window around them [2]. Thus, in addition to duplication, transfer, and loss events, the DTLOR model adds *origin* events to indicate that a gene is transferred from outside of the species tree and *rearrangement* events that account for changes in the syntenic regions of genes on the genome.

In the DTL model, reconciliation is generally performed using a maximum parsimony formulation. A positive cost is associated with each type of event and the objective is to find a reconciliation that minimizes the total cost of the incurred events. Efficient algorithms have been developed for maximum parsimony reconciliations (MPRs) in the DTL model [4–6], and several software tools implement these algorithms [7–10].

In earlier related work, Delabre et al. studied a related problem of reconciliation with synteny information in the Duplication-Loss model; horizontal transfer events were not considered in that work [11]. Szöllősi et al. [3] proposed an event called "transfer from the dead" to account for gene evolution that occurs outside of the species tree and Jacox et al. [10] described an extension of an existing DTL maximum parsimony reconciliation algorithm to compute most parsimonious reconciliations with this additional event. Our work differs in two significant ways from that prior work. First, while "transfer from the dead" allows gene lineages to transfer out of and back into the sampled species trees multiple times, the DTLOR model only permits transfer from the outside under the assumption that the species tree comprises closely-related species and, thus, transfers out of the species tree and back in are considered to be relatively rare. Second, the DTLOR model captures rearrangement events, which are not considered in conjunction with DTL events in previous models. Reconstructing rearrangement events is particularly important in identifying genomic islands in bacteria [2].

In summary, in this paper we extend the DTL model to allow origin (O) and rearrangement (R) events. We give an exact polynomial-time algorithm for maximum parsimony reconciliation in the DTLOR model. Since gene trees are often non-binary due to lack of signal in their sequence data, we show how maximum parsimony reconciliations can be found in fixed-parameter polynomial time for non-binary gene trees where the parameter is the maximum branching factor of a node. Finally, we describe preliminary results from the integration of the DTLOR MPR algorithm into the xenoGI tool [2] which may provide new insights into microbial evolution.

Definitions

An instance of the DTLOR reconciliation problem comprises undated rooted species and gene trees, S and G, respectively; a positive integer syntenic region number associated with each leaf vertex (extant gene) in G; and a mapping of the leaves of G to the leaves of S. We assume that both trees are binary, but consider the case that gene trees may be non-binary in Sect. 4. Some leaves of the gene tree may be in the same syntenic region while others may be in unique syntenic regions. The DTLOR model comprises the standard DTL events (duplication, transfer, and loss; described in detail below) [4] and two additional events called *origin* and *rearrangement*. Each of those five event types has an associated positive cost.

A syntenic region number is a positive integer from the set of syntenic region numbers of the leaves of the gene tree (called an *actual syntenic region*) or the special *unknown syntenic region* symbol $*$. When a gene vertex is labeled with $*$, that vertex is assumed to be evolving outside the species tree. When a gene vertex is assigned an actual syntenic region but its parent has unknown syntenic region, this means that the gene entered the species tree through transfer from outside, inducing an origin event. Rearrangement indicates a change in syntenic region that happens during the course of evolution within the species tree.

The rules for syntenic region numbering are as follows:

1. If a vertex u is labeled $*$ and v is a child of u, then v may be labeled with either $*$ or an actual syntenic region number.
2. If a vertex u is labeled with an actual syntenic region number, then its children must be labeled with actual syntenic region numbers. Note that this implies that any vertex labeled with an actual syntenic region number has the property that all of its descendants are also labeled with actual syntenic region numbers.

Constraint 1 ensures that genes can originate outside the species tree while constraint 2 ensures that once a gene is found in the species tree it continues evolving within the tree.

The DTL events in this model are analogous to those in the DTL model. O and R events are induced as follows:

1. If a vertex u is labeled $*$ and a child v is labeled with an actual syntenic region number, then vertex v induces an O event.
2. If a vertex u and its child v have actual syntenic region numbers and those two syntenic region numbers are different, then an R event is induced on the edge between u and v.

The objective of the DTLOR maximum parsimony reconciliation (DTLOR MPR) problem is to map the vertices and edges of the gene tree onto the species tree and to identify a syntenic region number with each internal vertex in the gene tree, minimizing the total cost of the induced events. Note that this model implicitly assumes that duplications are tandem or proximal duplications, and thus a duplication event by itself does not imply a change of syntenic region. A duplication that gives rise to a copy at a different syntenic region is modeled implicitly by a duplication and a rearrangement event. The model can be extended to permit other types of duplication events.

Notation

Let S and G denote a pair of undated species and gene trees, respectively. Throughout this and the next section, we assume that S and G are binary. In Sect. 4 we extend results to non-binary trees.

For a tree T, let root(T) be the root and Le(T) be the set of leaves or *tips*. For a non-root vertex v in the tree, $p(v)$ is the parent of v. For a non-leaf vertex v, v_1 and v_2 denote its two children. We assume that each tree T has an additional *handle* edge, namely an edge $(u, \text{root}(T))$. The handle of S is denoted e^S and the handle of G is denoted e^G. For a vertex v of T, we let $T(v)$ be the subtree of T rooted at v, including its own handle edge e^v from $p(v)$ to v. An edge of a tree T is said to be a *leaf edge* if its terminus is a leaf and is said to be an *internal edge* otherwise.

The DTLOR MPR problem

An instance of the DTLOR-MPR problem is a 10-tuple $(S, G, L, \phi, \gamma, \mathbf{D}, \mathbf{T}, \mathbf{L}, \mathbf{O}, \mathbf{R})$ where:

- $S = (V_S, E_S)$ and $G = (V_G, E_G)$ are binary species and gene trees, respectively;
- L is a finite set of syntenic regions which are represented by counting numbers;
- $\phi : \text{Le}(G) \rightarrow \text{Le}(S)$ is a mapping that associates each leaf of G with a leaf of S;
- $\gamma : \text{Le}(G) \rightarrow L$ is a surjective mapping that associates each leaf of G with a syntenic region;
- Parameters **D**, **T**, **L**, **O**, **R** are positive costs for duplication, transfer, loss, origin, and rearrangement events, described in detail below.

A *reconciliation* in the DTLOR model comprises a pair of mappings (Φ, Γ) that extend the mappings ϕ and γ. Specifically, $\Phi : V(G) \rightarrow V(S) \cup \{N\}$ maps the vertices of G to the vertices of S or the special N location representing a species that is not in the species tree S. The constraints on Φ are as follows:

1. $\Phi(g) = \phi(g)$ for each leaf g of G;
2. If g is an internal vertex of G and $\Phi(g) \neq N$ then the children of g, denoted g_1 and g_2, have the properties that

   (a) $\Phi(g_1) \neq N$ and $\Phi(g_2) \neq N$;
   (b) Neither $\Phi(g_1)$ nor $\Phi(g_2)$ is an ancestor of $\Phi(g)$; and
   (c) At least one of $\Phi(g_1)$ or $\Phi(g_2)$ is equal to or a descendant of $\Phi(g)$.

Constraint 1 ensures that the mapping Φ is consistent with the leaf mapping ϕ while constraint 2 ensures that for any gene vertex mapped to a species vertex, (a) the children of g are also mapped to species vertices, (b) the children are not mapped to species vertices that are ancestral to their parent, and (c) at most one child can transfer to a different clade.

Note that we assume that the trees are undated and it is therefore possible that a mapping that satisfies these constraints is, nonetheless, time-inconsistent in the sense that there is no ordering of the internal nodes of the species tree that is consistent with the set of duplication, transfer, and loss events. But time-inconsistencies in an MPR can be detected in polynomial-time [12, 13]. Moreover, the problem of finding MPRs that are guaranteed to be time-consistent is NP-hard [14].

Note also that unlike the DTL model, which requires every gene vertex to be mapped to a vertex in the species tree, the DTLOR model allows gene vertices to be mapped to the N location which is outside the sampled species tree.

The mapping Φ induces four types of events. For an internal gene tree vertex g, with children g_1 and g_2, and $\Phi(g) \neq N$, the events induced by Φ are as follows:

Speciation event: Vertex g induces a speciation event if one of $\Phi(g_1)$ and $\Phi(g_2)$ is in the left subtree and the other is in the right subtree of $\Phi(g)$.

Duplication event: Vertex g induces a duplication event if each of $\Phi(g_1)$ and $\Phi(g_2)$ is either equal to or a descendant of $\Phi(g)$ but does not satisfy the requirements for a speciation event.

Transfer event: Vertex g induces a transfer event if exactly one of $\Phi(g_1)$ and $\Phi(g_2)$ is either equal to or a descendant of $\Phi(g)$ and the other is neither an ancestor nor a descendant of $\Phi(g)$.

Loss events: Each non-root vertex g (including leaf vertices) may induce zero or more loss events as follows: If $\Phi(p(g)) \neq N$ is ancestral to $\Phi(g)$, then each species vertex s on the path from $\Phi(p(g))$ to $\Phi(g)$ induces a loss event, except for $\Phi(g)$ and also not $\Phi(p(g))$ if $p(g)$ induces a speciation event. For each loss induced by a vertex s on the path from $\Phi(p(g))$ to $\Phi(g)$, we say that g *passes* through s.

If $\Phi(g) = N$ then g induces none of these four types of events.

The mapping $\Gamma : V(G) \to L \cup \{*\}$ maps each vertex g in G to an element of L or the special syntenic region represented by $*$ indicating that it is in an unknown syntenic region because it occurs outside of the species tree. The constraints on Γ and its relationship to Φ are as follows:

1. $\Gamma(g) = \gamma(g)$ for each leaf g of G;
2. $\Phi(g) = N$ if and only if $\Gamma(g) = *$;
3. If $\Gamma(g) \neq *$ and g has children g_1 and g_2 then $\Gamma(g_1) \neq *$ and $\Gamma(g_2) \neq *$.

Constraint 1 ensures that the mapping Γ is consistent with the leaf mapping γ, constraint 2 ensures that if a gene vertex is mapped outside of the species tree then its syntenic region is not yet established, and constraint 3 ensures that once the syntenic region for a gene node is established, the syntenic regions of its children are also established. The mapping Γ induces events as follows:

Origin event: A non-root vertex g induces an origin event if $\Gamma(p(g)) = *$ and $\Gamma(g) \neq *$. The root vertex $\text{root}(G)$ induces an origin event if $\Gamma(\text{root}(G)) \neq *$.

Rearrangement event: A non-root vertex g induces a rearrangement event if $\Gamma(g) \neq *$, $\Gamma(p(g)) \neq *$, and $\Gamma(p(g)) \neq \Gamma(g)$.

The cost of a reconciliation is defined to be the sum of the number of duplication, transfer, loss, origin, and rearrangement events scaled by the event costs **D**, **T**, **L**, **O**, and **R**, respectively. Speciation events are assigned an implicit cost of zero because a gene is expected to diverge when the species that carries it diverges.

Methods

When a gene vertex g induces an origin event, all of the genes in the subtree $G(g)$ rooted at g must have actual syntenic regions (by rule 3 in the definition of Γ), and the genes in that subtree are mapped to species in S (by rule 2 in the definition of Γ), that is, $\Phi(g') \in V_S$ and $\Gamma(g') \in L$ for all $g' \in G(g)$. The mappings Φ and Γ are only related by the constraint that $\Phi(g) = N$ iff $\Gamma(g) = *$. Thus, if g induces an origin event, then the pair of mappings Φ and Γ restricted to the domain $G(g)$ are independent. Therefore, for an *origin subtree*, a subtree of G whose root induces an origin event, the process of finding an optimal species mapping Φ can be decoupled from the process of finding an optimal syntenic region mapping Γ. Further, by definition, vertices that induce origin events cannot be ancestrally related. Thus, in a reconciliation (Φ, Γ) where g', g'' induce origin events, the species and syntenic region mappings restricted to the origin subtree $G(g')$ are independent of the mappings restricted to the origin subtree $G(g'')$.

For binary gene trees, we use a dynamic programming algorithm to compute the optimal cost of a species mapping of each subtree of the gene tree. Then, we use a second dynamic programming algorithm to compute the optimal cost for the syntenic region mapping for each subtree. Finally, a third algorithm combines these results to find an optimal solution to the DTLOR MPR problem. For non-binary gene trees, this decoupling is no longer possible and a different (and less efficient) algorithm is presented in Sect. 9.

Computing the species map

Next, we give an efficient algorithm for computing an optimal species mapping for each origin subtree $G(g)$. The algorithm is similar to other DTL reconciliation algorithms [4], but the variant used here is useful in the extensions and generalizations in later sections.

For a species mapping Φ, or its restriction to an origin subtree of the gene tree, we say that a gene tree edge e_g is *placed* on species tree edge e_s if either $\Phi(g) = s$ or if the path from $\Phi(p(g))$ to $\Phi(g)$ includes vertex s, unless $p(g)$ is involved in a speciation event at s. As a special case, if g is the root of an origin subtree, then $\Phi(p(g)) = N$. In this case, there is no path from $\Phi(p(g))$ to $\Phi(g)$, so e_g is placed on e_s if and only if $\Phi(g) = s$. If

$\Phi(g) = s$ we say that e_g *terminates* on edge e_s and if $\Phi(g)$ is a descendant of s then a loss event is induced and we say that e_g *continues* on the corresponding child edge of e_s.

Let $C(g)$ denote the optimal cost for a species mapping restricted to the domain of $G(g)$ and let $C(e_g, e_s)$ denote the optimal cost for a species mapping of $G(g)$ such that e_g is placed on e_s. Then $C(g) = \min_{e_s \in E_S} C(e_g, e_s)$.

We now describe an algorithm for computing $C(e_g, e_s)$. The algorithm computes the C table by considering edges in the gene tree bottom-up (postorder): An edge e_g is considered if either g is a leaf or the children edges e_{g_1} and e_{g_2} have already been considered. For each edge e_g under consideration, we now consider each edge e_s in the species tree in postorder.

To compute $C(e_g, e_s)$, we enumerate the four possible cases:

- In the base case, if g and s are leaves, then:

$$C(e_g, e_s) = \begin{cases} 0 & \text{if } \phi(g) = s \\ \infty & \text{otherwise} \end{cases} \tag{1}$$

- If neither g nor s is a leaf, then either g is mapped to s or not. If g is not mapped to s, then it induces a loss at s by being mapped to one of its children. Otherwise, g is mapped to s, which induces either a speciation, duplication, or transfer event, which incurs a corresponding cost. Thus,

$$\begin{aligned} C(e_g, e_s) = \min\{&\text{Spec}(e_g, e_s), \text{Loss}(e_g, e_s) \\ &\text{Dup}(e_g, e_s), \text{Transfer}(e_g, e_s)\} \end{aligned} \tag{2}$$

  where the computation of Spec, Loss, Dup, and Transfer are described below.
- If g is not a leaf but s is a leaf, then speciation and loss at g are not possible, so:

$$C(e_g, e_s) = \min\{\text{Dup}(e_g, e_s), \text{Transfer}(e_g, e_s)\} \tag{3}$$

- If g is a leaf but s is not a leaf, then speciation, duplication and transfer at g are not possible, so:

$$C(e_g, e_s) = \text{Loss}(e_g, e_s) \tag{4}$$

The functions Spec(e_g, e_s), Loss(e_g, e_s), Dup(e_g, e_s), and Transfer(e_g, e_s) are computed as follows:

$$\begin{aligned} \text{Spec}(e_g, e_s) = \min\{&C(e_{g_1}, e_{s_1}) + C(e_{g_2}, e_{s_2}), \\ &C(e_{g_1}, e_{s_2}) + C(e_{g_2}, e_{s_1})\} \end{aligned} \tag{5}$$

$$\text{Loss}(e_g, e_s) = \mathbf{L} + \min\{C(e_g, e_{s_1}), C(e_g, e_{s_2})\} \tag{6}$$

$$\text{Dup}(e_g, e_s) = \mathbf{D} + C(e_{g_1}, e_s) + C(e_{g_2}, e_s) \tag{7}$$

$$\text{Transfer}(e_g, e_s) = \mathbf{T} + \min \begin{cases} C(e_{g_1}, e_s) + \text{Best-Transfer}(e_{g_2}, e_s), \\ C(e_{g_2}, e_s) + \text{Best-Transfer}(e_{g_1}, e_s) \end{cases} \quad (8)$$

The speciation term (5) considers both ways in which the children edges of e_g can be placed onto the children edges of e_s in a speciation event. The loss term (6) considers both ways in which edge e_g can continue, either on one child of e_s or the other. The duplication term (7) places both children edges of e_g on e_s. In the transfer term (8), we consider both ways of selecting the transferred child edge. The non-transferred child edge of e_g remains on e_s, but the transferred child edge is placed on a species edge determined by Best-Transfer; Best-Transfer(e_j, e_s) denotes the minimum cost of a mapping of the subtree $G(g_j)$ assuming that e_j is placed on a species edge that is neither ancestral nor descendant to e_s. In order to compute these values, the algorithm maintains another table called Best-Entry(e_g, e_s) which stores the minimum of $C(e_g, e_i)$ over all e_i in the subtree rooted at e_s. The algorithm is given in Algorithm 1.

```
for each gene edge e_g in postorder do
    for each species edge e_s in postorder do
        if g is not a leaf then
            if s is not a leaf then
                C(e_g, e_s) =
                  min{SPEC(e_g, e_s), LOSS(e_g, e_s)
                  DUP(e_g, e_s), TRANSFER(e_g, e_s)};
            else
                C(e_g, e_s) =
                  min{DUP(e_g, e_s), TRANSFER(e_g, e_s)};
            end
        else
            if s is not a leaf then
                C(e_g, e_s) = LOSS(e_g, e_s)
            else
                if φ(g) = s then C(e_g, e_s) = 0 else
                  C(e_g, e_s) = ∞;
            end
        end
        if s is a leaf then
            BEST-ENTRY(e_g, e_s) = C(e_g, e_s)
        else
            BEST-ENTRY(e_g, e_s) =
              min{C(e_g, e_s), BEST-ENTRY(e_g, e_s1),
              BEST-ENTRY(e_g, e_s2)};
        end
    end
    BEST-TRANSFER(e_g, e^S) = ∞ ;
    for each non-handle species edge e_s in preorder do
        BEST-TRANSFER(e_g, e_s1) =
          min{BEST-TRANSFER(e_g, e_s), BEST-ENTRY(e_g, e_s2)};

        BEST-TRANSFER(e_g, e_s2) =
          min{BEST-TRANSFER(e_g, e_s), BEST-ENTRY(e_g, e_s1)};

    end
    C(g) = min_{e_s ∈ E_S} C(e_g, e_s)
end
```

Computing the synteny map

We use another dynamic programming algorithm to find the optimal cost for a syntenic region mapping for each subtree $G(g)$. Let syn(g) denote the optimal cost for a syntenic region mapping of $G(g)$. Let $\text{syn}(g, \ell)$ denote the optimal cost for a syntenic region mapping of $G(g)$ such that g has the syntenic region ℓ. Then $\text{syn}(g) = \min_{\ell \in L} \text{syn}(g, \ell)$.

If g is a leaf, then $\Gamma(g) = \gamma(g)$ (by rule 1 in the definition of Γ). Thus,

$$\text{syn}(g, \ell) = \begin{cases} 0 & \text{if } \gamma(g) = \ell \\ \infty & \text{otherwise} \end{cases} \tag{9}$$

If g is not a leaf, then (recalling that g_1 and g_2 denote the children of g):

$$\begin{aligned} \text{syn}(g, \ell) = &\min\{\text{syn}(g_1, \ell), \mathbf{R} + \text{syn}(g_1)\} + \\ &\min\{\text{syn}(g_2, \ell), \mathbf{R} + \text{syn}(g_2)\} \end{aligned} \tag{10}$$

This accounts for each child g_1 and g_2 either remaining in the same syntenic region as g or potentially changing to a new region and incurring a cost of $\mathbf{R}$. The algorithm for computing $\text{syn}(g, \ell)$ is summarized in Algorithm 2.

```
for each gene vertex g in postorder do
    for each actual syntenic region ℓ ∈ L do
        if g is a leaf then
            syn(g, ℓ) = 0 if ℓ = γ(g)
            syn(g, ℓ) = ∞ otherwise
        else
            SYN(g, ℓ) =
              min{SYN(g1, ℓ), R + SYN(g1)} +
              min{SYN(g2, ℓ), R + SYN(g2)}
        end
    end
    SYN(g) = min_{ℓ∈L} SYN(g, ℓ)
end
```

Solving the DTLOR MPR problem

Let $\text{Origin}(g)$ denote the cost of reconciling an origin subtree $G(g)$ which, as noted earlier, can be computed as $\text{Origin}(g) = \mathbf{O} + C(g) + \text{syn}(g)$. To find a maximum parsimony reconciliation, we must therefore determine the optimal locations for origin events.

Let $\text{Null}(g)$ be the optimal cost of reconciling $G(g)$ such that g has the unknown syntenic region $*$. Since the given mapping γ of leaves to syntenic regions must be respected, g may not be assigned syntenic region $*$ if g is a leaf. Thus, $\text{Null}(g)$ is calculated as:

$$\text{Null}(g) = \begin{cases} \infty & \text{if } g \text{ is a leaf} \\ \min\{\text{Null}(g_1), \text{Origin}(g_1)\} + & \\ \min\{\text{Null}(g_2), \text{Origin}(g_2)\} & \text{otherwise} \end{cases} \tag{11}$$

The optimal cost for reconciling the entire gene tree G is given by:

$$\text{Opt} = \min\{\text{Null}(\text{root}(G)), \text{Origin}(\text{root}(G))\} \tag{12}$$

The algorithm for computing $\text{Null}(g)$ is summarized in Algorithm 3.

```
for each gene vertex g in postorder do
    ORIGIN(g) = O + SYN(g) + C(g)
    if g is a leaf then
        NULL(g) = ∞
    else
        NULL(g) =
          min{NULL(g_1), ORIGIN(g_1)} +
          min{NULL(g_2), ORIGIN(g_2)}
    end
end
return min{NULL(root(G)), ORIGIN(root(G))}
```

Note that if we wish to reconstruct an optimal solution, the DP tables C, syn, Null can be annotated in the standard way, allowing the solutions to be reconstructed by tracing through the table. We first trace through the Null table to find a set of origin events that produce an optimal solution. For any gene vertex not in any of the origin subtrees induced by these origin events, they are labeled with the unknown syntenic region $*$. For each origin subtree, we trace through the syn table to get an optimal syntenic region mapping, and then we trace through the C table to get an optimal species mapping. Because of loss events, there can be multiple $C(e_g, e_s)$ entries that involve the same gene vertex in an optimal solution. The mapping of that gene vertex corresponds to the lowest such e_s.

The proofs of the following are given in the Appendix:

Lemma 1 *Algorithm 1 correctly computes $C(g)$ for every gene vertex $g \in V(G)$.*

Lemma 2 *Algorithm 2 correctly computes* syn(g) *for every gene vertex $g \in V(G)$.*

Theorem 1 *Algorithm 3 correctly computes the optimal solution to the DTLOR-MPR problem.*

Time complexity

Computing each entry of the C table takes constant time, so the running time for computing the C table is $O(|G||S|)$. Computing the syn table takes $O(|G||L|)$ time, and computing the Origin and Null entries takes $O(|G|)$ time. In total, the asymptotic running time of this algorithm is $O(|G||S| + |G||L|)$.

Non-binary gene trees

While it is generally possible to construct accurate species trees using a variety of methods, gene trees are susceptible to ambiguity due to the relatively little information available in their sequence data. Consequently, phylogenetic trees for genes often have non-binary vertices, also known as *multifurcations* or *soft polytomies*, which correspond to an unknown ordering of the underlying sequence of divergences [15]. In this case, we wish to expand each multifurcation into a sequence of binary divergences, leading to a binary gene tree. Such an expansion is called a *resolution* or *binarization* of the non-binary tree. The DTLOR MPR problem for non-binary trees seeks to find the optimal reconciliation of G and S over all possible resolutions of G.

Unfortunately, the number of resolutions of a non-binary tree can be exponential in the number of vertices in the tree. It is, therefore, impractical to explicitly consider every resolution. However, Kordi and Bansal [15] and Jacox et al. [16] demonstrated the existence of fixed-parameter polynomial-time algorithms for finding maximum parsimony reconciliations for non-binary trees in the DTL model. These algorithms operate in polynomial time assuming the maximum number of children of any non-binary vertex is bounded by some constant k. More precisely, a fixed-parameter algorithm in this context runs in time $O(f(k)p(m, n))$ where m and n denote the sizes of the gene and species trees, k is the maximum branching factor of any gene vertex, $p(m, n)$ is a polynomial in m and n, and $f(k)$ is some function of k which may even be exponential in k. In particular, $f(k)$ in this context is the number of distinct binary resolutions of a tree comprising a root and k children times the size of such a binary resolution, which can be shown to be $f(k) = O(2^k(k-1)!)$. For any fixed k, this value is a fixed constant. Jacox et al. [16] offer an approach that results in an $f(k)$ which is smaller but still potentially exponential in k. Importantly, a fixed-parameter polynomial-time algorithm is much more efficient and practical than an exponential time algorithm such as the naive approach of enumerating all possible resolutions of a non-binary tree which would have running time $O((2^k(k-1)!)^n)$.

In this section, we describe a fixed-parameter polynomial time algorithm for the DTLOR MPR problem. Following the approach of Kordi and Bansal [15], our algorithm expands each individual non-binary vertex into every possible binary resolution but avoids enumerating all possible binary resolutions of the entire tree, thus resulting in a fixed-parameter polynomial-time algorithm rather than an exponential-time algorithm. While our algorithm leverages the important ideas for resolving non-binary vertices first proposed by Kordi and Bansal [15], it requires a new algorithm due to the advent of O and R events.

Each binary resolution of a non-binary gene tree implies a different topology for the gene tree, which induces potentially different costs for both the species mapping and the syntenic region mapping. Note that one resolution may be most favorable for minimizing the cost of the species mapping while a different one may admit the least expensive syntenic region mapping (Fig. 1). Thus, while in binary gene trees the species mapping and syntenic region mappings could be efficiently solved independently and then merged into an optimal solution for the DTLOR MPR problem, the situation is more complicated in the presence of non-binary gene trees. The algorithm presented here considers the species mapping and syntenic region mapping simultaneously as non-binary vertices are resolved one-by-one. Importantly, once the best

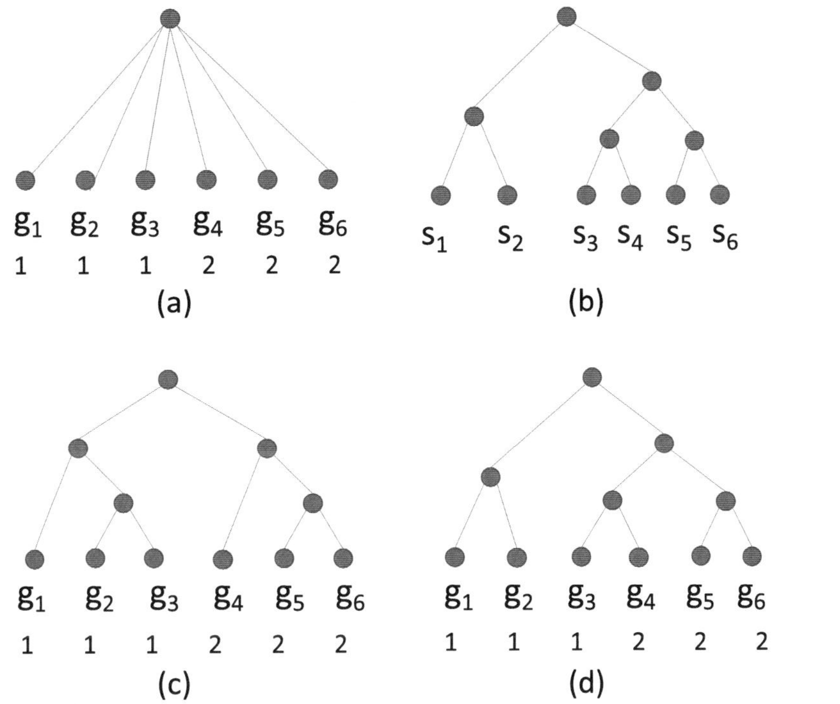

Fig. 1 An example showing that one resolution of a multifurcation can be optimal for species mapping while another may be optimal for syntenic region mapping. **a** A gene tree with six leaves labeled with their syntenic regions. **b** A species tree. The tip association is $\phi(g_i) = s_i, 1 \leq i \leq 6$. **c** A binary resolution of the gene tree in which the optimal species mapping cost is necessarily greater than the origin cost **O** since this tree is not isomorphic to the species tree; the optimal number of rearrangements in this case is 1. **d** Another binary resolution of the gene tree in which the optimal species mapping cost is just the origin cost **O** since this tree is isomorphic to the species tree; the optimal number of rearrangements in this case is 2

resolution is found for the subtree rooted at a given gene vertex, that value can be saved and used as the dynamic program processes the ancestors of g. For this reason, it is not necessary to consider complete resolutions of the gene tree but, instead, the non-binary vertices can be resolved one-at-a-time. This results in an algorithm that is additive, rather than multiplicative, in the number of resolutions of individual non-binary vertices.

The following definitions and equations assume that the terminus vertex g of edge e_g is either a leaf or has exactly two children. Later, we show how to apply these to non-binary trees. Let $C(e_g, e_s, \ell)$ denote the optimal cost of reconciling the subtree $G(g)$ with S such that e_g is placed on e_s and g has the syntenic region ℓ. Note that in contrast to $C(e_g, e_s)$ used in the previous section, $C(e_g, e_s, \ell)$ also encodes the constraint that g has syntenic region ℓ and the total cost includes the cost of rearrangement events in the subtree $G(g)$. We define Best-Entry(e_g, e_s, ℓ) and Best-Transfer(e_g, e_s, ℓ) analogously to Best-Entry(e_g, e_s) and

Best-Transfer(e_g, e_s), respectively, in the previous section. We define $C(e_g, e_s, L) = \min_{\ell \in L} C(e_g, e_s, \ell)$ and

$$\text{Best-Transfer}(e_g, e_s, L) = \min_{\ell \in L} \text{Best-Transfer}(e_g, e_s, \ell)$$

We compute $C(e_g, e_s, \ell)$ in postorder. There are four cases:

- In the base case, if g and s are leaves, then:

$$C(e_g, e_s, \ell) = \begin{cases} 0 & \text{if } \phi(g) = s \text{ and } \gamma(g) = \ell \\ \infty & \text{otherwise} \end{cases} \tag{13}$$

- If neither g nor s is a leaf, then:

$$\begin{aligned} C(e_g, e_s, \ell) = &\min\{\text{Spec}(e_g, e_s, \ell), \text{Loss}(e_g, e_s, \ell), \\ &\text{Dup}(e_g, e_s, \ell), \text{Transfer}(e_g, e_s, \ell)\} \end{aligned} \tag{14}$$

  where the computation of Spec, Loss, Dup, and Transfer are given below.

- If g is not a leaf but s is a leaf, then:

$$C(e_g, e_s, \ell) = \min\{\text{Dup}(e_g, e_s, \ell), \text{Transfer}(e_g, e_s, \ell)\} \tag{15}$$

- If g is a leaf but s is not a leaf, then:

$$C(e_g, e_s, \ell) = \text{Loss}(e_g, e_s, \ell) \tag{16}$$

The functions $\text{Spec}(e_g, e_s, \ell)$, $\text{Loss}(e_g, e_s, \ell)$, $\text{Dup}(e_g, e_s, \ell)$, and $\text{Transfer}(e_g, e_s, \ell)$ are computed as follows:

$$\begin{aligned} \text{Spec}(e_g, e_s, \ell) = \min\{ \\ &\min\{C(e_{g_1}, e_{s_1}, \ell), \mathbf{R} + C(e_{g_1}, e_{s_1}, L)\} + \\ &\quad \min\{C(e_{g_2}, e_{s_2}, \ell), \mathbf{R} + C(e_{g_2}, e_{s_2}, L)\}, \\ &\min\{C(e_{g_1}, e_{s_2}, \ell), \mathbf{R} + C(e_{g_1}, e_{s_2}, L)\} + \\ &\quad \min\{C(e_{g_2}, e_{s_1}, \ell), \mathbf{R} + C(e_{g_2}, e_{s_1}, L)\}\} \end{aligned} \tag{17}$$

$$\text{Loss}(e_g, e_s, \ell) = \mathbf{L} + \min\{C(e_g, e_{s_1}, \ell), C(e_g, e_{s_2}, \ell)\} \tag{18}$$

$$\begin{aligned} \text{Dup}(e_g, e_s, \ell) = \\ &\mathbf{D} + \min\{C(e_{g_1}, e_s, \ell), \mathbf{R} + C(e_{g_1}, e_s, L)\} \\ &\quad + \min\{C(e_{g_2}, e_s, \ell), \mathbf{R} + C(e_{g_2}, e_s, L)\} \end{aligned} \tag{19}$$

$$\begin{aligned} \text{Transfer}(e_g, e_s, \ell) = \mathbf{T} + \min\{ \\ &\min\{C(e_{g_1}, e_s, \ell), \mathbf{R} + C(e_{g_1}, e_s, L)\} + \\ &\quad \min\{\text{Best-Transfer}(e_{g_2}, e_s, \ell), \\ &\quad \mathbf{R} + \text{Best-Transfer}(e_{g_2}, e_s, L)\}, \\ &\quad \min\{C(e_{g_2}, e_s, \ell), \mathbf{R} + C(e_{g_2}, e_s, L)\} + \\ &\quad \min\{\text{Best-Transfer}(e_{g_1}, e_s, \ell), \\ &\mathbf{R} + \text{Best-Transfer}(e_{g_1}, e_s, L)\}\} \end{aligned} \tag{20}$$

In order to compute Best-Transfer, we compute Best-Entry(e_g, e_s, ℓ) as follows. If s is a leaf, then

$$\text{Best-Entry}(e_g, e_s, \ell) = C(e_g, e_s, \ell). \tag{21}$$

Otherwise,

$$\begin{aligned}\text{Best-Entry}(e_g, e_s, \ell) = \min\{&C(e_g, e_s, \ell),\\ &\text{Best-Entry}(e_g, e_{s_1}, \ell), \text{Best-Entry}(e_g, e_{s_2}, \ell)\}\end{aligned} \tag{22}$$

Best-Transfer(e_g, e_s, ℓ) is then computed in preorder: First, for the handle edge e^S

$$\text{Best-Transfer}(e_g, e^S, \ell) = \infty \tag{23}$$

For all other edges, e_s with child edges e_{s_1} and e_{s_2}

$$\begin{aligned}&\text{Best-Transfer}(e_g, e_{s_1}, \ell) = \min\{\\ &\quad \text{Best-Transfer}(e_g, e_s, \ell), \text{Best-Entry}(e_g, e_{s_2}, \ell)\}\end{aligned} \tag{24}$$

$$\begin{aligned}&\text{Best-Transfer}(e_g, e_{s_2}, \ell) = \min\{\\ &\quad \text{Best-Transfer}(e_g, e_s, \ell), \text{Best-Entry}(e_g, e_{s_1}, \ell)\}\end{aligned} \tag{25}$$

Now, we consider the case that each internal vertex g has an arbitrary number of children denoted $g_1, \ldots g_k$, $k \geq 2$. A *binary resolution for* g is defined to be a binary tree whose root is g and whose leaves are $g_1, g_2, \ldots, g_k$. Let $BR(g)$ denote the set of all binary resolutions for g. Note that if g has two children, then it has just one binary resolution. Note also that a binary resolution for a vertex g is different from a binary resolution for the entire gene tree; the former only resolves g into a binary subtree whereas the latter resolves all non-binary vertices in G.

Let Null(g) be the optimal cost of reconciling $G(g)$ with S such that g has the unknown syntenic region $*$. Let Origin(g) be the optimal cost of reconciling $G(g)$ with S such that g induces an origin event. Let H be a binary resolution for g and let $G^H(g)$ denote the subtree $G(g)$, along with its handle, such that g and its children have been replaced by H. Note that if g has exactly two children, then $G^H(g) = G(g)$. Let C^H, Best-Entry^H, Best-Transfer^H, Origin^H, Null^H correspond to C, Best-Entry, Best-Transfer, Origin, and Null (from the previous section) for $G^H(g)$.

Let e_h be an edge in H. If e_h is a leaf edge in H (and thus h is one of the children of g in G), then $C^H(e_h, e_s, \ell) = C(e_h, e_s, \ell)$ for all e_s, ℓ, Origin$^H(h)$ = Origin(h), and Null$^H(h)$ = Null(h). Thus, as the algorithm considers each gene edge e_g in postorder, it then considers each binary resolution H of g and the induced subtree $G^H(g)$. Within $G^H(g)$ it considers the edges of H in postorder to compute the optimal reconciliation for $G^H(g)$. Finally, the optimal reconciliation over all binary resolutions H of g yields the optimal reconciliation for $G(g)$. The algorithm is summarized in Algorithm 4. (Recall that h_1 and h_2 denote the two children of vertex h.)

```
Initialize all entries in tables used below to ∞;
for each gene edge e_g in postorder do
    if e_g is a leaf edge then
        ORIGIN(g) = O ;
        C(e_g, e_φ(g), γ(g)) = 0;
    end
    for each H ∈ BR(g) do
        for each leaf edge e_h in H do
            Initialize ORIGIN^H(h) = ORIGIN(h);
            Initialize NULL^H(h) = NULL(h);
            for each syntenic region ℓ do
                for each species edge e_s in postorder do
                    Initialize C^H(e_h, e_s, ℓ) = C(e_h, e_s, ℓ);
                end
            end
        end
        for each internal edge e_h in H in postorder do
            for each syntenic region ℓ do
                for each species edge e_s in postorder do
                    Compute C^H(e_h, e_s, ℓ) ;
                    Compute BEST-ENTRY^H(e_h, e_s, ℓ);
                end
                for each species edge e_s in preorder do
                    Compute
                      BEST-TRANSFER^H(e_h, e_s1, ℓ);
                    Compute
                      BEST-TRANSFER^H(e_h, e_s2, ℓ);
                end
            end
            ORIGIN^H(h) = min_{e_s,ℓ} C^H(e_h, e_s, ℓ) + O;
            NULL^H(h) =
              min{NULL^H(h_1), ORIGIN^H(h_1)} +
              min{NULL^H(h_2), ORIGIN^H(h_2)};
        end
        for each syntenic region ℓ do
            for each species edge e_s in postorder do
                C(e_g, e_s, ℓ) =
                  min{C^H(e_g, e_s, ℓ), C(e_g, e_s, ℓ)};
                C(e_g, e_s, L) =
                  min{C(e_g, e_s, ℓ), C(e_g, e_s, L)};
            end
        end
        ORIGIN(g) = min{ORIGIN^H(g), ORIGIN(g)}
         NULL(g) = min{NULL^H(g), NULL(g)}
    end
end
return min{ORIGIN(root(G)), NULL(root(G))}
```

Theorem 2 *Algorithm 4 correctly computes the optimal solution to the DTLOR-MPR problem for non-binary gene trees.*

The proof is summarized in the Appendix.

Time complexity

The algorithm first initializes Origin, Null, C entries for all the leaf edges of the gene tree, which takes $O(|G|)$ time. Then, for each internal gene edge e_g, the algorithm loops through all binary resolutions H at the gene vertex and computes C^H, Best-Entry^H, Best-Transfer^H, Origin^H, and Null^H, which takes $O(|H||S||L|) = O(k|S||L|)$ time. (Here $|H|$ is the size of any binary resolution at a gene vertex, which is bounded by $O(k)$ where k is the maximum degree of any vertex in the gene tree.) For each H, the algorithm updates all entries of $C(e_g, e_s, \ell)$ and $C(e_g, e_s, L)$, which takes $O(|L||S|)$ time. In total, the running time for computing all the DP entries for all gene edges and binary resolutions is $O(f(k)k|G||S||L|)$ time, where $f(k)$ upper bounds the number of binary resolutions at any gene vertex.

Results

The implementation of the DTLOR MPR Algorithm (Algorithm 3) was integrated into the xenoGI software package [2] which seeks to reconstruct the history of genome evolution in clades of microbes. xenoGI takes as input a set of sequenced genomes, identifies gene families within this set, and groups those families by common origin. The previous version of xenoGI created gene families in a species-tree aware way, but did not make use of reconciliations. It was able to map gene families onto the species tree and identify their point of origin. However, it was not able reconstruct events in the subsequent evolution of the gene family (e.g. losses or rearrangements). The integration of the DTLOR MPR algorithm allows xenoGI to reconstruct these subsequent events, providing potentially important new insights into microbial evolution.

Within the new DTLOR version of xenoGI, we construct a gene tree for every family, then reconcile it with the species tree. The resulting reconciliation can be used to refine the family (e.g. split it into multiple parts based on the placement of origin events) and to provide detailed information about the family's subsequent evolution.

Table 1 shows the running time for the DTLOR MPR Algorithm within xenoGI on all gene trees given inputs ranging from 4 to 15 bacterial genomes (species). Trees were constructed using FastTree [17] and MUSCLE [18] In each case, DTLOR was run on every binary gene tree with more than two leaves, with gene trees being rooted in all possible ways. These calculations were performed on a commodity server (50 AMD Opteron 6276 2.3 GHz processors, 503 GB RAM). The DTLOR costs were set at 1, 1, 1, 2, 2, respectively.

In one of our enteric bacterial test data sets (Dataset B in Table 1), we examined DTLOR output for a known genomic island, the Acid Fitness Island (AFI) [19]. The corresponding species tree is shown in Fig. 2. This island is thought to have originated with an insertion of 19 genes on branch s2 (the branch leading to the internal vertex s2) via horizontal transfer from outside the clade. It then evolved in the clade and was inherited in the four descendant strains, with the notable loss of nine genes along the branch leading to E. coli K12 [2]. For nearly all the gene families in this island, DTLOR produced reconciliations that place the origin of the family on the branch leading to s2, and it correctly recognized loss events on the K12 branch where those occurred. In a few cases,

Table 1 Runtimes for four different species trees

Species Tree	# Genomes	# Gene Trees	Median GT size	Total Time (s)
A	4	3288	4	4.3
B	6	4032	6	10.6
C	11	4750	9	128.8
D	15	5510	9	563.8

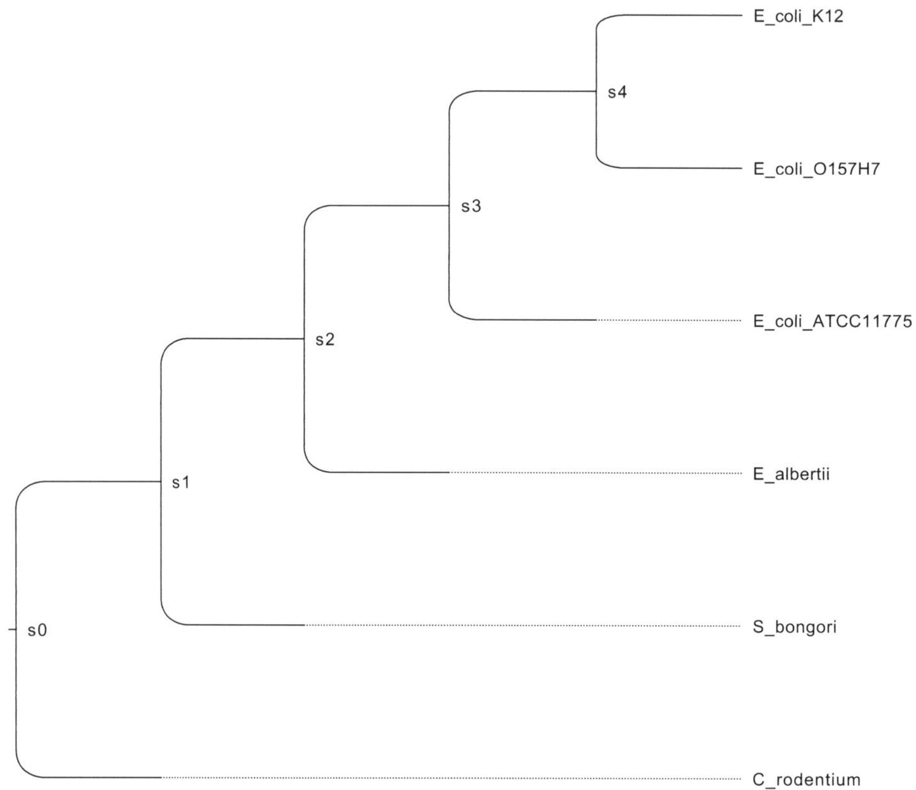

Fig. 2 The species tree for Dataset B

there were multiple most parsimonious reconciliations (MPRs), one of which agreed with the above scenario and was deemed correct, and the others did not agree. Finally there is one family (glutamate decarboxylase) with an evidently complicated post-insertion evolution that is not fully understood. In this case, none of the MPRs using the selected event costs appear to be correct. (The evolution of this family likely involved gene conversion, but the MPR lacks transfer events using the event costs that we used in this experiment.)

Discussion

Several important problems remain to be studied. First, the impact of event costs is not well-understood. Just as in the case of the DTL model, different event costs can give rise to different reconciliations which, in turn, can lead to different conclusions. We believe that the costscape algorithm developed for the DTL model [20] may be extendible to the DTLOR model, which would provide insights into the impact of event costs to the solution space.

Second, it is often the case that there are many distinct MPRs. In fact, even in the DTL model, the number of MPRs can be exponential in the size of the two trees [21]. A subset of these MPRs may be of particular interest because they contain certain evolutionary events that are strongly believed to have occurred (e.g., a horizontal transfer on a particular branch of the species tree). It is desirable, therefore, to efficiently filter the set of MPRs to only include those that contain a specified set of events, count the number of

MPRs in the filtered set, compute support values on the constituent events in that space of MPRs, and select representative reconciliations from this set.

Finally, further systematic studies are needed to determine the full impact of the DTLOR MPR algorithm on the analyses that can now be performed with the enhanced xenoGI tool, including the case of non-binary gene trees.

Conclusions

In this paper, we have described the DTLOR model which extends the well-known DTL model to include origin and rearrangement events. This model is particularly applicable to the evolution of microbes where the species tree is, in many cases, not fully sampled. Therefore, reconciliations must be able to account for transfer events from outside the sampled tree. In addition, the DTLOR model allows for syntenic rearrangement, which is also prevalent in microbial gene families.

We have described efficient algorithms for maximum parsimony reconciliation in the DTLOR model. In binary gene trees, our algorithm solves the DTL reconciliation problem and the sytnenic region problems independently and then combines the results of those two algorithms, resulting in a particularly efficient solution. When the gene tree is non-binary, the two subproblems can no longer be decoupled in this way, and our algorithm for this case considers all of the events simultaneously.

Appendix

This appendix contains proofs of results from the main paper.

Proof of Lemma 1

Proof

We prove that the algorithm correctly computes $C(e_g, \cdot)$ and Best-Transfer$(e_g, \cdot)$ *by structural induction on G. For the base case for $C(e_g, \cdot)$, consider a leaf edge e_g. We perform structural induction on S. In the base case where e_s is a leaf edge, g must map to $\phi(g)$, so $C(e_g, e_s)$ is computed correctly by equation 1. In the inductive step, consider a non-leaf edge e_s. In this case, g must induce a loss event at s since g is a leaf, so $C(e_g, e_s) = \text{Loss}(e_g, e_s)$ (by Eq. 4). By the inductive hypothesis, for each descendant branch $e_{s'}$ of e_s, $C(e_g, e_{s'})$ is computed correctly, so* Loss(e_g, e_s) *is computed correctly (by Eq. 6). This concludes the base case for $C(e_g, \cdot)$.*

In the base case for Best-Transfer$(e_g, \cdot)$, we consider a leaf edge e_g. Since its correctness relies on Best-Entry$(e_g, \cdot)$, we first use structural induction on S to prove that Best-Entry$(e_g, \cdot)$ is correctly computed. In the base case, e_s is a leaf edge, so the only choice for e_g to enter the subtree rooted at e_s is at e_s. Since $C(e_g, e_s)$ is correctly computed, Best-Entry(e_g, e_s) is also correctly computed on line 17. In the inductive step, consider a non-leaf edge e_s. By the inductive hypothesis, Best-Entry$(e_g, e_{s'})$ is computed correctly for each descendant edge $e_{s'}$ of e_s. The ways for e_g to enter the subtree rooted at e_s are at e_s, the left subtree of e_s, or the right subtree of e_s, thus Best-Entry(e_g, e_s) is computed correctly on line 19.

Now we prove the base case for Best-Transfer$(e_g, \cdot)$ using structural induction on S from the handle edge e^S. In the base case where $e_s = e^S$, all edges in S is a descendent of e^S, so there is no valid species edge for the child edge of e_g to transfer to. Thus Best-Transfer(e_g, e^S) is computed correctly on line 22. In the inductive step, we consider a non-root edge e_{s_1}, which has a sibling edge e_{s_2}. By the inductive hypothesis Best-Transfer$(e_g, p(e_{s_1}))$ is computed correctly and by the inductive proof on Best-Entry$(e_g, \cdot)$, Best-Entry(e_g, e_{s_2}) is computed correctly. Since the species edges that e_g are allowed to transfer to from e_{s_1} not only include the same edges if e_g were to transfer from $p(e_{s_1})$, but also edges in the subtree rooted at e_{s_2}, and the optimal cost of placing e_g inside the subtree rooted at e_{s_2} is given by Best-Entry(e_g, e_{s_2}), Best-Transfer(e_g, e_s) is computed correctly. This concludes the base case for Best-Transfer$(e_g, \cdot)$.

To conclude the proof of correctness of $C(e_g, \cdot)$, we consider a non-leaf edge e_g. We use structural induction on S. In the base case, e_s is a leaf edge; the only two possibilities are e_g duplicates on e_s or transfers on e_s. The correctness of Dup(e_g, e_s) is guaranteed by the correctness of $C(e_{g'}, \cdot)$, while the correctness of Transfer(e_g, e_s) is guaranteed by the correctness of both $C(e_{g'}, \cdot)$ and Best-Transfer$(e_{g'}, \cdot)$ for each descendant edge $e_{g'}$ of e_g . In the inductive step, e_s is not a leaf edge, then $C(e_g, e_s)$ is, by definition, the minimum of Spec(e_g, e_s), Dup(e_g, e_s), Transfer(e_g, e_s), and Loss(e_g, e_s) (see Eq. 2). Again, Spec(e_g, e_s), Dup(e_g, e_s) and

Transfer(e_g, e_s) are computed correctly by the correctness of $C(e_{g'}, \cdot)$ and Best-Transfer$(e_{g'}, \cdot)$. The correctness of Loss(e_g, e_s) is guaranteed by the inductive hypothesis on the correctness of $C(e_g, e_{s'})$ for every descendant edge $e_{s'}$ of e_s. This concludes the inductive step for $C(e_g, e_s)$. The inductive step for Best-Transfer$(e_g, \cdot)$ is analogous to the proof for the base case.

Finally, $C(g)$ is the optimal cost for a species mapping of $G(g)$ such that g can be mapped to any species in S. By definition, $C(g) = \min_s \mathcal{C}(g, s)$ where $\mathcal{C}(g, s)$ denotes the optimal cost for a species mapping of $G(g)$ in which g is mapped to s.

Consider $\min_{e_s} C(e_g, e_s)$. Since $C(e_g, e_s)$ is the optimal cost of a species mapping where e_g is placed on e_s, this implies a mapping of g to s or one of its descendants. If g is mapped to s, then $C(e_g, e_s) = \mathcal{C}(g, s)$. If $C(e_g, e_s)$ involves a mapping of g to s' which is a descendant of s, then it induces loss events. Because loss events have a non-negative cost, $C(e_g, e_s') \le C(e_g, e_s)$, and thus $\min_{e_s} C(e_g, e_s)$ only includes those entries of $C(e_g, e_s)$ where g is mapped to s. Thus, $C(g) = \min_s \mathcal{C}(g, s) = \min_{e_s} C(e_g, e_s)$.

□

Proof of Lemma 2

Proof

We prove the correctness of syn(g, ℓ) *and* syn(g) *by induction on* g. *In the base case,* g *is a leaf, then the syntenic region of* g *must be* $\gamma(g)$, *so both* syn(g, ℓ) *and* syn(g) *are computed correctly on line 5. In the inductive step, consider an internal vertex* g. *By the inductive hypothesis,* syn(g_i, ℓ') *and* syn(g_i) *is computed correctly for* $g_i \in \{g_1, g_2\}$ *and* $\ell' \in L$. *The syntenic regions for the left and the right child are chosen independently, and the total cost for* syn(g, ℓ) *is the sum of the costs for choosing syntenic regions for the left subtree and the right subtree of* g. *If* g_i *is a child of* g *with syntenic region* ℓ, *then no* R *event is induced and the cost is* syn(g_i, ℓ). *If an* R *event is induced by a change in syntenic region, then it is optimal to choose a syntenic region that minimizes the total cost of choosing syntenic regions for* $G(g_i)$. *Thus, the cost for a syntenic region mapping of* $G(g_i)$ *with an* R *event is* syn$(g_i) + \mathbf{R}$. *The cost for a syntenic region mapping of* $G(g_i)$ *as a whole is the minimum of these two possibilities, and this is the same for both children of* g. *Thus,* syn(g, ℓ) *is computed correctly on line 7. Then, by definition,* syn(g) *is also computed correctly by taking the minimum of* syn(g, ℓ) *over all* $\ell \in L$. □

Proof of Theorem 1

Proof

First we prove the correctness of Origin(g). *Since a species mapping for an origin subtree* $G(g)$ *is independent of a syntenic region mapping of the same subtree, the optimal cost of a reconciliation of* $G(g)$ *is the optimal cost* $C(g)$ *for a species mapping and the optimal cost* syn(g) *for a syntenic region mapping and the cost of the origin event at* g. *Thus* Origin(g) *is computed correctly on line 2.*

Now we prove Null(g) is computed correctly by an induction on g. In the base case, g is a leaf, so g must have a known syntenic region $\gamma(g)$. Thus Null$(g) = \infty$ as computed on line 4. In the inductive case, consider an internal vertex g. By the inductive hypothesis, Null(g_i) is computed correctly for any $g_i \in \{g_1, g_2\}$. There are two cases we need to consider for each child since it can either induce an origin event or remain unassigned to any real syntenic region. Since we already know Origin(g_i) and Null(g_i) are computed correctly, taking the minimum over the two cases yield the optimal cost for assigning each child and the sum of the costs for both children is the optimal cost for assigning g to the unknown syntenic region $*$. Thus Null(g) is computed correctly on line 6.

Since the root of the gene tree can either be mapped to the unknown syntenic region or some actual syntenic region, we minimize over the two cases and get the optimal cost for reconciling the entire gene tree. Using standard DP traceback techniques, we can also obtain the mappings and events involved in an optimal solution. □

Proof of Theorem 2

The correctness of Algorithm 4 is a direct extension of the proof of correctness of Algorithm 1, now using triple induction to account for the third argument, the syntenic region, in the DP table. The correctness for non-binary vertices then extends analogously to the proof in [15] for the DTL model.

Abbreviations
DTLOR: Duplication–transfer–loss–origin–rearrangement; MPR: Maximum parsimony reconciliation.

Acknowledgements
The authors thank Yi-Chieh Wu for valuable discussions and the anonymous reviewers for their comments and suggestions.

About this supplement
This article has been published as part of BMC Bioinformatics Volume 22 Supplement 10 2021: Selected articles from the 19th Asia Pacific Bioinformatics Conference (APBC 2021): bioinformatics. The full contents of the supplement are available at https://bmcbioinformatics.biomedcentral.com/articles/supplements/volume-22-supplement-10.

Authors' contributions
RLH and EB conceived this research. SS contributed to the model. RLH and NL wrote the first implementation of the algorithm. JL and RM made contributions to the development and analysis of the algorithms, analysis, and proofs of correctness. JL, RM, EB, and RLH wrote the paper. All authors read and approved the final manuscript

Funding
This material is based upon work supported by the National Science Foundation under IIS-1905885 to RLH. Publication costs are funded by support from Harvey Mudd College. The funding agency did not play any role in the design of the methodology, creation of the algorithms, analysis, and interpretation of data, or in writing the manuscript

Availability of data and materials
The DTLOR algorithm for binary trees has been implemented in Python 3 and is available for download at https://github.com/aremath/DTLOR.

Declarations

Competing Interests
The authors declare that they have no competing interests.

Author details
[1]Department of Computer Science, Harvey Mudd College, Claremont, CA, USA. [2]Department of Biology, Harvey Mudd College, Claremont, CA, USA. [3]Department of Computer Science and Engineering, University of California Santa Cruz, Santa Cruz, CA, USA.

Received: 13 June 2021 Accepted: 6 July 2021
Published online: 04 August 2021

References
1. Ochman H, Lawrence JG, Groisman EA. Lateral gene transfer and the nature of bacterial innovation. Nature. 2000;405(6784):299–304.
2. Bush EC, Clark AE, DeRanek CA, Eng A, Forman J, Heath K, Lee AB, Stoebel DM, Wang Z, Wilber M, Wu H. Xenogi: reconstructing the history of genomic island insertions in clades of closely related bacteria. BMC Bioinform. 2018;19(1):32–13211. https://doi.org/10.1186/s12859-018-2038-0.
3. Szöllősi GJ, Tannier E, Lartillot N, Daubin V. Lateral gene transfer from the dead. Syst Biol. 2013;62(3):386–97.
4. Bansal MS, Alm EJ, Kellis M. Efficient algorithms for the reconciliation problem with gene duplication, horizontal transfer and loss. Bioinformatics. 2012;28(12):283–91.
5. Doyon J-P, Scornavacca C, Gorbunov KY, Szöllosi J G, Ranwez V, Berry V. An efficient algorithm for gene/species trees parsimonious reconciliation with losses, duplications and transfers. Comp Genomics. 2011;6398:93–108.
6. Tofigh A. Using trees to capture reticulate evolution: lateral gene transfers and cancer progression. PhD thesis, KTH Royal Institute of Technology (2009).
7. Bansal MS, Kellis M, Kordi M, Kundu S. RANGER-DTL 2.0: rigorous reconstruction of gene-family evolution by duplication, transfer and loss. Bioinformatics 2018;34(18):3214–3216 . https://doi.org/10.1093/bioinformatics/bty314.
8. Chen K, Durand D, Farach-Colton M. NOTUNG: A program for dating gene duplications and optimizing gene family trees. J Comput Biol. 2000;7(3–4):429–47. https://doi.org/10.1089/106652700750050871.
9. Conow C, Fielder D, Ovadia Y, Libeskind-Hadas R. Jane: A new tool for cophylogeny reconstruction problem. Algorithms Mol. Biol. 2010;5(16).
10. Jacox E, Chauve C, Szöllősi GJ, Ponty Y, Scornavacca C. ecceTERA: comprehensive gene tree-species tree reconciliation using parsimony. Bioinformatics. 2016;32(13):2056–8. https://doi.org/10.1093/bioinformatics/btw105.

11. Delabre M, El-Mabrouk N, Huber K, Lafond M, Moulton V, Noutahi E, Castellanos M. Evolution through segmental duplications and losses: a super-reconciliation approach. Algorithms Mol Biol AMB. 2020;15.
12. Ma W, Smirnov D, Libeskind-Hadas R. DTL reconciliation repair. BMC Bioinform. 2017;18(3):76. https://doi.org/10.1186/s12859-017-1463-9.
13. Stolzer M, Lai H, Xu M, Sathaye D, Vernot B, Durand D. Inferring duplications, losses, transfers and incomplete lineage sorting with nonbinary species trees. Bioinformatics. 2012;28(18):409–15.
14. Ovadia Y, Fielder D, Conow C, Libeskind-Hadas R. The cophylogeny reconstruction problem is NP-complete. J Comput Biol. 2011;18(1):59–65.
15. Kordi M, Bansal MS. Exact algorithms for duplication-transfer-loss reconciliation with non-binary gene trees. IEEE Trans Comput Biol Bioinform. 2017. https://doi.org/10.1109/TCBB.2017.2710342.
16. Jacox E, Weller M, Tannier E, Scornavacca C. Resolution and reconciliation of non-binary gene trees with transfers, duplications and losses. Bioinformatics. 2017;33(7):980–7. https://doi.org/10.1093/bioinformatics/btw778.
17. Price MN, Dehal PS, Arkin AP. Fasttree 2 -approximately maximum-likelihood trees for large alignments. PLoS ONE. 2010;5(3):1–10. https://doi.org/10.1371/journal.pone.0009490.
18. Edgar RC. MUSCLE: multiple sequence alignment with high accuracy and high throughput. Nucl Acids Res. 2004;32(5):1792–7. https://doi.org/10.1093/nar/gkh340.
19. Hommais F, Evelyne K, Coppée J-Y, Lacroix C, Yeramian E, Danchin A, Bertin P. Gade (yhie): a novel activator involved in the response to acid environment in *Escherichia coli*. Microbiology (Reading, England). 2004;150:61–72. https://doi.org/10.1099/mic.0.26659-0.
20. Libeskind-Hadas R, Wu Y-C, Bansal MS, Kellis M. Pareto-optimal phylogenetic tree reconciliation. Bioinformatics. 2014;30(12):87–95.
21. Bansal MS, Alm EJ, Kellis M. Reconciliation revisited: handling multiple optima when reconciling with duplication, transfer, and loss. J Comput Biol. 2013;20(10):738–54. https://doi.org/10.1089/cmb.2013.0073.

Wang and Zheng *BMC Bioinformatics* (2021) 22:419
https://doi.org/10.1186/s12859-021-04330-1

BMC Bioinformatics

METHODOLOGY Open Access

Velo-Predictor: an ensemble learning pipeline for RNA velocity prediction

Xin Wang and Jie Zheng*

From The 19th Asia Pacific Bioinformatics Conference (APBC 2021) Tainan, Taiwan. 3-5 February 2021

*Correspondence: zhengjie@shanghaitech.edu.cn
School of Information Science and Technology, ShanghaiTech University, 393 Middle Huaxia Road, Pudong District, 201210 Shanghai, China

Abstract

Background: RNA velocity is a novel and powerful concept which enables the inference of dynamical cell state changes from seemingly static single-cell RNA sequencing (scRNA-seq) data. However, accurate estimation of RNA velocity is still a challenging problem, and the underlying kinetic mechanisms of transcriptional and splicing regulations are not fully clear. Moreover, scRNA-seq data tend to be sparse compared with possible cell states, and a given dataset of estimated RNA velocities needs imputation for some cell states not yet covered.

Results: We formulate RNA velocity prediction as a supervised learning problem of classification for the first time, where a cell state space is divided into equal-sized segments by directions as classes, and the estimated RNA velocity vectors are considered as ground truth. We propose Velo-Predictor, an ensemble learning pipeline for predicting RNA velocities from scRNA-seq data. We test different models on two real datasets, Velo-Predictor exhibits good performance, especially when XGBoost was used as the base predictor. Parameter analysis and visualization also show that the method is robust and able to make biologically meaningful predictions.

Conclusion: The accurate result shows that Velo-Predictor can effectively simplify the procedure by learning a predictive model from gene expression data, which could help to construct a continous landscape and give biologists an intuitive picture about the trend of cellular dynamics.

Keywords: RNA velocity, Single cell, Ensemble learning, Landscape

Background

Recent advances in high-throughput RNA sequencing technologies [1] have enabled analysis of transcription at single-cell level [2], which has provided immense opportunities to unravel the underlying mechanisms of gene expression regulation. However, in many cases, dynamical information of cell state transition is limited. When the sequencing is completed, the expression data provide only a snapshot of a cell [3]. Currently, trajectory inference (including pseudotime analysis) is a primary task to identify cells in various states of differentiation [4]. In general, trajectory inference methods need to construct graphs. There are various approaches to trajectory reconstrcution, e.g. SCUBA [5] is based on bifurcation analysis, SCENT [6] and scEpath [7] use a measurement of entropy of cell states. HopLand [8] and Topslam [9] project cells to a landscape with optimized parameters.

A major limitation of most trajectory inference methods [10] is they do not connect data to underlying molecular kinetics. La Manno et al. found that spliced and unspliced mRNAs can be distinguished in standard single-cell RNA-seq protocols [11], and the timescale of differentiation during development is comparable to the typical half-life of an mRNA. Hence, we can use the abundances of mRNAs to estimate splicing rate and degradation rate. They proposed a simple kinetic framework for estimating changes in mRNA levels of individual cells. This framework is based on the central dogma of molecular biology. Gorini and Maas proposed a first-order differential equation to model this biological process [12], to which Zeisel et al. added intermediate steps [13].

The original steady-state model for RNA velocity proposed by La Manno et al. assumes that transcriptional phases endure long enough to reach a steady state equilibrium, and the equilibrium mRNA levels can be approximated with a linear regression by simplification with a common splicing rate. Recently, to relax this assumption, Volker Bergen et al. proposed an algorithm called "scVelo" [14], which includes a stochastic model and a dynamical model in addition to the steady-state model. The stochastic model treats the transcription, splicing and degradation as probabilistic events, which means steady-state levels are approximated not only from mRNA levels, but also from intrinsic expression variability. The dynamical model considers non-stationary populations and different splicing rates across genes, and the dynamics is solved in a maximum likelihood framework using the expectation maximization (EM) algorithm. The dynamical model is slower but can provide more consistent velocity estimation and better identification of transcriptional states.

The concept of RNA velocity and its associated algorithms and models have become very popular in single-cell biology. However, this technique needs the support of RNA sequencing protocols. Moreover, to get splicing information we need to run a complex preprocessing pipeline which involves the issues of file format and is time consuming. More importantly, the data of estimated RNA velocities are still sparse compared with the size of the uncovered cell state space. Here we propose an ensemble learning pipeline for the prediction of RNA velocities, which can skip the complex procedures for splicing analysis, etc. When we have a new data sample from the same biological context, we can predict the direction of RNA velocity from a state unkown in the traning data. This is similar to the pedestrian prediction [15] in a driverless transportation system, or the prediction of next movements of basketball players on the court [16]. It is possible to further combine all the transient movements into long trajectories of cells. Inspired by the concept of Waddington's epigenetic landscape, which is a classical metaphor for cell differentiation, we can treat cells as balls rolling down through a potential surface. Based on the predicted RNA velocities and cell trajectories, we can reconstruct the landscape, as an intuitive platform for single-cell data visualization.

Methods

Velocity estimation

Velocyto CLI or loompy/kallisto was used to obtain spliced/unspliced reads annotations. We filter the genes with counts number (both spliced and unspliced) smaller than the threshold, keep the top high variability genes. Then normalize in cell level and did logarithm transform. On Euclidean distances PCA space of counts matrix, a nearest neighbor graph was computed, first and second moments were obtained for each cell. According to the basic reaction kinetics:

$$\frac{dU(t)}{dt} = \alpha_k(t) - \beta \cdot U(t), \tag{1}$$

$$\frac{dS(t)}{dt} = \beta \cdot U(t) - \gamma \cdot S(t), \tag{2}$$

where *S(t)* represents mature mRNA abundance over time, *U(t)* represents pre-mRNA abundance over time, α is the rate of transcription, β is the rate of splicing, and γ is the rate of degradation. k and t are cell-specific latent variables, where k represents discrete transcriptional state, and t represents latent time.

RNA velocity is termed as the time derivative of mature spliced mRNA $v(t) = \frac{dS(t)}{dt}$. Three approaches are provided in scVelo to do velocity estimation: steady state model, stochastic model and dynamical model. The basic difference between them is that the assumptions about the parameters are different. The data preprocessing steps are shown in Algorithm 1. For the sake of completeness and readers' convenience, we have rephrased their description of methods for RNA velocity estimation into the pseudocode. After velocity estimation we can get a multi-dimensional RNA velocity vector V for each transcriptional state of a single cell. Combining this information we can further inference cell future state of an individual cell. The movements can be UMAP projected into a lower dimensional embedding D to visualize.

Algorithm 1: Target preparation

Input: Anndata format data with two count matrices (n x m) of unspliced and spliced abundances {n represents cell number and m represents gene number}
Output: velocity vector $V = (v_1, v_2, ..., v_n)$ for every cell

filter genes according to detection and dispersion level;
normalize and logarithmize data;
compute first and second order moments;
if *model == 'steady state'* **then**
  $\beta \leftarrow 1$;
  $\gamma' \leftarrow \frac{\gamma}{\beta}$ solved via least square fit;
  for $i \leftarrow 1...n$ **do**
    $v_i \leftarrow u_i - \gamma' s_i$
if *model == 'dynamic model'* **then**
  integrate Eq.1 and Eq.2, $\theta \leftarrow (\alpha(k), \beta, \gamma)$, $\hat{x}(t) \leftarrow (\hat{u}(t), \hat{s}(t))$;
  construct negative log-likelihood to minimize;
  while *not converge* **do**
    E-step: Assign a latent time t_i to the observed value ${x_i}^{obs}$ by minimizing the distance to the phase trajectory $(\hat{x}(t|\theta))_t$ in each transcriptional state;
    M-step: Update θ to maximize the log-likelihood
  Substitute params to get V;
if *model == 'stochastic model'* **then**
  add higher order moments and treat transcription, splicing and degradation as probabilistic events;
  solve γ via generalized least square fit;
  get V
return V;

Problem statement

Our goal is to predict the RNA velocity vector of each cell based on its gene expression data. As illustrated in a 2D space, we formulate it as a classification problem through an equal division of a 2D circle into d equal-sized segments, as shown in Fig. 1. If the predicted and the original target directions fall in the same segment, we count it as a true positive, etc. A slightly more realistic formulation of the problem could be the regression of angles of the RNA velocities from a fixed direction. But we will leave that as a future work.

Input preprocessing

Figure 2 shows the whole pipeline of our work. We start our supervised learning task from the gene expression matrix. Single-cell genomic data may be sparse and suffer from technical noise and bias. Therefore, to improve the behavior we need to do a de-noising step, or called feature engineering step. There are several ways to do such as scVI [17], scVAE [18] and DCA [19]. They basically use auto-encoder to find the hidden layer with minimum reconstruction error. After comparison we do feature selection based on gene ranking by ScVelo. ScVelo ranking is based on cluster-specific t-test to find genes with significantly higher/lower differential velocity, and we select the top k genes from each cluster as the model features. We use the lower-dimensional embedding $D \in \mathbf{R}^{2n}$ obtained from the velocity estimation step as our ground truth.

After the division, the distribution of labels is not balanced. Therefore, we provide several ways to rescue: over-sampling, down-sampling and combine-sampling. We first test different sampling ways on different base models. For oversampling, we test adaptive synthetic (ADASYN) [20], synthetic minority oversampling technique (SMOTE) [21] and some variants of SMOTE, such as border line smote (BLS) [22] and svm-smote which uses support vector machine (SVM) [23]. For down-sampling, cluster centroid (CC) [24], random under sampler (RUS), NearMiss [25], repeated edited nearest neighbours (RENN) [26], neighbourhood cleaning rule (NCR) [27] and one side selection (OSS) [28] are used. Then, we further test combine-sampling methods of SMOTETomek [29] and SMOTEENN [30].

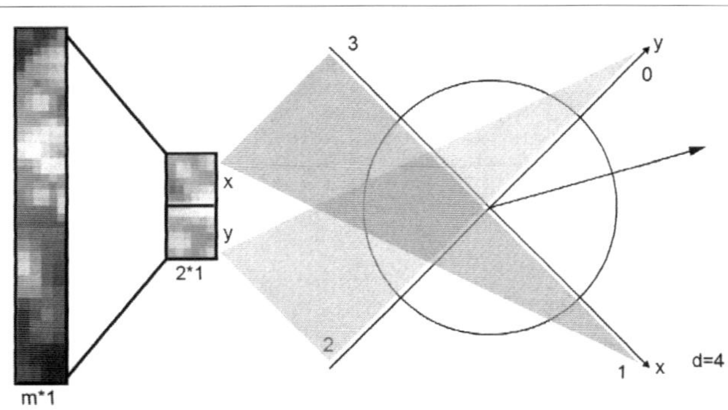

Fig. 1 Circle partition. Assign d classification labels through equal division of a two dimensional circular plane. Here $d = 4$

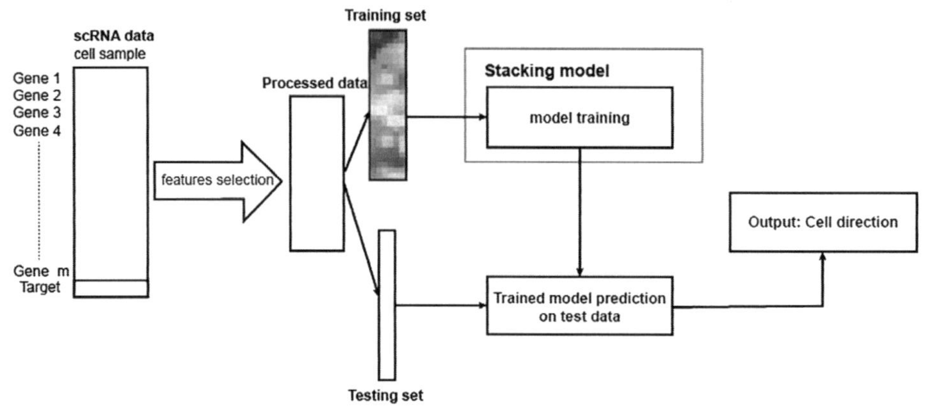

Fig. 2 Overall pipeline of Velo-Predictor. The pipeline includes stages of feature selection, model training and prediction

Model training

We divide the sample data into a training set and test set. The training set is for model training and test set is for model evaluation. For training, the parameters are saved and can be directly used for prediction in testing part. We adapt a stacking structure model. Figure 3 and Algorithm 2 show the detail. We use mlxtend [31] and Scikit-learn [32] packages for implementation. We choose random forest (RF), GBDT, extra tree classifier (ET), adaboost (ADA) and XGBoost model to be the first layer, and the second layer is a simple Logistic regression classifier. To avoid over-fitting, we use cross validation concept to divide the training set to K subsets, where $K-1$ subsets are used to fit the first layer of classifiers. Then in each round, the unused subset will be predicted by the fitted classifier, and all the resulting predictions are stacked to feed into the second layer.

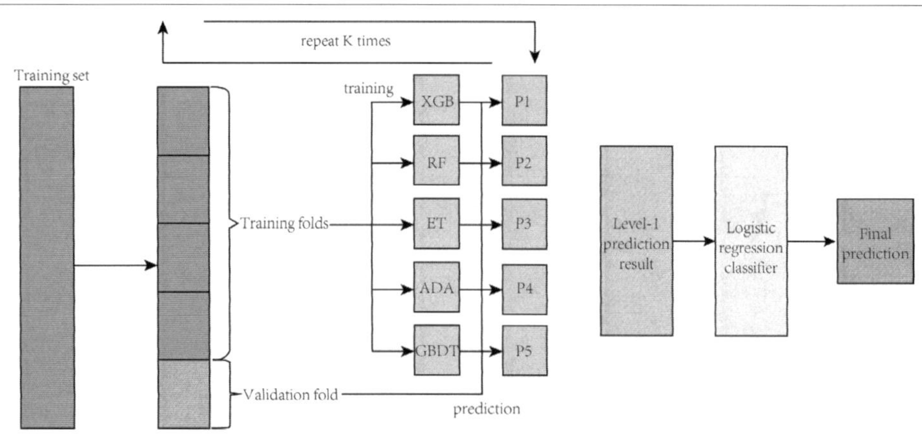

Fig. 3 Architecture of the stacking model. Stacking with K-fold cross validation, where the first layer includes XGBoost, Random Forest, etc. as the base models. The second layer (as the meta classifier) is logistic regression classifier

Algorithm 2: Stacking model with K-fold

Input: Training data $X = \{x_i, y_i\}_{i=1}^{n} (x_i \in R^m)$, test data T
Output: Velocity prediction on test data
split X to K equal-sized subsets: $X = \{X_1, X_2, ..., X_K\}$;
for $k \leftarrow 1...K$ **do**
  for $j \leftarrow 1...5$ **do**
    $X \backslash X_i$ to learn first level classifier h_{kj} ;
  for $x_i \in X_k$ **do**
    Get $\{x_i^{'}, y_i\}$, where $x_i^{'} = \{h_{k1}, h_{k2}...h_{k5}\}$;
learn logistic regression classifier $h^{'}$ from $\{x_i^{'}, y_i\}$;
for $j \leftarrow 1...5$ **do**
  relearn first level classifiers h_j on X ;
$H(x) = h^{'}(h_1(x), h_2(x)...h_5(x))$ to predict on test data ;
return $H(T)$

Results

Data sets

We train and test the models on two single-cell RNA-seq datasets. One is the Mouse hippocampal dentate gyrus neurogenesis (DGN) dataset [33] available from NCBI Gene Expression Omnibus (GEO) under accession ID GSE95753. It consists of RNA-seq data of 13,913 genes and 2930 cells from multiple lineages. The other dataset is Pancreatic endocrinogenesis (PE) [34] also available from NCBI GEO under accession ID GSE132188, which comprises the transcriptional levels of 27,998 genes of 3,696 pancreatic epithelial and Ngn3-Venus fusion cells sampled from mouse embryonic day 15.5. The number of cells and the numbers of genes (with different values of *k*, the number of top genes selected from each cluster as features) are shown in Table 1.

To test the generalization ability of our models, we randomly divide the cell samples into two disjoint sets with ratio of 7:3, 7 for training and 3 for testing. Figure 4a shows the proportion of labels in the DG datasets.

Class imbalance issue

To illustrate how to address the class imbalance issue, we take the DG dataset as an example. Figure 4a shows the label proportion of the DG dataset. Figure 4b–d shows the ROC curve and corresponding AUC score of different sampling strategies. The AUC score is not enough for imbalanced data, thus we also consider the precision on each of the four classes and the balanced score as metrics. The metrics can be calculated according to the following equations:

$$
\begin{aligned}
PRE(precision) &= \frac{TP}{TP + FP} \\
Balanced\ Score &= \frac{TPR + TNR}{2}
\end{aligned}
\tag{3}
$$

Table 2 shows the most representative performance of each methods. We can see the down-sampling methods perform poorly because of loss of information. The over-sampling methods are better but may introduce some biases. The best way is to combine them. Therefore we choose SMOTETomek as our final choice for the DG dataset.

Table 1 Data statistics

	DGN	PE
Cell number	2930	3696
Gene number	13,913	27,998
Gene number (top k = 3)	41	24
Gene number (top k = 5)	63	38

Table 2 Performance of Random Forest on the DG dataset with different sampling methods

	PRE-0	PRE-1	PRE-2	PRE-3	Balanced score
Origin	0.65	0.68	0.57	0.66	0.52
SMOTE	0.8	0.75	0.69	0.67	0.73
NCR	0.67	0.72	0.67	0.75	0.55
ADASYN	0.8	0.76	0.7	0.68	0.73
NearMiss	0.67	0.61	0.46	0.46	0.53
SMOTETomek	0.8	0.77	0.71	0.7	0.75
SMOTEENN	0.83	0.81	0.74	0.4	0.64

Functional analysis

After parameters fine tuning through grid search, Fig. 5 visualizes the performances of base models and stacking model. Figure 6a shows the loss curve on first fold, the behaviors of the other folds are similar. XGBoost model can also provide the log loss curve and the most important genes learning from the data (Fig. 6b). The impact of hyper parameters k the number of top genes and d the number classes is shown in Fig. 7b. Parameter k controls the feature selection part. The curve rises first and after k reaches 20, it starts to oscillate. In the previous experiment we set k to 3, although we can increase k to get better performance. Parameter d controls the granularity of prediction, and the result shows that the number of divisions is 8. When we continue to increase d, the task becomes more difficult so that the score will decay. Figure 7a shows the best performance of stacking model on DG dataset with $d = 8$, $k = 20$. Table 3 shows the comparison of base models and final stacking model. Stacking model's performance is even slightly worse than XGBoost. We think that is probably due to the following reasons. First, the dataset is too small because stacking is not so strong when the data set is not big enough. Secondly, we can increase the model diversity of the first layer, and add more models to improve the performance. Thirdly, we did not use cross validation technique in the random forest model. In conclusion, we think the score difference is small and users can choose different provided models according to their size of data set.

Visualization

We use the UMAP toolkit of scVelo [14] to map cells and velocity vectors into two dimensional space, and we assume the incoming new data has the same distribution with our training data. We projected the new data in the same way as previous embedding, and give them a small red arrow which indicates the prediction of our velocity

Table 3 Performance of base and stacking models

	PRE-0	PRE-1	PRE-2	PRE-3	Balanced score
XGBoost	0.85	0.81	0.75	0.73	0.79
RF	0.83	0.8	0.72	0.66	0.75
GBDT	0.82	0.7	0.64	0.63	0.70
ET	0.87	0.79	0.72	0.73	0.76
ADA	0.66	0.58	0.55	0.58	0.60
stacking	0.83	0.8	0.74	0.7	0.78

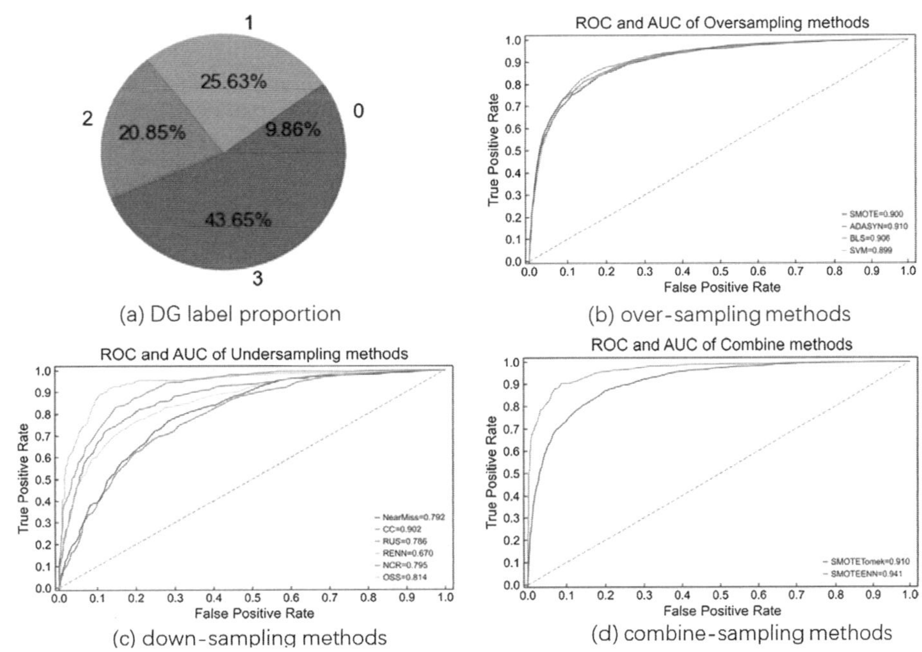

Fig. 4 Methods comparison. **a** Direction label proportion of the DG dataset. **b–d** Performance of different methods to solve the class imbalance issue on the DG dataset

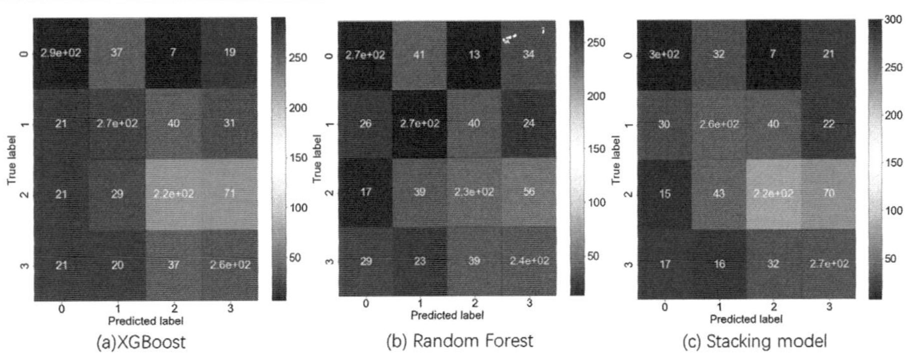

Fig. 5 Confusion matrix of different models. Here shows the case when the number of classes *d* equals 4 on the DG dataset. **a** The performance of the XGBoost model, where each square corresponds to a predicted label and a true label for direction of RNA velocity, and the value in each square is the number of cells meeting the condition (color becomes darker as the number increases). We can see that the diagnals have darker colors which means that some samples are correctly classified. **b** The performance of the Random Forest model. **c** The performance of the Stacking model

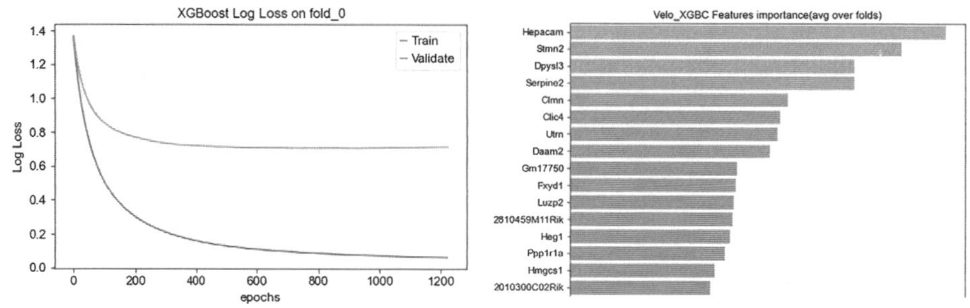

(a) XGBoost loss curve on fold 0 (b) Visualization of gene Importance

Fig. 6 Loss curve and most important genes of XGBoost. **a** Training and validation log loss curve along epoches of XGBoost on DG dataset. **b** Gene importance ranking of XGBoost

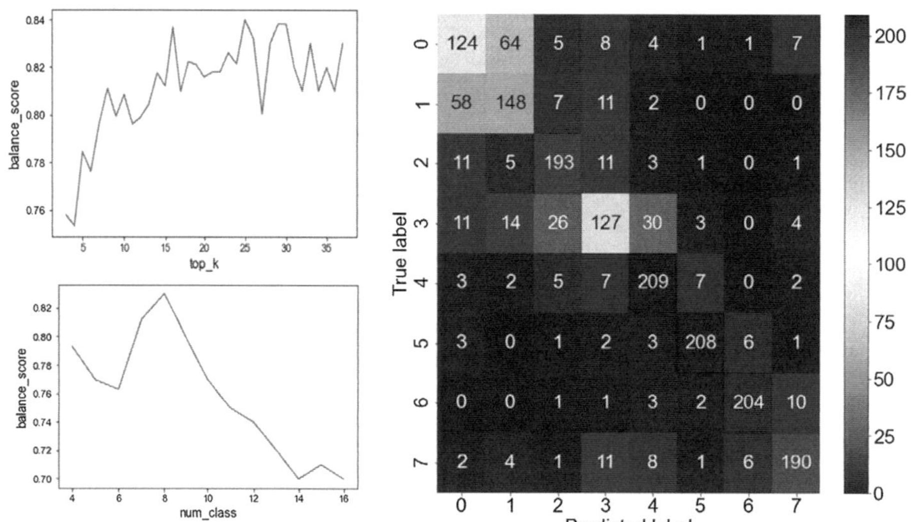

(a) Score curve of hyperparameters (b) Confusion matrix of Stacking model (k = 20, d = 8)

Fig. 7 The impact of hyperparameters on DG dataset. **a** Balance score curve with different hyperparameters. The upper one shows the balance score increases with the increasing top gene number k. After k reaches 20, the balance score begins to oscillate which means 20 is enough for a good performance. The bottom one shows the performance drops when the number of classes exceeds 8. **b** Confusion matrix of the Stacking model when the number of genes k equals 20 and the number of classes d equals to 8. Each square corresponds to predicted label and a true label, and the value in each square is the number of cells satisfying the condition (color is darker as the number increases). The numbers along the diagnal show the performance of the Stacking model on DG dataset

direction information. Each point in the figure represents a cell, the arrow is the velocity information of cells. It gives us an intuitive instruction of which way where a specific cell goes to. In Biology, it will tell us the differential path of a cell, we can see it performs well. Figure 8 shows the result on dataset PE, different colors indicate different clusters, above figure is the ground truth. In below figure, red arrow is our prediction outcome. Comparing with the same location in ground truth figure, we can see that the outcome is consistent with the ground truth. For detail, the orange dots represent pre-endocrine cells, and the perple dots represent epsilon cells. Through the zoom-in window, the comparision shows clearly that cell movements on 2D space are well captured.

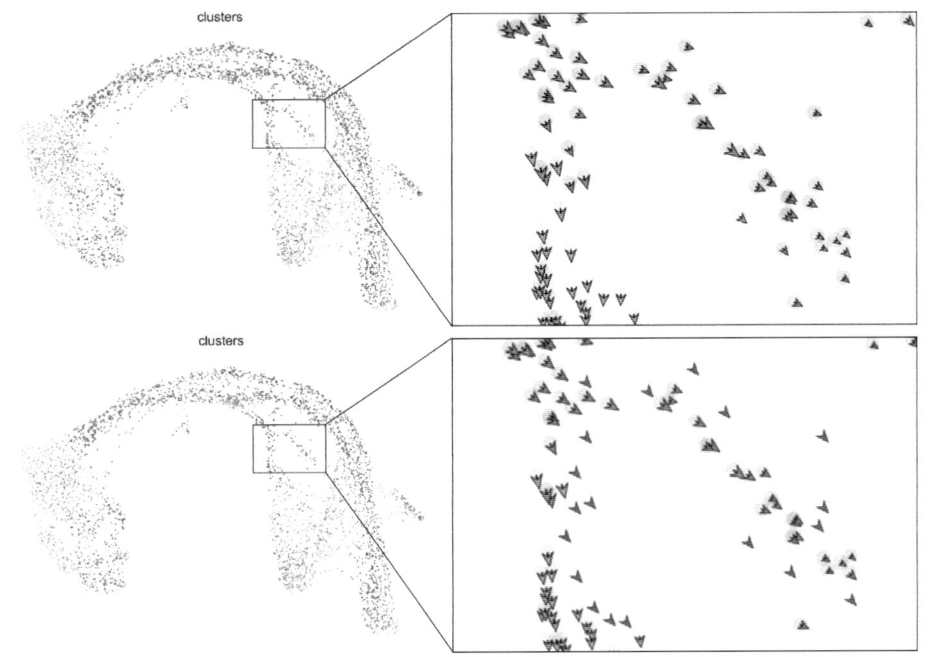

Fig. 8 Schematic diagram of movement prediction on PE dataset. The leftside figure shows the velocity distribution of PE dataset, where different colors indicate different clusters. In the zoom-in figures on the right side, the upper one shows cells and velocity directions with original labels (training set), and in the bottom figure, the red arrows without halos represent cells and velocity directions with predicted labels (test set)

Discussion

One limitation of Velo-Predictor is that its performance may depend on data. Although we have used the ensemble learning framework to balance the results from different baseline models for different sample-feature ratios, it is still an empirical approach. When the data distribution is not so complete and balanced, the prediction may be less accurate. The lowest input data size depends on the properties of data (e.g. in terms of samples and features) and biological scenarios (e.g. types of data). We have tested the case ($k = 5$, $d = 4$) on the DG dataset by adjusting the testing data size, and when it exceeds 40%, the number of errors will increase to hundreds. In general, more complete data that cover the dynamical processes in the biological scenarios under study would be much preferred. Besides, the interpretability of the model is still a challenge.

Our work provides a prediction-based approach for study of cell differentiation mechanisms. Such predictions can also help impute the state space not yet covered by the scRNA-seq data, and the interpolation can help construct a continuous landscape surface. In the future, we can use single-cell multi-omic data to learn the bifucation point of cell differentiation more precisely. With the imputed direction information we can further do trajectory inference. Combined with an energy function such as that in the Hopfield network model used in our previous work [8], the predicted RNA velocities can be used to model the Waddington's epigenetic landscape. Therefore, the prediction of RNA velocity can give biologists an intuitive picture about the trend of cellular dynamics, which is informative for their research.

Conclusion

In this paper, we described Velo-Predictor, an ensemble learning pipeline for RNA velocity prediction. While RNA velocity estimation is not straightforward, our pipeline can simplify the procedure by learning a predictive model from gene expression data. The results showed that our pipeline can predict the directions of cell state transitions accurately.

Abbreviations

scRNA-seq: Single-cell RNA sequencing; EM: Expectation maximization; ADASYN: Adaptive synthetic; SMOTE: Synthetic minority over sampling technique; BLS: Border line smote; SVM: Support vector machine variant of smote; CC: Cluster centroid; RUS: Random under sampler; RENN: Repeated edited nearest neighbours; NCR: Neighbourhood cleaning rule; OSS: One side selection; RF: Random forest; GBDT: Gradient boosting decision tree; ET: Extra tree classifier; ADA: Adaboost; DGN: Dentate gyrus neurogenesis; GEO: Gene Expression Omnibus; PE: Pancreatic endocrinogenesis.

Acknowledgements

Not applicable.

About this supplement

This article has been published as part of BMC Bioinformatics Volume 22 Supplement 10 2021: Selected articles from the 19th Asia Pacific Bioinformatics Conference (APBC 2021): bioinformatics. The full contents of the supplement are available at https://bmcbioinformatics.biomedcentral.com/articles/supplements/volume-22-supplement-10.

Authors' contributions

XW and JZ conceived the project. XW developed the method and did the experiments. XW and JZ wrote the manuscript. Both authors read and approved the final manuscript.

Funding

This research was supported by a startup grant from the ShanghaiTech University. Publication costs are funded by the same startup grant. The funding body had no role in the design of the study and collection, analysis, and interpretation of data and in writing the manuscript.

Availability of data and materials

The single cell RNAseq data are publicly available in Gene Expression Omnibus, accession numbers GSE95753 and GSE132188. The code is available at https://github.com/clay001/Velo-Predictor.

Declarations

Ethics approval and consent to participate

Not applicable.

Consent to publish

Not applicable.

Competing interests

The authors declare that they have no competing interests.

Received: 16 August 2021 Accepted: 23 August 2021
Published online: 03 September 2021

References

1. Zheng GX, Terry JM, Belgrader P, Ryvkin P, Bent ZW, Wilson R, Ziraldo SB, Wheeler TD, McDermott GP, Zhu J, et al. Massively parallel digital transcriptional profiling of single cells. Nat Commun. 2017;8(1):1–12.
2. Svensson V, Vento-Tormo R, Teichmann SA. Exponential scaling of single-cell RNA-seq in the past decade. Nat Protoc. 2018;13(4):599–604.
3. Weinreb C, Wolock S, Tusi BK, Socolovsky M, Klein AM. Fundamental limits on dynamic inference from single-cell snapshots. Proc Natl Acad Sci. 2018;115(10):2467–76.
4. Saelens W, Cannoodt R, Todorov H, Saeys Y. A comparison of single-cell trajectory inference methods. Nat Biotechnol. 2019;37(5):547–54.
5. Marco E, Karp RL, Guo G, Robson P, Hart AH, Trippa L, Yuan G-C. Bifurcation analysis of single-cell gene expression data reveals epigenetic landscape. Proc Natl Acad Sci. 2014;111(52):5643–50.
6. Teschendorff AE, Enver T. Single-cell entropy for accurate estimation of differentiation potency from a cell's transcriptome. Nat Commun. 2017;8(1):1–15.
7. Jin S, MacLean AL, Peng T, Nie Q. scEpath: energy landscape-based inference of transition probabilities and cellular trajectories from single-cell transcriptomic data. Bioinformatics. 2018;34(12):2077–86.

8. Guo J, Zheng J. HopLand: single-cell pseudotime recovery using continuous hopfield network-based modeling of Waddington's epigenetic landscape. Bioinformatics. 2017;33(14):102–9.
9. Zwiessele M, Lawrence ND. Topslam: Waddington landscape recovery for single cell experiments. BioRxiv. 2016;057778.
10. Trapnell C, Cacchiarelli D, Grimsby J, Pokharel P, Li S, Morse M, Lennon NJ, Livak KJ, Mikkelsen TS, Rinn JL. The dynamics and regulators of cell fate decisions are revealed by pseudotemporal ordering of single cells. Nat Biotechnol. 2014;32(4):381.
11. La Manno G, Soldatov R, Zeisel A, Braun E, Hochgerner H, Petukhov V, Lidschreiber K, Kastriti ME, Lönnerberg P, Furlan A, et al. RNA velocity of single cells. Nature. 2018;560(7719):494–8.
12. Gorini L, Maas WK. The potential for the formation of a biosynthetic enzyme in *Escherichia coli*. Biochim Biophys Acta. 1957;25(1):208.
13. Zeisel A, Köstler WJ, Molotski N, Tsai JM, Krauthgamer R, Jacob-Hirsch J, Rechavi G, Soen Y, Jung S, Yarden Y, et al. Coupled pre-mRNA and mRNA dynamics unveil operational strategies underlying transcriptional responses to stimuli. Mol Syst Biol. 2011;7(1):529.
14. Bergen V, Lange M, Peidli S, Wolf FA, Theis FJ. Generalizing RNA velocity to transient cell states through dynamical modeling. Nat Biotechnol. 2020;38:1408–14.
15. Alahi A, Goel K, Ramanathan V, Robicquet A, Fei-Fei L, Savarese S. Social lstm: human trajectory prediction in crowded spaces. In: Proceedings of the IEEE conference on computer vision and pattern recognition; 2016. p. 961–71.
16. Felsen P, Lucey P, Ganguly S. Where will they go? Predicting fine-grained adversarial multi-agent motion using conditional variational autoencoders. In: Proceedings of the European conference on computer vision (ECCV); 2018. p. 732–47.
17. Lopez R, Regier J, Cole MB, Jordan MI, Yosef N. Deep generative modeling for single-cell transcriptomics. Nat Methods. 2018;15(12):1053–8.
18. Grønbech CH, Vording MF, Timshel PN, Sønderby CK, Pers TH, Winther O. scVAE: variational auto-encoders for single-cell gene expression datas. BioRxiv; 2018:318295.
19. Eraslan G, Simon LM, Mircea M, Mueller NS, Theis FJ. Single-cell RNA-seq denoising using a deep count autoencoder. Nat Commun. 2019;10(1):1–14.
20. He H, Bai Y, Garcia EA, Li S. ADASYN: adaptive synthetic sampling approach for imbalanced learning. In: 2008 IEEE international joint conference on neural networks (IEEE world congress on computational intelligence). IEEE; 2008. p. 1322–28.
21. Chawla NV, Bowyer KW, Hall LO, Kegelmeyer WP. SMOTE: synthetic minority over-sampling technique. J Artif Intell Res. 2002;16:321–57.
22. Han H, Wang W-Y, Mao B-H. Borderline-smote: a new over-sampling method in imbalanced data sets learning. In: International conference on intelligent computing. Springer; 2005. p. 878–887.
23. Nguyen HM, Cooper EW, Kamei K. Borderline over-sampling for imbalanced data classification. Int J Knowl Eng Soft Data Paradigms. 2011;3(1):4–21.
24. Yuwono M, Su SW, Moulton B, Nguyen H. Fast unsupervised learning method for rapid estimation of cluster centroids. In: 2012 IEEE congress on evolutionary computation. IEEE; 2012. p. 1–8 .
25. Mani I, Zhang I. kNN approach to unbalanced data distributions: a case study involving information extraction. In: Proceedings of workshop on learning from imbalanced datasets. vol 126; 2003.
26. Tomek I, et al. An experiment with the edited nearest-nieghbor rule; 1976.
27. Laurikkala J. Improving identification of difficult small classes by balancing class distribution. In: Conference on artificial intelligence in medicine in Europe. Springer; 2001, p. 63–6.
28. Kubat M, Matwin S, et al. Addressing the curse of imbalanced training sets: one-sided selection. In: Icml, vol. x97; 1997. p. 179–86. Citeseer
29. Batista GE, Prati RC, Monard MC. A study of the behavior of several methods for balancing machine learning training data. ACM SIGKDD Explor Newsl. 2004;6(1):20–9.
30. Batista GE, Bazzan AL, Monard MC. Balancing training data for automated annotation of keywords: a case study. In: WOB; 2003. p. 10–18.
31. Mlxtend Raschka S. providing machine learning and data science utilities and extensions to python's scientific computing stack. J Open Source Softw. 2018;3(24):638 (https://doi.org/10.21105/joss.00638).
32. Pedregosa F, Varoquaux G, Gramfort A, Michel V, Thirion B, Grisel O, Blondel M, Prettenhofer P, Weiss R, Dubourg V, Vanderplas J, Passos A, Cournapeau D, Brucher M, Perrot M, Duchesnay E. Scikit-learn: machine learning in Python. J Mach Learn Res. 2011;12:2825–30.
33. Hochgerner H, Zeisel A, Lönnerberg P, Linnarsson S. Conserved properties of dentate gyrus neurogenesis across postnatal development revealed by single-cell RNA sequencing. Nature Neurosci. 2018;21(2):290–9.
34. Bastidas-Ponce A, Tritschler S, Dony L, Scheibner K, Tarquis-Medina M, Salinno C, Schirge S, Burtscher I, Böttcher A, Theis FJ, et al. Comprehensive single cell mRNA profiling reveals a detailed roadmap for pancreatic endocrinogenesis. Development. 2019;146(12):dev173849.

Publisher's Note

Bourgeais et al. BMC Bioinformatics (2021) 22:455
https://doi.org/10.1186/s12859-021-04370-7

BMC Bioinformatics

RESEARCH

Open Access

Deep GONet: self-explainable deep neural network based on Gene Ontology for phenotype prediction from gene expression data

Victoria Bourgeais*, Farida Zehraoui, Mohamed Ben Hamdoune and Blaise Hanczar

From The 19th Asia Pacific Bioinformatics Conference (APBC 2021) Tainan, Taiwan. 3-5 February 2021

*Correspondence:
victoria.bourgeais@univ-evry.fr
IBISC, Univ Evry,
Université Paris-Saclay,
91020 Évry-Courcouronnes,
France

Abstract

Background: With the rapid advancement of genomic sequencing techniques, massive production of gene expression data is becoming possible, which prompts the development of precision medicine. Deep learning is a promising approach for phenotype prediction (clinical diagnosis, prognosis, and drug response) based on gene expression profile. Existing deep learning models are usually considered as black-boxes that provide accurate predictions but are not interpretable. However, accuracy and interpretation are both essential for precision medicine. In addition, most models do not integrate the knowledge of the domain. Hence, making deep learning models interpretable for medical applications using prior biological knowledge is the main focus of this paper.

Results: In this paper, we propose a new self-explainable deep learning model, called Deep GONet, integrating the Gene Ontology into the hierarchical architecture of the neural network. This model is based on a fully-connected architecture constrained by the Gene Ontology annotations, such that each neuron represents a biological function. The experiments on cancer diagnosis datasets demonstrate that Deep GONet is both easily interpretable and highly performant to discriminate cancer and non-cancer samples.

Conclusions: Our model provides an explanation to its predictions by identifying the most important neurons and associating them with biological functions, making the model understandable for biologists and physicians.

Keywords: Gene expression, Phenotype prediction, Model interpretation, Deep learning, Gene Ontology

Background

With the rapid advances of data acquisition technologies, collecting large amounts of different-type data (images, ECG, genomics...) becomes simpler in the medical field. It inspires a new form of this field, i.e., precision medicine, which takes advantage of these available data to improve profoundly diagnosis, prognosis, or therapeutic decision. Precision medicine has access to detect in advance a disease, such as cancer, anticipate the progression of the disease, and adapt the therapy according to the characteristics of patients. Among these data, genomic data and especially gene expression data play a key role in the development of precision medicine. Gene expression profile is known to be an indicator of the cellular state and allows the study of complex diseases.

For many years, machine learning has been used on transcriptomic data to construct classifiers predicting phenotypes (diagnosis, prognosis, treatment) [1]. In the last decade, deep learning has become the source of the most impressive improvements in machine learning [2]. It shows its superiority in many domains such as image analysis or natural language processing. Its main advantage is that it constructs high levels of data abstraction by stacking multiple linear and non-linear processing units. Deep learning has recently been applied to classification based on gene expression problems. Unlike images or texts, gene expression data have no structure. The architectures used in the literature are, therefore, mainly autoencoders and multilayer perceptrons (MLP) [3]. For instance, *Stacked Denoising Autoencoders* [4, 5] are exploited to extract a lower dimension from the data, then a classifier (such as support vector machine (SVM) or MLP) is applied to perform classification. MLP are used in [6, 7] to predict directly diseases without dimension reduction. Despite promising first results, deep learning has not made a breakthrough in gene expression classification yet because of the often small size of the available training sets. Deep learning is especially good with large training sets. In the next years, with the increasing production of transcriptomic data, it is highly likely that deep learning will play a major role to solve these problems.

One of the main concerns of the application of deep learning in the medical field is its lack of interpretability. Indeed, the neural networks are black-box models, which means that the model cannot provide an explanation to its decision. The interpretation of machine learning algorithms is one of the most essential topics nowadays, especially in the case of medical application for three main reasons. First, both the physician and his patient must understand why the model predicts a given phenotype. Particularly, it can influence later decisions such as the choice of the treatment. Second, it is important to ensure that the model bases its predictions on a reliable representation of the data and does not focus on irrelevant artifacts. This will highly impact the trust of the physicians toward the predictions regardless of the performances of the model. Finally, the model with high-accurate predictions may have identified interesting patterns that biologists would like to investigate.

We can distinguish two main approaches for interpreting the black-boxes: the post-hoc methods and the self-explainable models [8]. In a post-hoc method, the black-box model is first learned and then an interpretation method is used to explain the predictions. Several post-hoc methods with different purposes are proposed in the literature. Among them, proxy methods, which approximate a black-box model by an interpretable model, can help interpret the general behavior of the model. For example, Ribeiro *et al.* [9] propose a linear proxy method, *Local Interpretable Model-Agnostic Explanations* (LIME), to approximate any black-box model. Interpretation methods specific to deep learning have been recently proposed, namely gradient-based methods [10]. The model prediction is explained by backpropagating the signal from the output to the input. This type of method enables the identification of the most relevant features and neurons involved in the decision making. Several gradient-based methods are proposed in the literature including *Layerwise Relevance Propagation* (LRP) [11, 12], *Integrated Gradients* [13], and *DeepLift* [14]. In [15], the authors show that among these methods, DeepLift and LRP are better aligned with human intuition since they satisfy some desirable properties. The self-explainable models are inherently interpretable models. By definition, they include the decision trees, rules systems, sparse linear models. However, these three models generally do not perform well on high-dimensional complex data. Few works on self-explainability have been proposed for deep learning. Melis and Jaakkola [16] introduce a built-in interpretable model, *Self-Explainable Neural Network*, that behaves locally as a linear model.

A general opinion is that the black-boxes are more accurate than the self-explainable models. The capacity of interpretability is often viewed as a constraint of the model that decreases its performance. There would be a trade-off between performance and interpretability. However, recently a part of the machine learning community claims that performance and interpretability are not exclusive. Rudin [17] explains why black-box models should be avoided for crucial decisions, like in medical applications, even with the use of post-hoc interpretation. For example, the proxy methods create a new model that approximates the decision process of black-box models, leading to an imperfect fidelity in explanation. In addition, different explanations can be obtained for the same prediction using different interpretation methods or the same interpretation method with different parameters [18, 19]. Rudin, therefore, promotes the development of high-accurate self-explainable models. Self-explainable deep learning model is one of the solutions.

All of these methods, post-hoc and self-explainable, consider that the interpretation of a model consists of the identification of the inputs and the part of the model, in case of deep learning the set of neurons, that support the predictions. In the context of phenotype prediction from gene expression, these methods generally do not provide an understandable explanation. The explanation must be completed with knowledge of the domain. For example, we have to explain which biological functions are represented in the model and which ones are used to compute the patient outcomes.

Few works have been published on the construction of self-explainable neural networks based on gene expression data using prior biological knowledge. Prior knowledge comes from the ontologies such as Gene Ontology (GO) [20], Kyoto Encyclopedia of Genes and Genomes (KEGG) [21], Reactome [22], or Search Tool for the Retrieval of Interacting Genes/Proteins (STRING) [23]. Among them, the closest literature to our work [24] incorporates GO into a neural network, called *Gene Ontology Neural Network* (GONN). They replace one hidden layer by one level of the GO subontologies and connect the input features by partial connections according to the annotations with the ontologies. In this way, some input features cannot be included if there are not connected to the ontologies. Similarly, *Gene-Pathway-Disease* (GPD) [25],

Pathway-Associated Sparse Deep Neural Network (PASNet) [26], and *Gene Regulatory network-based Regularized Artificial Neural Network* (GRRANN) [27] integrate respectively biological pathways and regulators from protein-protein/protein-gene interactions in the first layer. These architectures contain at most two hidden layers. For example, PASNet tries to capture nonlinear interactions between pathways in a second hidden layer. However, deep neural networks allow deeper representations of hierarchical relations between gene expression and biological objects. By using prior knowledge, these works first attempt to boost the model performances on their target tasks. Yet, they do not clearly show whether the neurons correspond to the associated biological concept or not. Integrating knowledge may not be so beneficial for learning.

In this paper, we propose a self-explainable deep fully-connected neural network, called *Deep GONet*, based on gene expression data. This model is constrained by prior biological knowledge from GO, which is widely used in the bioinformatics community. The architecture represents different levels of the ontology preserving the hierarchical relationships between the GO terms by using sparse regularization. Our objective is to build an accurate and relevant interpretable model for cancer detection. Each neuron is associated with a GO function and the links between these functions are represented by the network connections. A prediction of the network can, therefore, be directly explained by the set of biological functions.

The paper is organized as follows. We first describe the proposed novel model, Deep GONet, for biological interpretation. Then, the model is evaluated on two public datasets and compared with other approaches. We also provide how to obtain the explanations of outcomes and their biological significations at three levels (disease, subdisease, and patient). The conclusions to this paper and some future research directions are finally presented.

Methods

We propose a new neural network model, *Deep GONet*, that is self-explainable and embeds the biological knowledge contained in GO. Our model is based on a MLP constrained by the GO structure. The constraints are introduced into the network using an adaptive regularization term.

The architecture of Deep GONet

Our model takes in the input layer the gene expression profile of a patient and returns in the output layer the prediction of a phenotype of this patient. The architecture of the hidden layers represents the structure of GO. GO gathers three ontologies that respectively describe the following categories: biological process (GO-BP), molecular function (GO-MF), and cellular component (GO-CC). We chose to base the architecture of the hidden layers on the GO-BP since it provides larger processes implied by the activity of the genes, which can be more useful for phenotype prediction. However, it is possible to implement the GO-MF or GO-CC in the network architecture with the same method.

GO-BP is structured as a directed acyclic graph (DAG) containing 11991 nodes (version of October 2019) annotated with the input layer and distributed over 19 levels as illustrated in the top of Fig. 1. Each node is a GO term representing a biological function. Two GO terms are linked if their biological functions are related and the

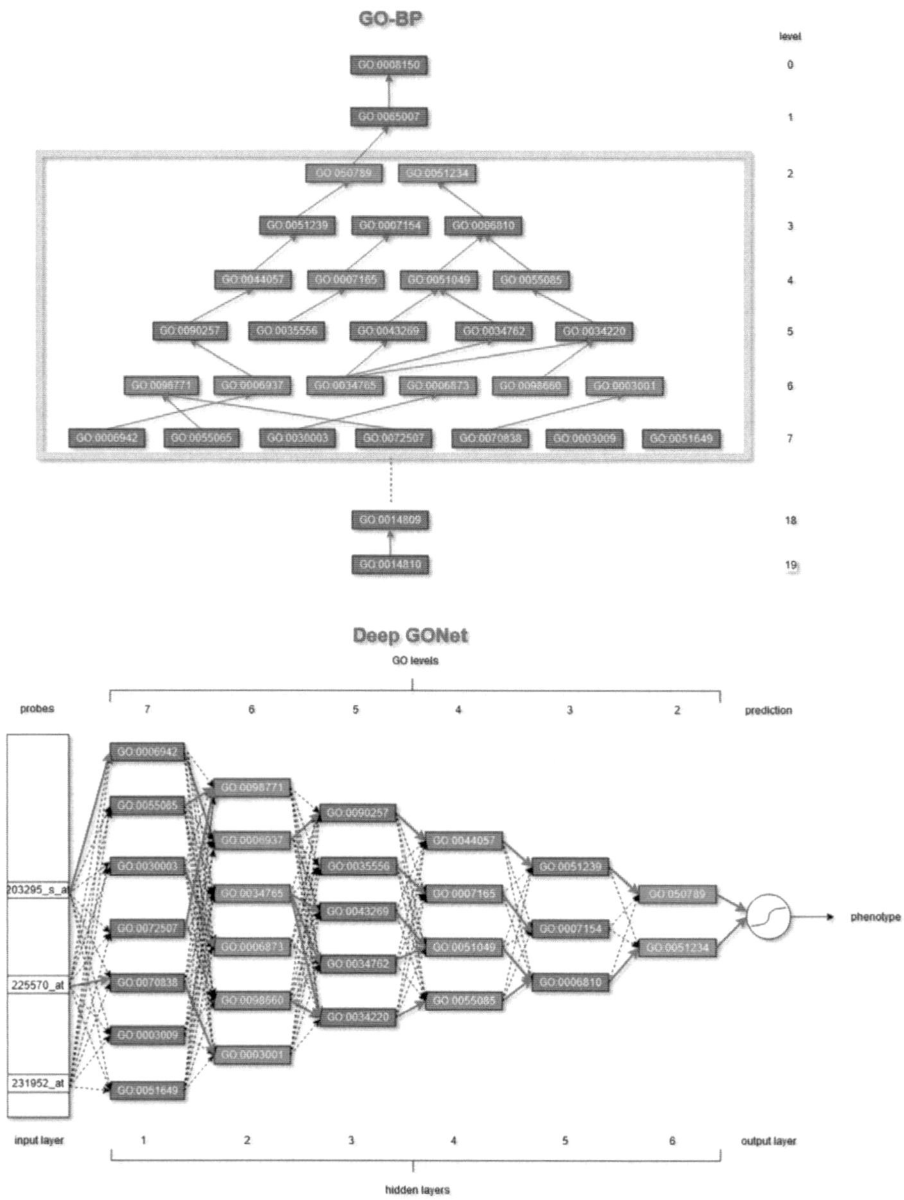

Fig. 1 A subset of GO-BP (top) and the corresponding Deep GONet architecture (down). The green box represents the GO levels implemented in Deep GONet. The red and black dashed arrows represent respectively the GO and noGO connections

majority of these relations are "is a" relations. The GO terms are connected respecting a hierarchical bottom-up orientation. A GO term is assigned to a dedicated level according to its longest path to the root (i.e., GO:0008150). The GO terms in lower levels correspond to more specific functions, like *positive regulation of skeletal* (GO:0014810 at the 19th level), whereas the nodes in upper levels are more general functions such as the root function GO:0008150. The GO terms are also linked to genes via GO annotations. A parent GO term (i.e., destination of incoming connections) inherits, therefore, the set of genes from its children (i.e., origins of incoming connections).

Our neural network architecture represents the GO-BP, i.e., each hidden layer l represents a GO level h, each neuron a GO term, and each input variable a gene. Since the lowest levels of GO contain few very specific terms and the highest levels are very general, it seems not useful to implement the whole GO in our architecture. The selection of the levels is, therefore, part of the hyperparameters of the model to determine. In our experiments, the level 7 to level 2 have been selected as illustrated by the green box in the Fig. 1.

Our model is based on a fully-connected MLP that consists of an input layer, L hidden layers, and an output layer for phenotype prediction. The input layer is composed of genes or gene products (e.g., probes). A probe is a short DNA sequence targeting a region of one or several genes. It is the measure used in microarray data. Each neuron is connected to all neurons of the previous layer and all neurons of the next layer. Each hidden layer corresponds to a level in GO-BP and its neurons match all the GO terms of the target level. Note that the incorporation of the knowledge must respect the goal of the neural network to construct an abstract representation of the data through its hierarchical architecture. The first hidden layer of a neural network extracts the low-level features from the input layer, it corresponds to the lowest selected level of GO containing more specific biological functions. In the last hidden layers, the high-level features represent the most general biological functions of the highest GO levels. The bottom of the Fig. 1 illustrates where the levels 7 to 2 of GO are implemented in the architecture of the neural network. The activation of the i-th neuron of the hidden layer l can be expressed as: $a_i^{(l)} = f\left(\sum_{j=1}^{N_{l-1}} a_j^{(l-1)} w_{ji}^{(l)} + b_i^{(l)}\right) = f\left(z_i^{(l)}\right)$, where $w_{ji}^{(l)}$ is the weight of the connection from the j-th neuron of the layer $l-1$ to the i-th neuron of the layer l, $b_i^{(l)}$ is the bias of the i-th neuron of the layer l, N_{l-1} the number of neurons in the layer $l-1$, $z_i^{(l)}$ is the sum of signals received from the previous layer $l-1$, and f is the rectified linear unit function (ReLU) defined as $f(x) = \max(0, x)$. The output layer finally estimates the probability to belong to each class. For binary problem, the output layer contains only one neuron with a sigmoid function, given by $a^{(L)} = \frac{1}{1-\exp(z^{(L)})}$, returning the probability to predict the positive class. Note that for multiclass problem, the output layer should contain one neuron for each class k with a softmax function, defined as $a_k^{(L)} = \frac{\exp(z_k^{(L)})}{\sum_{j=1}^{K} \exp(z_j^{(L)})}$ where K represents the number of classes, to get a probability of belonging to each class.

In our fully-connected architecture, we identify two types of connections :

- connections corresponding to links in GO-BP (colored in red in Fig. 1), called GO connections;
- connections between two nodes that are not linked in GO-BP (marked by dashed arrows in Fig. 1), called noGO connections.

A probe in the input layer in Fig. 1 is connected to the neurons of the first hidden layer via a GO connection if it is associated with the corresponding GO term in the lowest chosen level of GO-BP (i.e., level 7 in Fig. 1), or via a noGO connection otherwise. Note that the neurons of the next hidden layers (i.e., 2 to 6 in Fig. 1) are not directly connected to the probes. These neurons can be indirectly connected to their probes by the propagation of gene expression through the GO connections of the previous layers. If we want to represent exactly the GO-BP, we can cut all noGO connections and keep only the GO connections in our architecture. However, GO only represents the current knowledge we have on biology. The ontologies change continuously with the outcoming of new scientific discoveries. Some links can be missing or wrong, and many genes are not associated with their right corresponding GO term. 33% of the probes from the microarray HG-U133Plus2 used in our experiments have no GO annotations (such as the probe 231952_at in Fig. 1). This means that these probes would not be connected to the neural network if we use only the GO connections. They would not be used to compute the prediction even if they bring relevant information. This situation could impact negatively the accuracy of the neural network. To deal with the errors and the incompleteness of the knowledge represented in GO, we keep all connections in our architecture (both GO and noGO connections). However, the noGO connections are penalized to favor the use of GO connections to compute the predictions.

Learning and regularization of the network

The model is constrained by a customized regularization term, named L_{GO}, to favor the GO connections and penalize the noGO connections. This regularization term is defined as follows:

$$L_{GO} = \sum_{l=1}^{L} \| W^{(l)} \otimes (1 - C^{(l)}) \|_2^2 \tag{1}$$

where L is the number of hidden layers of the neural network, $W^{(l)}$ is the weight matrix of the layer l, and $\otimes$ is the pointwise product. $C^{(l)}$ is the adjacency matrix that encodes the connections between the GO terms of the layer $l-1$ and l (i.e., the corresponding levels $h+1$ and h in GO-BP). More precisely, if a GO term i at the corresponding level h in GO-BP is a parent of GO term j from the level $h+1$, then $c_{j,i}^{(l)} = 1$ else $c_{j,i}^{(l)} = 0$. For the output layer, $C^{(L)}$ is a matrix of ones. The loss of our model is the sum of the common cross-entropy loss and our regularization term:

$$L = \sum_{i=1}^{N} \sum_{k=1}^{K} (-y_{i,k} \log \hat{y}_{i,k}) + \alpha L_{GO} \tag{2}$$

where N and K are respectively the number of samples in the training set and the number of classes. $y_{i,k}$ is the indicator of the true class, i.e., $y_{i,k} = 1$ when the i-th sample belongs to the class k, or 0 otherwise. Note that each sample only belongs to one class. $\hat{y}_{i,k}$ is the probability that the i-th sample belongs to the class k computed by the neural network. During the inference phase, we select the class with the highest probability to get the final prediction. Finally, α is a hyperparameter that weights the regularization term. With α close to 0, the regularization term vanishes, our model becomes a classical MLP without interpretation capacity. With a high value of α, the learning algorithm focuses on the regularization term and ignores the cross-entropy. The resulting neural network represents perfectly the GO connections by cutting the noGO connections, but it has a weak prediction capacity. α is a crucial hyperparameter that controls the trade-off between the minimization of the cross-entropy and the loss L_{GO}.

Results

Dataset

We validate our model on two datasets. The first one comes from a cross-experimental study compiling microarray data of over 40.000 publicity available Affymetrix HG-U133Plus2 chip arrays [28]. These arrays were produced under different experimental protocols and concerned seventeen types of tissue. The dataset contains 54675 probes for 22309 samples whose 14749 (66.11%) are cancer and 7560 (33.89%) are non-cancer. We standardize it to a mean of zero and a standard deviation of one, and split it into a training set of 17847 examples (11799 cancer, 6048 non-cancer) and a test set of 4462 examples (2950 cancer, 1512 non-cancer). Note that the original proportions of cancer and non-cancer samples are preserved in each set. The training set is used to train predictive models and the test set to evaluate their performances.

The second dataset is RNA-Seq data from The Cancer Genome Atlas (TCGA) combining 5982 samples of 11 cancer types and 482 normal samples from different tissues. Table 1 lists the number of cancer samples by cancer type. The number of input features is 56602. Before standardization, the data have been pre-normalized with FPKM (fragments per kilobase per million mapped reads) and transformed using $\log_2$. 80% of the dataset goes into a training set and the remaining 20% into the test set.

Performances and sensitivity analysis

In this first experiment, we compare the performances of Deep GONet on cancer prediction from gene expression profile with the state-of-the-art. Binary classification is evaluated on the microarray dataset with a sigmoid function in the output layer whereas multiclass classification is performed on the RNA-Seq data with a softmax.

Deep GONet model is learned from the training set using a standard learning procedure. The number of layers and nodes are determined by the levels chosen in GO-BP. Different levels of GO-BP have been tested. For both datasets, we fix the architecture with levels 7 to 2 of GO-BP since it gives us the best performance. We test different values of the training hyperparameters and select the following settings. The weights and biases are initialized with He initializer. On the microarray dataset, dropout layers with a ratio of 0.6 are added after each hidden layer to reduce overfitting. The network parameters are optimized using adam with an adaptive learning rate of 0.001. On the RNA-Seq dataset, we choose the stochastic gradient descent with a momentum equal to 0.9. The number of epochs of the training is set up to 600. Different values of the hyperparameter α, controlling the regularization term L_{GO}, are tested in the interval $[0, 10^1]$. The accuracy of the model is estimated from the test set according to the value of α to investigate the impact of this hyperparameter on the performance of the model. To reduce the variability of the results coming from the random initialization of the model parameters, 10

Table 1 Number of samples by cancer types in the TCGA dataset

Cancer type	BRCA	HNSC	KIRC	LGG	LIHC	LUAD	LUSC	OV	PRAD	THCA	UCEC
#patients	1102	500	538	511	371	533	502	374	498	502	551
Frequency (%)	17.05	7.74	8.32	7.91	5.74	8.25	7.77	5.79	7.71	7.77	8.53

models for each value of α are learned with different random seeds for the initialization of the parameters.

Our method is compared with classical fully-connected networks using L_2 or L_1 regularization terms. These regularization terms apply a penalty on all the connections regardless of the type (GO or noGO). L_2 is the squared magnitude of the weights $L_2 = \sum_{l=1}^{L} \|W^{(l)}\|_2^2$, and L_1 is the absolute value of the magnitude of the weights $L_1 = \sum_{l=1}^{L} |W^{(l)}|$. These regularization terms are also controlled by a hyperparameter α. In addition, a model without any regularization is tested for comparison at $\alpha = 0$. Note that all these models use the same baseline described in Fig. 1. They are trained and tested with the same procedure used for Deep GONet. The accuracy of each model is estimated from the test set. All the experiments have been executed on a GPU RTX 2080Ti using Tensorflow 1.12.

In what follows, the results on the microarray dataset (Fig. 2a–c) are commented, but similar results are observed on the TCGA dataset (Fig. 2d–f).

Figure 2a (resp. Fig. 2d) plots the average and the standard deviation of the accuracy of a model with a L_1, L_2, and L_{GO} penalty according to α. The three curves begin at the same point since $\alpha = 0$ corresponds to a model without regularization. We can see that the model without regularization and the one with L_{GO} and L_2 at $\alpha = 10^{-5}$ achieve the best accuracy (0.945). Note that L_{GO} and L_2 outperform L_1. We also test classical machine learning methods with scikit-learn python package: Random Forest (Gini criterion, number of trees=100), SVM (linear kernel, C=1.0), XGboost (number of trees=100, learning rate=0.1), and MLP (three layers with respectively 1000, 500 and 200 nodes). Metrics performance for each method are detailed in Table 2 (resp. Table 3 for TCGA). Similar performances are obtained. These results show that our method obtains the same accuracy with the state-of-the-art algorithms, which are not self-explainable. For all models, the average accuracy decreases for high value of α. The accuracy drops to 0.66 which corresponds to the proportion of the majority class, meaning that the models learn nothing and associate all examples to the cancer class. In this case, the regularization term takes too much importance relative to the cross-entropy. We note some special points, at $\alpha = 10^{-1}$ (for L_{GO} and L_2) and at $\alpha = 10^{-3}$ (for L_1), with high variability. At these values of α, some models fail to learn with an accuracy of 0.66, whereas others succeed by reaching an accuracy around 0.9. That's why the average is between these two extremes.

In the following, we analyze the behavior of GO and noGO connections in Deep GONet and standard MLP. Figure 2b (resp. Fig. 2e) displays the ratio between the absolute-value norms of the GO (Eq. 3) and noGO (Eq. 4) connections, defined respectively as:

$$\frac{1}{L}\frac{1}{\#GO}\sum_{i=1}^{L}|W^{(l)} \otimes (C^{(l)})|, \tag{3}$$

$$\frac{1}{L}\frac{1}{\#noGO}\sum_{i=1}^{L}|W^{(l)} \otimes (1 - C^{(l)})|. \tag{4}$$

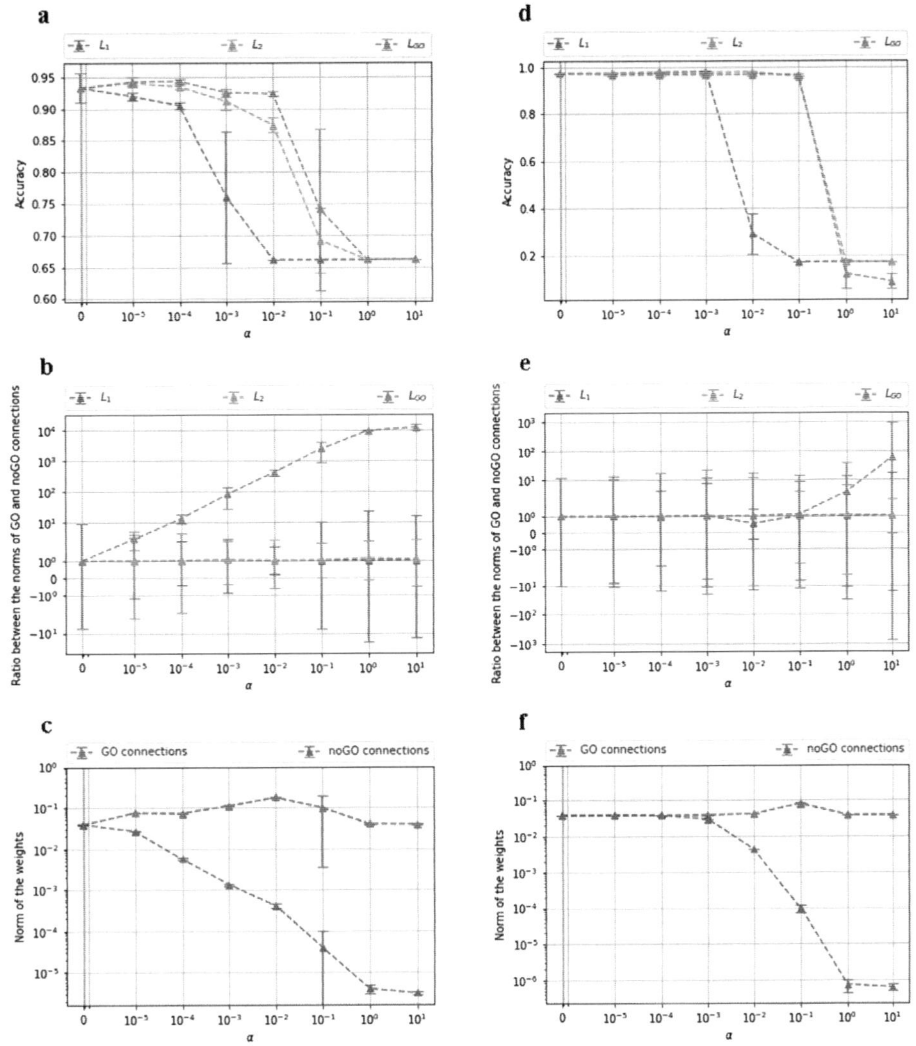

Fig. 2 Results on the microarray dataset (left column) and the TCGA dataset (right column). **a–d** Accuracy of the models according to α. **b–e** Ratio between GO and noGO connections weights according to α. **c–f** Absolute-value norms of the GO and noGO connections from L_{GO} models according to α

Table 2 Comparison of the performances of the models on the microarray dataset

Model	Accuracy	Precision	Recall	F1-Score	MCC	AUC
RF	0.904	0.932	0.921	0.927	0.786	0.895
SVM	0.948	0.964	0.957	0.961	0.885	0.944
XGBoost	0.936	0.954	0.948	0.951	0.857	0.930
MLP	0.951	0.974	0.952	0.963	0.893	0.986
Deep GONet	0.925	0.943	0.943	0.943	0.832	0.916

For L_2 and L_1, the ratio is stuck to 1 whatever the value of α. As expected, no distinction is made between the two types of connections. On the opposite, the ratio of the model with L_{GO} regularization increases along with the growth of α and finally reaches its

Table 3 Comparison of the performances of the models on the RNA-Seq dataset

Model	Accuracy	Precision	Recall	F1-Score	MCC	AUC
RF	0.968	0.967	0.965	0.966	0.964	0.999
SVM	0.977	0.977	0.975	0.976	0.974	1.000
XGBoost	0.974	0.972	0.972	0.972	0.971	1.000
MLP	0.962	0.961	0.960	0.960	0.958	0.998
Deep GONet	0.970	0.970	0.967	0.968	0.967	0.998

highest value of 10^4. For this model, Fig. 2c (resp. Fig. 2f) shows the average of the absolute-value norms of the GO (Eq. 3) and noGO (Eq. 4) connections. Note that the green line in Fig. 2b (resp. Fig. 2e) is obtained from the division of the red line by the green line from Fig. 2c (resp. Fig. 2f). We can observe that the average norm of the GO connections remains between 10^{-2} and 10^{-1}. In contrast, the average norm of noGO connections decreases with α, following the accuracy trend. With $\alpha = 0$ and $\alpha = 10^{-5}$, the average norm of the noGO connections is very close to the one of the GO connections. The ratio between the two norms, illustrated in Fig. 2b, is below 10^1. From 10^{-4} to 10^1, the gap between the two norms becomes larger. The norm of noGO connections is converging almost to 0, leading to a ratio of 10^1 at $\alpha = 10^{-4}$ and the highest ratio of 10^4 at $\alpha = 10^1$. As a consequence, L_{GO} seems to penalize well the noGO connections with high value of α. At $\alpha = 10^1$, the model is equivalent to a model containing only GO connections, all noGO connections are set to 0, respecting the hierarchy of GO scrupulously. However, the accuracy curves in Fig. 2a show that with a large value of α, the model is not able to learn anymore. It means that some noGO connections are necessary for accurate predictions. In particular, the flexibility brought by the fully-connected architecture makes it possible. This advantage will be further inspected in the next sections.

In summary, imposing a number of layers and neurons is not enough to make the model interpretable. An appropriate regularization term should be added to the loss function to constrain it along with biological knowledge. If the regularization term is not customized, the GO and noGO connections will be considered identically like with a L_2 or L_1 regularization. This results in a non-interpretable model without any prior knowledge. Our model Deep GONet reaches similar prediction performances than the state-of-the-art, in both (i) penalizing properly the noGO connections, and (ii) privileging enough the GO connections to let the major information flow by them.

On the microarray dataset, the models at $\alpha = 10^{-2}$ achieve an average accuracy around 0.92 and an average ratio of 10^3. Since they represent a good trade-off between noGO connections penalization and accuracy, we analyze in-depth and interpret biologically one of the models learned with $\alpha = 10^{-2}$ in the rest of this paper. The study will focus on the microarray dataset, but similar analyses can be conducted on the other dataset.

Analysis of the Deep GONet architecture

The first part of this analysis is to check that the architecture of the model chosen in the previous section is very close to the subhierarchy of GO-BP. This model has been learned with $\alpha = 10^{-2}$ and reaches an accuracy of 0.925 (reported in Table 2) as well as a ratio between GO and noGO connections around 10^3. Table 4 presents in detail its architecture. The first two rows summarize the corresponding levels from GO-BP and the number of neurons from the input layer to the output one (see Fig. 1). The last two rows give for each layer the number of incoming connections (GO and noGO) and the number of incoming GO connections. Note that the total number of connections plus the number of neurons constitute the number of parameters of the model (i.e., around 90.105M). The number of connections decreases through the layers because the number of neurons by layer becomes smaller. This table shows that the large majority of the connections are noGO connections, only 0.05% are GO connections (i.e., around 48K).

Figure 3 displays for each layer the sorting of the incoming connections according to the absolute value of their weight. The incoming GO (resp. noGO) connections are colored in red (resp. green). We first note that the connection matrices are very sparse, few connections have their weight significantly different from 0. This means that the gene expression is not uniformly propagated through the entire network and only a small part of the network is useful for the prediction. For all hidden layers, most of the GO connections are ranked before the noGO connections. Some of the GO connections can have a very high weight (around 10^0). The high-weighted incoming GO connections of a neuron promote the activation of its corresponding function. The value of the noGO connections is close to 0 as expected by the application of the L_{GO} penalization. Some GO connections are ordered at the bottom of the rank. For example, the 43 505th GO connection of the first layer is ranked 33 041 190th. The GO connections, which do not seem to be useful for the network, get a very low value (7.10^{-6} for our example). On the opposite, despite the application of the L_{GO} penalty on noGO connections, few of them have higher weight than GO connections as illustrated in the figure of the second hidden layer. These results show that the architecture of our model is very close to the GO-BP architecture since most of the weights of noGO connections are set to 0. The rare noGO connections with high weight are interesting. It represents links that the network has to build to compute accurate predictions. It would be interesting to investigate which GO terms or probes that have been connected by these noGO connections.

The next analyses of our network will be based on two sets of values: the neurons activation and neurons relevance. The activation $a_i^{(l)}$ of a neuron i from layer l gives information about how much signal $z_i^{(l)}$ flows from this neuron. However, high activation doesn't necessarily mean that the neuron contributes highly to the prediction. A neuron highly activated by a given sample may have outcoming connections with very low weight. In this case, it will contribute a few to the prediction. Therefore, to identify which neurons and connections are used to compute the predictions, we employ a gradient-based method, *Layerwise Relevance Propagation* (LRP) [11, 12]. The aim of LRP is to

Table 4 Details about the architecture of Deep GONet

Layer	Input	1	2	3	4	5	6	Output	Total
Level GO-BP	–	7	6	5	4	3	2	–	6
#neurons	54675	1574	1386	951	515	255	90	1	4772
#connections	–	86M	2.2M	1.3M	490K	131K	23K	90	90.1M
#GO connections	–	43504	1709	1585	1010	491	175	–	48K

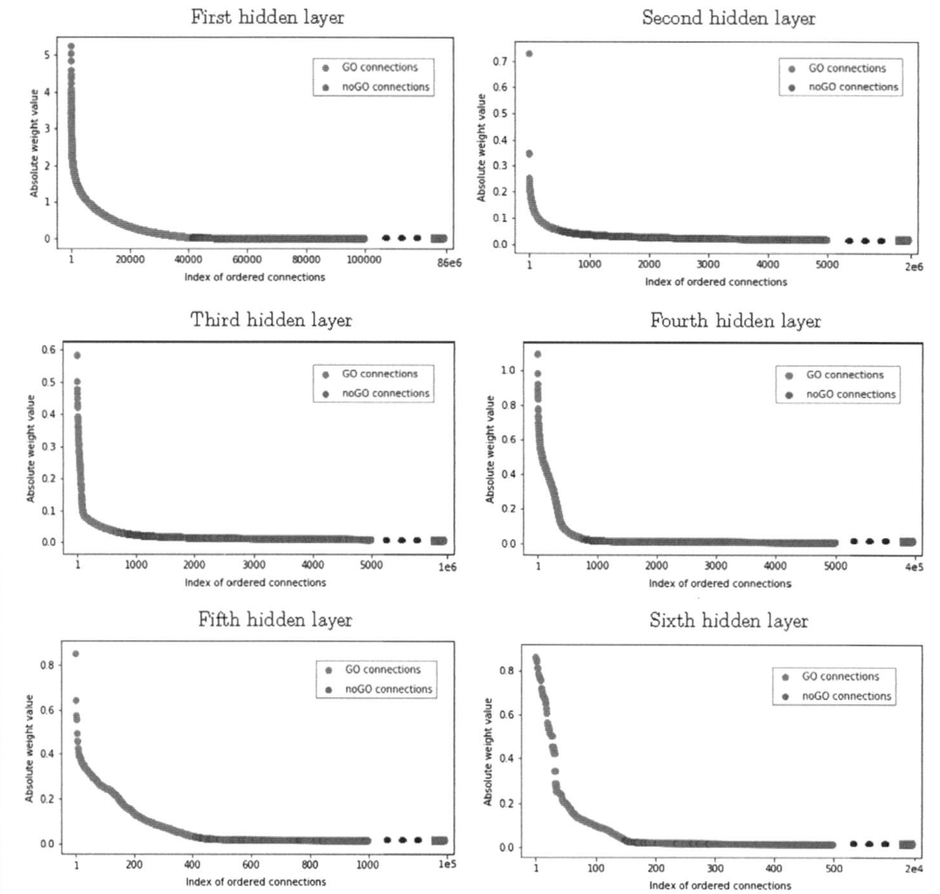

Fig. 3 Sorting of incoming connections from each layer according to their absolute weight value

retropropagate the output signal of one sample from the upper hidden layer to the input layer. A relevance score assigned to a neuron i of a layer l is given by

$$R_i^{(l)} = \sum_{j=0}^{N_{l+1}} \frac{a_i^{(l)} w_{i,j}}{\sum_k a_k^{(l)} w_{k,j} + \epsilon} R_j^{(l+1)} \qquad (5)$$

where ϵ is a factor of stabilization (equals to 10^{-7} in our experiments), $R_j^{(l+1)}$ is the relevance of a neuron j of the upper layer $l+1$, and $R^{(L)} = z^{(L)}$. This score represents the proportion of the output signal passing through the neuron and its outcoming connections. The relevance of a neuron represents its importance in the computation of the prediction. For each patient, we can get a relevance (resp. activation) profile by layer composed of neurons relevance (resp. activation). An analysis of the neurons relevance of each layer confirms the fact that only a small subset of neurons is important to compute a given prediction.

Biological significance of the neurons

In this section, we check that the neurons of our network actually represent their corresponding GO term, i.e., the activation of a given neuron represents the expression of the corresponding biological function. For that, we use the fact that each GO term in GO is associated with a set of probes. If a given neuron really represents its corresponding GO term, the set of probes associated with this GO term should activate the neuron more than any other set of probes. We propose a procedure illustrated in Fig. 4 based on this principle to test the biological significance of the neurons and evaluate the relationship with the importance of the neurons by using LRP with the package innvestigate [29]. We detail in the following only the analysis of the first hidden layer. However, we can apply similar analyses to the other layers.

The first hidden layer contains 1574 neurons connected to the input layer. Each GO term is connected to a set of probes (median: 8, max: 1357, min: 1). Regarding this information, the target mask of a neuron is defined as follows:

- all the probes of the input layer, which are not connected to the GO term, are set to be 0;
- the values of the remaining probes in the set are unchanged.

In total, we have 1574 masks because none of the neurons has the same target mask. For every neuron, all these masks are applied to the input layer to identify whether the neuron is activated more by its target mask than the other masks. This can be measured by the rank of the target mask. The following procedure, illustrated in the top of Fig. 4, details how to get the rank of the target mask for each neuron in a layer l:

- *Step 1*: For each sample x from the full test set, the activation $a_{i,p}^{(l)}(x)$ of each neuron i for a given mask $m_p^{(l)}$ is calculated where $p = 1, \ldots, i, \ldots, N_l$. As a neuron and its target mask share the same index, the activation of a neuron i for its target mask is $a_{i,i}^{(l)}$. Note that there is no bias due to the length of the mask.

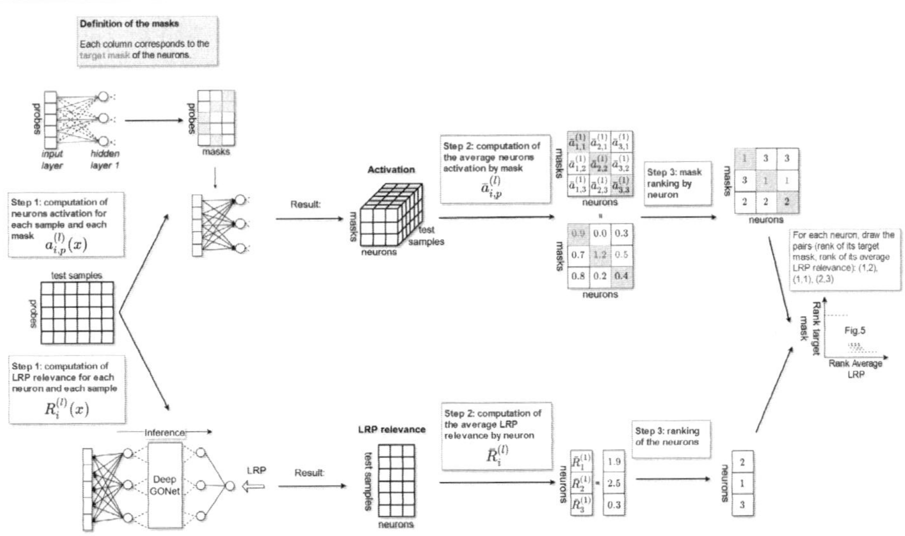

Fig. 4 Illustration of the procedure to evaluate the biological significance of neurons from the first layer. The upper part shows how to calculate the rank of the target mask of each neuron. The lower part shows how to compute the rank of the neurons according to their LRP relevance. The relationship between these two metrics is evaluated through a final figure (e.g., Fig. 5)

- *Step 2*: Then, the average value of these activations $\bar{a}_{i,p}^{(l)}$ is considered. For example, assuming that there are 3 neurons (3 masks) in the first hidden layer, for neuron 1, we obtain $\bar{a}_{1,1}^{(1)} = 0.9, \bar{a}_{1,2}^{(1)} = 0.7$, and $\bar{a}_{1,3}^{(1)} = 0.8$.
- *Step 3*: For each neuron, its activation values of all the masks are ordered in a decreasing way. Then, we have the rank $\bar{a}_{1,1}^{(1)}, \bar{a}_{1,3}^{(1)}, \bar{a}_{1,2}^{(1)}$. It means that the neuron 1 embodies the corresponding GO term because the rank of its target mask is 1.

We compare the rank of the target mask of the neurons with the rank of the neurons according to their LRP relevance. The computation of this rank, described in the bottom of the Fig. 4, follows the steps 1 to 3 without considering the masks. For each sample x and all the neurons i in a layer l, $R_i^{(l)}(x)$ is computed, then the average across the samples $\bar{R}_i^{(l)}$ is calculated. For example, we acquire $\bar{R}_1^{(1)} = 1.9$, $\bar{R}_2^{(1)} = 2.5$, and $\bar{R}_3^{(1)} = 0.3$ respectively for the neurons 1, 2, and 3. According to Step 3, the relevance scores are ordered in the following sequence $\bar{R}_2^{(1)}, \bar{R}_1^{(1)}, \bar{R}_3^{(1)}$. Then, based on this sequence, a rank to each neuron is attributed: neuron 2 gets the rank 1, and so on. Figure 5 plots the rank of the target masks of the neurons along y-axis and the rank of the neurons according to their LRP relevance along x-axis. Note that the value of the ranks is up to the total number of neurons (i.e., 1574).

For the y-axis, a rank can get a NULL value or a discrete value in the range [1,16]. On the one hand, a NULL rank means that the activation of a neuron for its target mask is zero, which concerns 603 neurons (i.e., 38.31% of the 1574 neurons). Specifically, in total 591 neurons have a zero activation value for any mask and generally a LRP rank above the 1000-th order (colored in orange in Fig. 5). The rest 12 neurons are activated by at least one another mask, and their LRP rank is below 1000-th order (colored in green in Fig. 5). On the other hand, there exist 971 neurons (61.69%) that have a positive activation for their target mask and show higher ranks, below order 1000. Among the 971 neurons, the target masks of 850 neurons rank 1, the other 121 neurons rank between 2 and 16. In conclusion, most of the neurons, which contribute highly to the prediction (LRP rank below order 1000), are well ranked for their target

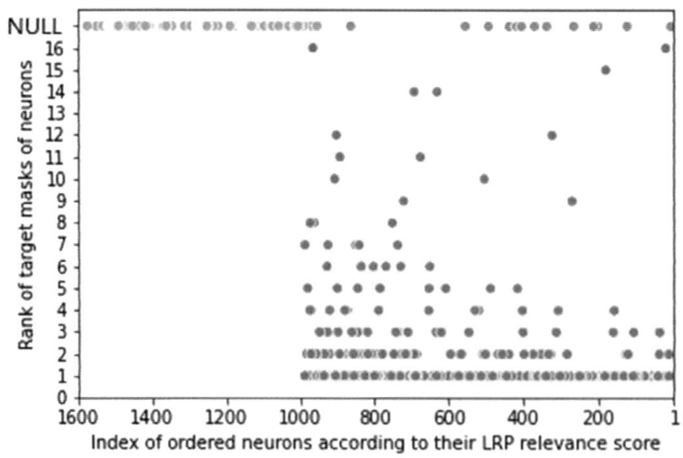

Fig. 5 Sorting of neurons from the first hidden layer according to the rank of their target mask (*y*-axis) and their LRP rank (*x*-axis)

mask. This means that the important neurons for the prediction mainly match with their corresponding GO term.

Concerning the neurons with a NULL rank for their target mask, the major part has low LRP relevance. These neurons are not important for the predictions, they will not be, therefore, used in the interpretation. The associated GO terms can be ignored. However, the few neurons that have a high LRP relevance and a NULL rank are much more interesting (colored in orange in Fig. 5). For instance, the neuron associated with GO:0071644 (*negative regulation of chemokine (C-C motif) ligand 4 production*) has a LRP rank of order 15, but it is not activated by its target mask. Its target mask is composed of 2 probes, linked by GO connections weighted 0.1 and 0.04 respectively. On the opposite, 890 of the 1000 first noGO connections from the input layer, which have the same value with GO connections of norm-1 0.01, point this neuron. Since these neurons are not activated by their target mask, we cannot conclude that they are associated with their corresponding GO term. We note that a large part of the noGO connections with high weight is connected to these neurons. Moreover, these noGO connections connect mainly probes with no annotation in GO, i.e., probes that have only noGO connections. We can assume that the network distorts these neurons from their primary use to propagate the information of probes without GO annotations via noGO connections. These neurons do not represent anymore their corresponding GO terms but an unknown biological information useful for the predictions.

Biological interpretation of the results

In this section, we show how to propose relevant biological interpretations of the model Deep GONet and its predictions. We provide three levels of interpretation: the disease level, the subtype of disease level, and the patient level. First, we study how our model detects cancer from heterogeneous samples basing on the neurons activation. Then, we analyze independently a subtype of cancer by extracting a subnetwork associated to it from the relevance scores computed by LRP. We finally present how the individual prediction of a patient can be explained.

Model interpretation at disease level

In this subsection, we study the clustering of samples correctly predicted as cancer according to their activation profiles. For each sample, an activation profile constituted of the activation of all neurons is computed during the forward pass. For each layer, we define an activation matrix of size (N, N_l) containing the activation of all neurons of this layer for all samples, where N is the number of samples, and N_l the number of neurons in layer l. From these activation matrices, we perform hierarchical clustering using the average linkage and the euclidean distance. The dendrograms of each layer are plotted in Fig. 6. The colors on the dendrogram represent the type of tissue of the samples. In the dendrogram of the first hidden layer, we see that the patients from the same tissues tend to be grouped into the same clusters. It is especially the case for bone (colored in orange), blood (colored in red), and lymph node (colored in cyan). Tissues of the same type tend to share the same activation profiles, meaning that some neurons and their corresponding GO terms are dedicated to one tissue. This clustering according to the tissue is still present in layer two although it is less significant. From the third hidden

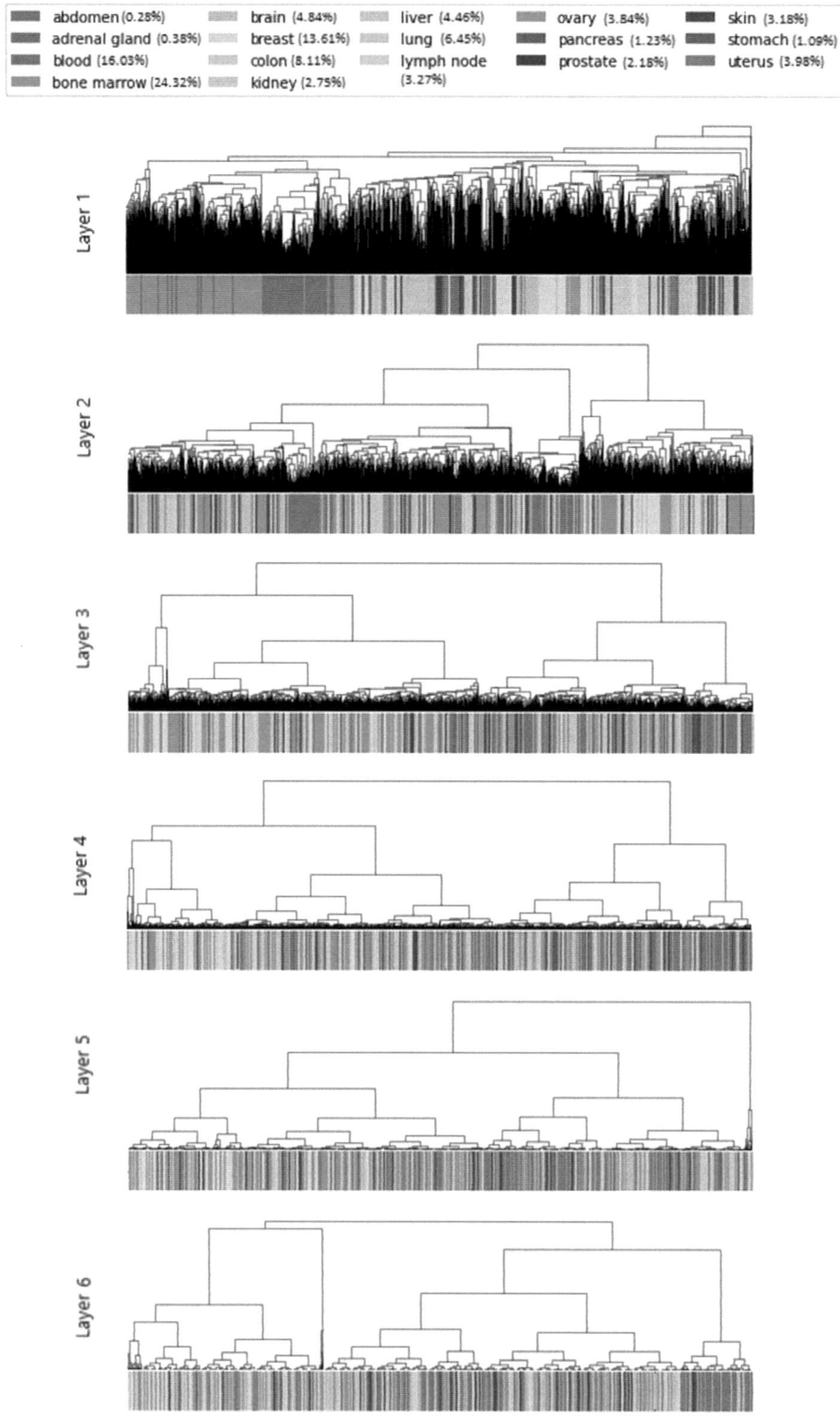

Fig. 6 Hierarchical clustering of test samples correctly predicted as cancer based on their activation profiles. Dendrograms are displayed by layer from the first hidden layer (top) to the sixth hidden layer (bottom)

layer, the clustering of samples from the same tissues becomes less clear. From layer four to six, the clusters contain samples from different tissues. This means that the same GO terms are activated for cancer prediction regardless of the characteristics of the inputs (such as the localization of the cancer). A signature of cancer shared by all tissues has been learned in the last layers of the network. In conclusion, according to the way our architecture is structured, the lower hidden layers gather more specific GO terms. The associated neurons are responsible to extract cancer features particular to a type of tissue. The model being more general incrementally, the upper hidden layers are instead in charge of extracting common cancer features to any type of tissue. It shows that our classifier is universal and able to extract the features shared by various cancers through common GO terms. In the last hidden layer, the existence of several clusters indicates that different neurons are activated to provide the same prediction since the signal from the input layer to the output layer propagates along different paths. Note that this capacity of extracting specific patterns in the first layers and generic patterns in the last layers is well-known in the deep learning models, which has been widely studied in the convolutional neural networks for image analysis [30].

Through the activation profiles, we see how the information is flowing in the general cancer network. In the next analyses, we will focus on neuron importance by using the relevance score. It's a better indicator for more granular results (e.g., at subdisease and patient level) to assess exactly which neurons contribute the most to individual sample's outcome.

Model interpretation at subdisease level: breast cancer

In this subsection, we show how to interpret biologically our model for a specific subtype of cancer.

To this aim, we propose a tool that points out the main biological functions used for cancer predictions and quantify their contribution. Similarly as shown in Fig. 4, we first compute the average LRP relevance of each neuron across a type of cancer samples from one tissue of interest. Then, for each layer, the neurons are sorted according to their relevance score and the most important ones are returned with their corresponding GO term and biological function. In Fig. 7, we give an example on the breast cancer. For each hidden layer, the five most important biological functions are reported with their LRP relevance. Note that this tool can also be used to determine which probes or genes are the most involved in the predictions.

This interpretation tool can be completed by a manual search in the literature in order to identify the links between the returned functions and the predicted phenotype. Biological and medical experts can, therefore, judge the relevance of the prediction based on this final interpretation. Among all GO terms supporting the prediction of cancer in this subnetwork, some of them are known to be related to cancer. The first hidden layer shows two terms (GO:0015031, GO:0006468) linked to protein activities that can highlight a high activity of protein disorder. In the second layer, both GO:0071420 and GO:1901258 reflect the immune activity response to cancer [31, 32]. The macrophage colony-stimulating factor is among one of the growth factors overexpressed in many tumors. Two additional terms related to protein (GO:0044257, GO:0006464) are present. On the third hidden layer, we have

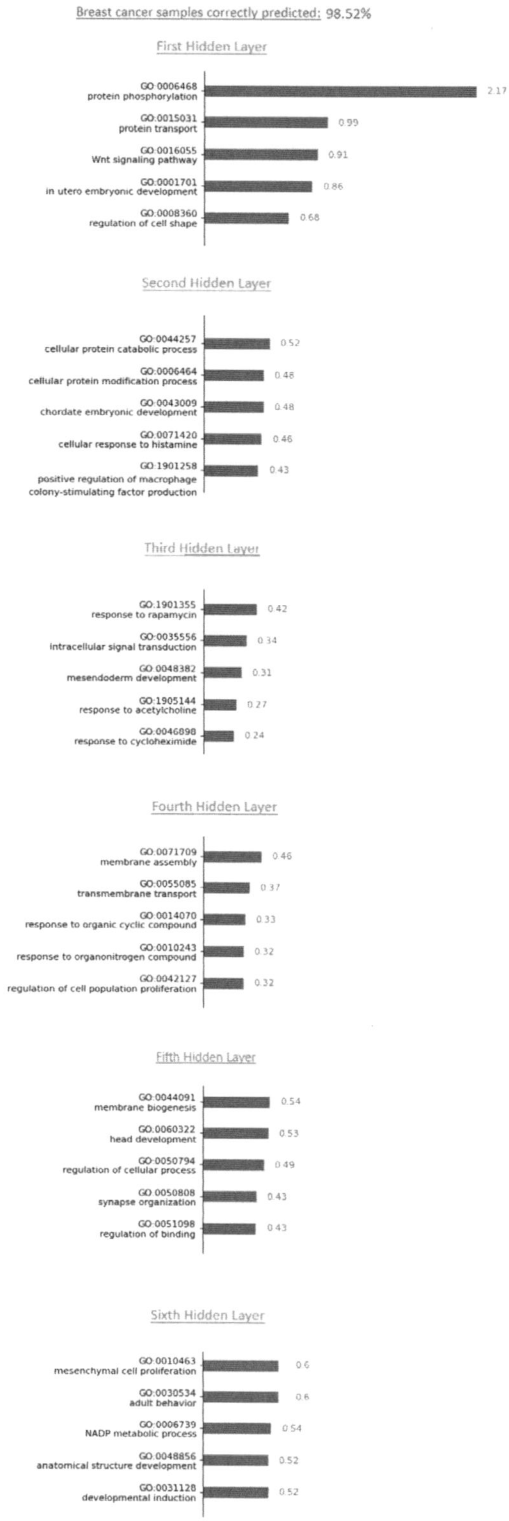

Fig. 7 Interpretation of a subnetwork for breast cancer. The GO terms are ordered according to their relevance score for each layer. The top-5 GO terms are given with literature support

GO:0035556 encoding the biological function *intracellular signal transduction*, part of cell communication. Intracellular signal transduction is a chain of biochemical reactions transmitting signals from the cell surface to receptors of various components within the cell. It finally ends with a cellular response as a cell state change, i.e., cell growth and many other processes. It was found that hyperactivity of these signal pathways can increase the proliferation of abnormal cells [33]. In the fourth hidden layer, GO:0042127 introduces the uncontrolled proliferation well known in tumor spread. Between the fourth and fifth hidden layers, different GO terms refer to membrane activity (GO:0071709, GO:0055085, GO:0044091). Many alterations of the membranes of tumor cells have been detected such as depolarisation [34]. GO:0050794 can point to the deregulation of cellular processes. Finally, concerning GO:0006739 in the last hidden layer, studies show that the quantity of this molecule can be much higher in cancer cells [35].

Note that if our objective had been to predict a subtype of cancer, the interpretation and the subnetworks extracted would be different. The interpretation of a model depends strongly on the phenotype prediction problem.

Prediction interpretation of a given patient

In this subsection, we show how to provide a biological interpretation of the predicted outcome of one single patient. The objective is to propose a tool to the physicians and scientists that makes understandable the prediction computed by the model for a patient. After our model predicts the outcome of a patient with a probability score, the LRP relevance of each neuron is computed. We can then apply the tool previously presented to obtain a rank of neurons by layer. Figure 8 shows an example of the biological interpretation that we propose. In this example, we explain the prediction of the patient 24509 predicted cancer by Deep GONet with a probability of 0.99. Note that this patient is from the breast cancer previous subset. As in Fig. 7, the top-5 important neurons are reported with their relevance score.

In the example of Fig. 8, in the first hidden layer, the term GO:0030335 can highlight the phenomenon of cancer cell invasion into surrounding tissues, which characterizes the beginning of tumor metastasis [36]. In the second hidden layer, GO:0010737 coding for *protein kinase A signaling* can refer to some dysregulations or mutations of the protein specie which contribute to all stages of cancer development [37]. In the third layer, the top-5th term (GO:0048864) can inform the production of cancer stem cells that have similar characteristics with normal stem cells. For the next layers, we find the same relevant GO terms from the previous biological interpretation at subdisease level. We can notice that especially in the first layers (one to three), there can be some differences in the most important GO terms for the prediction between one patient and patients from the same subdisease. In this way, we can identify patients that have distinct characteristics from the average.

We also observe that less than 1% of non-cancer samples have LRP relevances higher than those of the neurons in Fig. 8. It confirms that these neurons extract patterns characteristic of cancer in relation with the biological functions (tumor cell proliferation, protein disorder...).

Fig. 8 Interpretation of the prediction of the sample 24509. The GO terms are ordered according to their relevance score for each layer. The top-5 GO terms are given with literature support

Discussion

We point out that the goal of the interpretation is to explain how the model works and not how the biology works. Sometimes, there are no obvious relations between the biological functions, returned by the interpretation, and predicted phenotype. This does not necessarily mean that the predictions are not reliable. We remind that a model looks for correlations between the output and the input and not for causalities. When a function, which seems not related to the phenotype, is returned, it is possible that this function either has an indirect correlation or is linked by an unknown causality relation with the phenotype. However, the more biological functions returned by the interpretation are coherent with the phenotype, the more we can trust the model predictions. If the most part of the interpretation is incoherent with the current biological knowledge, the reliability of the model should be interrogated. The model may overfit or be mislead by a bias in the training set.

Although the model interpretation is not a tool for biological discovery, some parts of our neural network could be investigated in this way. We refer, in particular, on the high-weighted noGO connections and neurons diverted from their GO term. These elements connect to the network the probes that have not annotations. It could be interesting to understand why these probes have been used for prediction, they should be related to the phenotype. We could also investigate how the expression of these probes is combined into the hidden layer. The probes connected to the same neuron could have close biological functions related to the predicted phenotype. Our model can, therefore, help enrich GO by raising new hypotheses that have to be validated with further biological experiments.

Conclusion

In this paper, we propose, Deep GONet, a new self-explainable deep learning model for phenotype prediction based on gene expression data. We demonstrate that its prediction performances are equivalent to classical deep learning and machine learning methods. The whole architecture of Deep GONet is interpretable and easy-understandable by biologists since it reflects the knowledge that they usually employ. Each layer of Deep GONet corresponds to one level of GO and each neuron to a GO term. The addition of a customized regularization L_{GO} helps the model to better respect this knowledge by focusing on the real connections between the biological objects. The experiments presented on cancer detection show how to provide easily an interpretation of the model and its predictions, understandable by physicians and biologists. In this paper, the architecture of Deep GONet is based on GO-BP, but any other ontologies structured as a DAG (such as GO-CC, GO-MF) can be implemented in the neural network with the same approach. In addition, the model can be applied to other gene expression datasets, or other prediction tasks such as predicting the type of cancer or the prognostic, but it requires a retraining of the model.

In future works, we plan to improve Deep GONet by adding neurons to deal with the genes without GO annotations and a second branch representing the pathways to enrich the biological interpretations. We will finally investigate the links between the activation of a neuron and the activation of the corresponding biological function.

Abbreviations

BRCA: Breast invasive carcinoma; DAG: directed acyclic graph; GO: Gene Ontology; GO-BP: Biological Process Ontology; GO-CC: Cellular Component Ontology; GO-MF: Molecular Function Ontology; HNSC: Head and neck squamous cell carcinoma; KIRC: Kidney renal clear cell carcinoma; LGG: Brain lower grade glioma; LIHC: Liver hepatocellular carcinoma; LRP: Layerwise Relevance Propagation; LUAD: Lung adenocarcinoma; LUSC: Lung squamous cell carcinoma; MLP: multilayer perceptrons; OV: Ovarian serous cystadenocarcinoma; PRAD: Prostate adenocarcinoma; SVM: support vector machine; TCGA: The Cancer Genome Atlas; THCA: Thyroid carcinoma; UCEC: Uterine corpus endometrial carcinoma.

Acknowledgements

Not applicable.

About this supplement

This article has been published as part of BMC Bioinformatics Volume 22 Supplement 10 2021: Selected articles from the 19th Asia Pacific Bioinformatics Conference (APBC 2021): bioinformatics. The full contents of the supplement are available at https://bmcbioinformatics.biomedcentral.com/articles/supplements/volume-22-supplement-10.

Authors' contributions

VB and MBH designed the model Deep GONet. VB tested the model, performed the analysis, and edited the manuscript. FZ and BH supervised the research equally and reviewed the manuscript. All the authors approved the final version of the manuscript.

Funding

We received no funding for this work. Publication costs are funded by IBISC laboratory.

Availability of data and materials

The microarray dataset is accessible from the ArrayExpress database under the identifier E-MTAB-3732. The TCGA datasets can be downloaded from the Genomic Data Commons (GDC) data portal. Deep GONet is implemented in Python using the Tensorflow framework. The code is available at https://forge.ibisc.univ-evry.fr/vbourgeais/DeepGONet.git.

Declarations

Ethics approval and consent to participate

Not applicable.

Consent for publication

Not applicable.

Competing interests

The authors declare that they have no competing interests.

Received: 31 August 2021 Accepted: 8 September 2021
Published online: 22 September 2021

References

1. Kourou K, Exarchos TP, Exarchos KP, Karamouzis MV, Fotiadis DI. Machine learning applications in cancer prognosis and prediction. Comput Struct Biotechnol J. 2015;13:8–17.
2. Goodfellow I, Bengio Y, Courville A, Bengio Y. Deep learning, vol. 1. Cambridge: MIT Press; 2016.
3. Zou J, Huss M, Abid A, Mohammadi P, Torkamani A, Telenti A. A primer on deep learning in genomics. Nat Genet. 2019;51(1):12–8.
4. Fakoor R, Ladhak F, Nazi A, Huber M. Using deep learning to enhance cancer diagnosis and classification. In: Proceedings of the ICML workshop on the role of machine learning in transforming healthcare, vol. 28; 2013. .
5. Danaee P, Ghaeini R, Hendrix D. A deep learning approach for cancer detection and relevant gene identification. Pac Symp Biocomput. 2017;22:219–29.
6. Guo W, Xu Y, Feng X. DeepMetabolism: a deep learning system to predict phenotype from genome sequencing. arXiv e-prints. 2017. arXiv:1705.03094.
7. Hanczar B, Henriette M, Ratovomanana T, Zehraoui F. Phenotypes prediction from gene expression data with deep multilayer perceptron and unsupervised pre-training. Int J Biosci Biochem Bioinform. 2018;8:125–31.
8. Adadi A, Berrada M. Peeking inside the black-box: a survey on explainable artificial intelligence (XAI). IEEE Access. 2018;6:52138–60.
9. Ribeiro MT, Singh S, Guestrin C. "Why Should I Trust You?": Explaining the predictions of any classifier. In: Proceedings of the 22nd ACM SIGKDD international conference on knowledge discovery and data mining; 2016. p. 1135–1144.
10. Ancona M, Ceolini E, Öztireli C, Gross M. Gradient-based attribution methods. In: Explainable AI: interpreting, explaining and visualizing deep learning. Springer; 2019. p. 169–191.
11. Bach S, Binder A, Montavon G, Klauschen F, Müller KR, Samek W. On pixel-wise explanations for non-linear classifier decisions by layer-wise relevance propagation. PLoS ONE. 2015;10:1–46.
12. Montavon G, Samek W, Müller KR. Methods for interpreting and understanding deep neural networks. Digit Signal Process. 2017;73:1–15.
13. Sundararajan M, Taly A, Yan Q. Axiomatic attribution for deep networks. In: Proceedings of the 34th international conference on machine learning. vol. 70; 2017. p. 3319–3328.

14. Shrikumar A, Greenside P, Kundaje A. Learning important features through propagating activation differences. In: Proceedings of the 34th international conference on machine learning. vol. 70; 2017. p. 3145–3153.
15. Lundberg SM, Lee SI. A unified approach to interpreting model predictions. In: Advances in neural information processing systems; 2017. p. 4765–4774.
16. Melis DA, Jaakkola T. Towards robust interpretability with self-explaining neural networks. In: Advances in neural information processing systems; 2018. p. 7775–7784.
17. Rudin C. Stop explaining black box machine learning models for high stakes decisions and use interpretable models instead. Nat Mach Intell. 2019;1(5):206–15.
18. Ross AS, Hughes MC, Doshi-Velez F. Right for the right reasons: training differentiable models by constraining their explanations. In: Proceedings of the 26th international joint conference on artificial intelligence; 2017. p. 2662–2670.
19. Lissack M. Dealing with ambiguity "The 'Black Box' as a design choice. SheJi (forthcoming). 2016.
20. Consortium GO. The gene ontology (GO) database and informatics resource. Nucleic Acids Res. 2004;32(1):D258–61.
21. Kanehisa M, Goto S. KEGG: kyoto encyclopedia of genes and genomes. Nucleic Acids Res. 2000;28(1):27–30.
22. Fabregat A, Jupe S, Matthews L, Sidiropoulos K, Gillespie M, Garapati P, et al. The reactome pathway knowledgebase. Nucleic Acids Res. 2018;46(D1):D649–55.
23. Snel B, Lehmann G, Bork P, Huynen MA. STRING: a web-server to retrieve and display the repeatedly occurring neighbourhood of a gene. Nucleic Acids Res. 2000;28(18):3442–4.
24. Peng J, Wang X, Shang X. Combining gene ontology with deep neural networks to enhance the clustering of single cell RNA-Seq data. BMC Bioinform. 2019;20(8):284.
25. Gaudelet T, Malod-Dognin N, Sánchez-Valle J, Pancaldi V, Valencia A, Pržulj N. Unveiling new disease, pathway, and gene associations via multi-scale neural network. PLoS ONE. 2020;15(4):e0231059.
26. Hao J, Kim Y, Kim TK, Kang M. PASNet: pathway-associated sparse deep neural network for prognosis prediction from high-throughput data. BMC Bioinform. 2018;19:510.
27. Kang T, Ding W, Zhang L, Ziemek D, Zarringhalam K. A biological network-based regularized artificial neural network model for robust phenotype prediction from gene expression data. BMC Bioinform. 2017;18:565.
28. Torrente A, Lukk M, Xue V, Parkinson H, Rung J, Brazma A. Identification of cancer related genes using a comprehensive map of human gene expression. PLoS ONE. 2016;11(6):e1057484.
29. Alber M, Lapuschkin S, Seegerer P, Hägele M, Schütt KT, Montavon G, et al. Investigate neural networks! J Mach Learn Res. 2019;20(93):1–8.
30. Rawat W, Wang Z. Deep convolutional neural networks for image classification: a comprehensive review. Neural Comput. 2017;29(9):2352–449.
31. Medina VA, Rivera ES. Histamine receptors and cancer pharmacology. Br J Pharmacol. 2010;161(4):755–67.
32. Chockalingam S, Ghosh SS. Macrophage colony-stimulating factor and cancer: a review. Tumor Biol. 2014;35(11):10635–44.
33. Sever R, Brugge JS. Signal transduction in cancer. Cold Spring Harb Perspect Med. 2015;5(4):a006098.
34. Yang M, Brackenbury WJ. Membrane potential and cancer progression. Front Physiol. 2013;4:185.
35. Ciccarese F, Ciminale V. Escaping death: mitochondrial redox homeostasis in cancer cells. Front Oncol. 2017;7:117.
36. Yamaguchi H, Wyckoff J, Condeelis J. Cell migration in tumors. Curr Opin Cell Biol. 2005;17(5):559–64.
37. Bhullar KS, Lagarón NO, McGowan EM, Parmar I, Jha A, Hubbard BP, et al. Kinase-targeted cancer therapies: progress, challenges and future directions. Mol Cancer. 2018;17:48.

Publisher's Note

Springer Nature remains neutral with regard to jurisdictional claims in published maps and institutional affiliations.

Chen *et al. BMC Bioinformatics* (2021) 22:494
https://doi.org/10.1186/s12859-021-04403-1

BMC Bioinformatics

METHODOLOGY

Open Access

CirPred, the first structure modeling and linker design system for circularly permuted proteins

Teng-Ruei Chen[1,2†], Yen-Cheng Lin[3,4†], Yu-Wei Huang[1,2], Chih-Chieh Chen[5] and Wei-Cheng Lo[1,2,3,4,6*]

From The 19th Asia Pacific Bioinformatics Conference (APBC 2021) Tainan, Taiwan. 3-5 February 2021

*Correspondence:
WadeLo@nctu.edu.tw
†Teng-Ruei Chen and Yen-Cheng Lin have contributed equally to this work
[1] Institute of Bioinformatics and Systems Biology, National Chiao Tung University, Hsinchu, Taiwan
Full list of author information is available at the end of the article

Abstract

Background: This work aims to help develop new protein engineering techniques based on a structural rearrangement phenomenon called circular permutation (CP), equivalent to connecting the native termini of a protein followed by creating new termini at another site. Although CP has been applied in many fields, its implementation is still costly because of inevitable trials and errors.

Results: Here we present CirPred, a structure modeling and termini linker design method for circularly permuted proteins. Compared with state-of-the-art protein structure modeling methods, CirPred is the only one fully capable of both circularly-permuted modeling and traditional co-linear modeling. CirPred performs well when the permutant shares low sequence identity with the native protein and even when the permutant adopts a different conformation from the native protein because of three-dimensional (3D) domain swapping. Linker redesign experiments demonstrated that the linker design algorithm of CirPred achieved subangstrom accuracy.

Conclusions: The CirPred system is capable of (1) predicting the structure of circular permutants, (2) designing termini linkers, (3) performing traditional co-linear protein structure modeling, and (4) identifying the CP-induced occurrence of 3D domain swapping. This method is supposed helpful for broadening the application of CP, and its web server is available at http://10.life.nctu.edu.tw/CirPred/ and http://lo.life.nctu.edu.tw/CirPred/.

Keywords: Circular permutation, Protein engineering, Protein structure modeling, Protein structure prediction

Background

To facilitate the utilization of circular permutation (CP) as a protein engineering technique, we carried out this study. CP, a polypeptide backbone rearrangement, could be considered as if the native termini of a protein were linked and a new opening created elsewhere [1–3]. Studies on natural cases concluded that circular permutants (CPMs) usually retain their native structures and functions, sometimes with increased functional diversity or activity [1–3]. This property makes CP promising for bioengineering. By artificially creating CPMs, CP has been applied in various fields, such as studying the folding and function of proteins [4, 5], improving the stability, solubility, substrate affinity, substrate specificity, and activity of proteins [6–9]. It can be used to create biosensors, molecular switches, and novel bifunctional proteins [10–12]. Recently, it is also utilized to create split inteins [13, 14].

Despite being powerful, the implementation of CP poses challenges. First, CP is much more difficult, expensive, and time-consuming than traditional mutagenesis. Second, not every position is permissive for CP [7, 15]. Third, when the termini of a protein are distant, a peptide linker should be designed to connect them, or the CPMs are unlikely viable [9]. Forth, conventional modeling algorithms are inadequate for predicting the structure of circularly permuted proteins. We have previously developed a viable CP cutting site predictor [16, 17]. Nevertheless, there is still a lack of a 3D structure predictor and linker design algorithm for CP. Because of the rearrangement nature, when modeling a CPM, even state-of-the-art comparative modeling systems like the SWISS-MODEL [18], RaptorX [19], Robetta [20], and our work $(PS)^2$ [21] usually generate a partially modeled structure. So far, uneconomic trials and errors are inevitable for CP bioengineering.

In this work, we have developed the first CP structure modeling and linker design method named CirPred (Circularly-permuted protein structure Predictor), which integrates several algorithms of protein structural computation, machine learning, and molecular dynamics (MD) simulations in a "circularly-permuted" fashion (see Fig. 1 and "Methods" section). As tested with experimentally-verified CPMs of the dihydrofolate reductase (DHFR) [15], CirPred was the only comparative modeling method capable of producing complete models. Evaluated with ~ 1600 pairs of CPMs from literature and the Circular Permutation DataBase (CPDB) [22], the average alignment ratio and root-mean-square distance (RMSD) between CirPred-modeled and actual structures were better than 90% and 2.5 Å, respectively, even for CPMs sharing ~ 20% sequence identities. On average, linkers designed by CirPred possessed 70.2% sequence similarities with native linkers. For proteins with amino (N)- and carboxyl (C)-termini closer than 10 Å, the linkers designed by CirPred achieved an accuracy of 0.26 Å. Interestingly, we found CirPred capable of detecting protein 3D domain swapping (DS) [23]. When CP and DS co-occurred, CirPred provided accurate predictions of the structure and orientation of domains.

To fulfill the aims of this work, we implemented the CirPred into a rapid web server. With the assistance of this effective structure predictor, we hope that the time and cost of CP implementation can be significantly decreased, letting this powerful technique open protein and enzyme engineering to new possibilities.

Results

Comparison of structural models generated by CirPred and conventional modeling methods

Since proteins related by CP have different start points of the polypeptide sequence, conventional structure modeling methods typically meet difficulties in finding

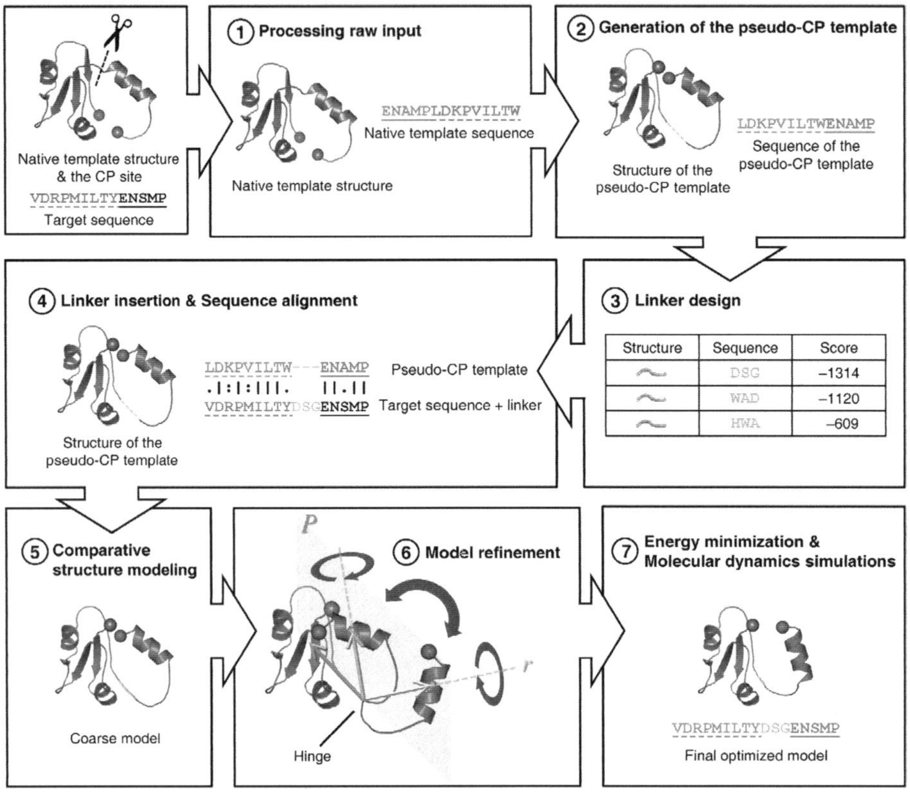

Fig. 1 Flowchart of the CirPred method. After obtaining the input data inclusive of a native template structure, a CP site, and a target sequence, seven major steps are performed. (1) The native template is processed to restore missing atoms and obtain the native amino acid sequence. (2) A pseudo-CP template structure and sequence are created according to the CP site. (3) A linker design procedure is carried out when necessary. (4) Sequence alignment between the pseudo-CP template and the target sequence is performed with native termini connected by the designed linker. (5) A coarse model is produced by comparative structure modeling. (6) The coarse model is refined by an algorithm that uses the CP site as a hinge to find the optimal orientation of domains (see "Methods" section). Each arrow represents a vector from the hinge to the center of mass of a domain. (7) Energy minimization and MD simulations are carried out to make the optimized model. In the web implementation, the linker design algorithm is activated only when the user does not provide the target sequence. In this situation, the target sequence is directly obtained from the template structure according to the CP site. The blue and red colors represent the N- and C-terminal proportions of a protein, respectively. The solid and dotted underlines indicate sequence fragments corresponding to the N- and C-terminal proportions of the native protein

suitable templates when processing a circularly permuted target protein. Even if the native protein of the target is provided as the template, they may still fail to build a full-length model. It is commonly observed that, delimited by the CP site, a part of the model is missing or predicted as extended loops/coils. In other cases, the model is predicted to possess two well-folded domains, but the predicted orientation of domains is incorrect.

The entire polypeptide of the DHFR had been scanned by CP to identify viable CP sites [15]. According to those CP sites, we generated all the permutant sequences and submitted them to state-of-the-art comparative modeling systems, including SWISS-MODEL [18], RaptorX [19], and Robetta [20] (see Additional file 1). When CP occurred at positions close to the center of the polypeptide, the quality of models built by these methods was generally good. As the CP site moved close to the N- or C-terminus, the missing, coiled, or incorrect-orientation modeling problems became increasingly serious. If a modeled permutant and the native DHFR were structurally aligned, the worst results (the lowest alignment ratio or highest RMSD) usually occurred when the CP site is situated around 1/4 or 3/4 of the sequence. The same permutant sequences were processed with the proposed CirPred method, and all the produced CPM models aligned well with the native protein. Take CP site residue 55 for example, the CPM model built by CirPred retained the correct conformation of DHFR (compare Fig. 2a, b). As for the model constructed by SWISS-MODEL, the proportion corresponding to residues 1–55 of the native DHFR was missing (Fig. 2c). The same proportion was predicted as a long coil by RaptorX (Fig. 2d). Robetta successfully modeled the two proportions delimited by the CP site, but the predicted orientation was wrong (Fig. 2e).

CirPred was the only one capable of correctly modeling all viable DHFR CPMs among the assessed methods, demonstrating its specificity to circularly permuted proteins. Additionally, it is noteworthy that CirPred is applicable to conventional co-linear modeling. By setting the CP site to be residue 1 of a protein, CirPred would create a co-linear model with comparable quality to models constructed by state-of-the-art modeling

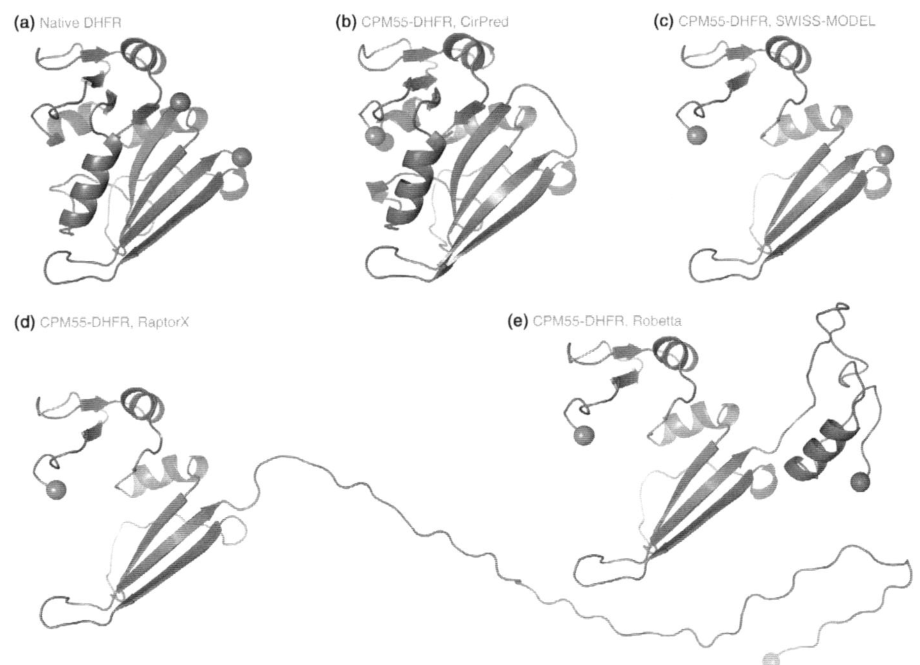

Fig. 2 Circularly-permuted protein structure models constructed by CirPred and conventional modeling methods. The CPM55 of DHFR has been reported viable (CP site on the native DHFR: Pro55) [15]. Several comparative structure modeling methods were applied to construct the model of this CPM. Conventional methods failed to model part of the CPM (panel **c** and **d**) or produced incorrect domain orientations (panel **e**). The proposed CirPred method correctly constructed the full-length model (panel **b**). Structures shown in this figure include **a** the native DHFR (PDB 1rx4A), and the model of CPM55 constructed by **b** CirPred **c** SWISS-MODEL [18], **d** RaptorX [19], and **e** Robetta [20]. Structures are drawn in the same orientation and scale. The blue and red colors indicate the N- and C-terminal proportions, respectively, of the protein. The α-carbon atoms of termini residues are shown as spheres

systems. See Additional file 2 for results of co-linear protein structure modeling performed using CirPred.

Comparison of models generated by CirPred and conventional modeling methods for proteins requiring termini linkers

When the native termini of a protein for CP were close, short poly-glycine or glycine/serine-rich linkers were frequently used [9]. However, longer linkers may be required for proteins with distant termini to be successfully engineered by CP [9]. We proposed a linker design protocol (see "Methods" section), which further differentiated the CirPred from traditional modeling systems. Here a linker redesign experiment was conducted to demonstrate how the proposed protocol could design a long peptide linker for proteins having distant termini.

The 1,3–1,4-β-glucanase from Bacillus with Protein Data Bank (PDB) entry 2ayhA (Fig. 3a) had been engineered by CP using Phe59 as the CP site, and the permuted

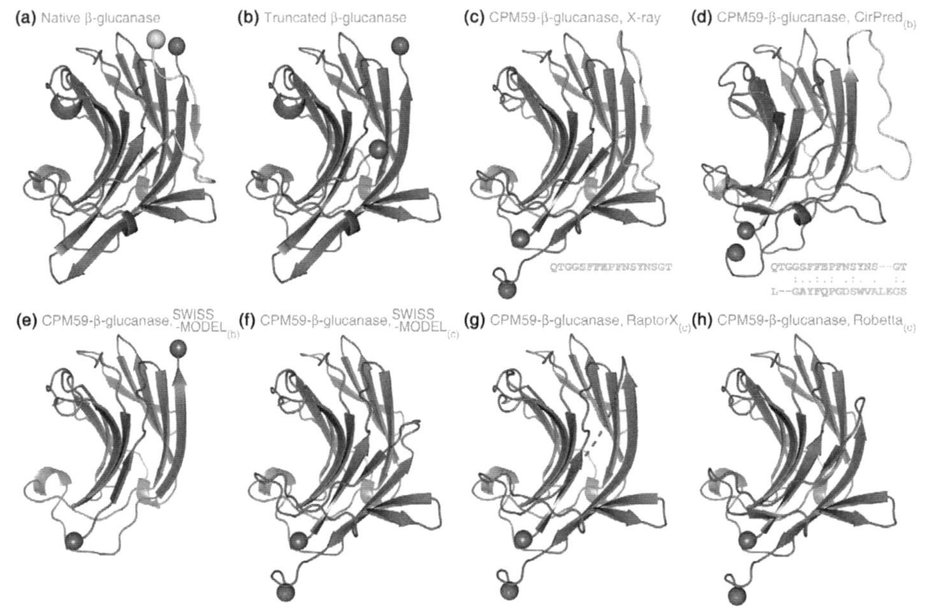

Fig. 3 Circularly-permuted models constructed by several modeling methods for proteins requiring a long termini linker. The CPM structure of a β-glucanase has been determined [24]. In this experiment, the 17 N-terminal residues of the native β-glucanase were removed to test the linker design capability of CirPred. The model constructed by CirPred was complete, and the designed linker was similar to the removed 17-residue fragment both in structure and sequence. Conventional modeling methods either twisted the structure to connect the native termini or failed to fill the gap in between. **a** Crystal structure of the native 1,3–1,4-β-glucanase (PDB 2ayhA). The yellow fragment corresponds to the 17 N-terminal residues for truncation. **b** The N-terminus truncated β-glucanase (denoted β-glucanase$_{\Delta 1-17}$). **c** Crystal structure of the CPM59 of the β-glucanase (PDB 1cpmA). The yellow fragment (and sequence) is equivalent to the truncated part of the native protein and was the target linker to be redesigned. **d** Model constructed by CirPred using **b** as the template. The yellow fragment is the redesigned linker, whose sequence similarity to the native linker is 70.6% (aligned by Stretcher [32]). **e** Model constructed by SWISS-MODEL using **b** as the template. The C-terminal proportion of the structure was missing. **f** Model constructed by SWISS-MODEL using **c** as the template. The longest main-frame β-strand was twisted to connect the native termini. **g** Model constructed by RaptorX using **c** as the template. As the dotted line indicates, an unreasonable gap remained between the native termini. **h** Model constructed by Robetta using **c** as the template. The direct connection between the native termini twisted the longest main-frame β-strand

structure had been determined (denoted as CPM59-β-glucanase; Fig. 3c) [24]. We deleted residues 1–17 from the native β-glucanase structure 2ayhA to create a new N-terminus 21.97 Å away from the C-terminus. The truncated β-glucanase (denoted as β-glucanase$_{\Delta 1-17}$; Fig. 3b) simulated a protein that possessed distant termini, and a long linker (17 residues) should be added for CP bioengineering. With β-glucanase$_{\Delta 1-17}$ as the template, we constructed the model of CPM59-β-glucanase using several methods. If a method could properly design a linker for β-glucanase$_{\Delta 1-17}$, the designed linker should likely be similar to the fragment in the actual structure of CPM59-β-glucanase (the yellow part in Fig. 3c) that is corresponding to the deleted residues in the native β-glucanase (the yellow part in Fig. 3a). The result showed that CirPred successfully designed a linker similar to the deleted residues in structure and sequence (see Fig. 3d). The last 58 residues of CPM59-β-glucanase were missing in the model constructed by SWISS-MODEL (Fig. 3e). As for RaptorX and Robetta, the long-coil and incorrect-orientation predictions exhibited in Fig. 2 persisted. We had then utilized the actual structure of CPM59-β-glucanase as the template to construct the model of CPM59 itself using SWISS-MODEL, RaptorX, and Robetta. These algorithms either directly connected the native termini of β-glucanase by a twist (Fig. 3f, h) or simply formed a big gap between the termini (Fig. 3g). It should be noted that only CirPred was equipped with a linker design protocol; therefore, the above comparisons were not made to compete with conventional modeling methods for performance but only to indicate the novelty of CirPred. See "Performance of linker design" section for large-scale evaluations of the proposed linker design protocol.

Performance on engineered CPs

Artificially engineered circular permutations are suitable materials for assessing the performance of a CP structure predictor because their parent proteins, CP sites, and additional mutations were well defined. The engineered CPMs collected in [25] were used to assess CirPred. The sequence of each engineered CPM was the target, which was subjected to CirPred modeling with its parent protein used as the template. Since the structure of all applied CPMs had been known, the performance of CirPred could be evaluated by aligning the known structures with constructed models. As shown in Table 1, the structural alignment qualities were high. The average alignment ratio and RMSD were 99.2% and 1.59 Å, respectively.

Large-scale assessments for various sequence identity levels

To examine how the modeling quality of CirPred would be influenced by the sequence identity between the target CPM and the template, we conducted an extensive test using CPDB, the largest dataset of manually-verified pairs of CPMs [22]. The 1568 CP pairs of CPDB with sequence identities $\geq$ 10% (Additional file 3) were grouped into subsets of decreasing identities. For each pair of CPMs, one was used as the template for modeling the other. The results are listed in Table 2. The quality of models remained high until the identity was lower than 20%. For those CP pairs with identities $\geq$ 20%, the average alignment ratios between models and actual structures were all > 90%, and the average RMSDs were generally < 2.50 Å. Even for those with identities < 20%, the average alignment ratio and RMSD still reached 74.0% and 3.90 Å,

Table 1 Structural alignment qualities between models constructed by CirPred and the actual structures of engineered CPMs

Target: engineered CPM (size: residues)	Template: native protein (size: residues)	CP site on the template[a]	Identity[b] (%)	Alignment ratio[c] (%)	RMSD[c] (Å)
1ajkA (212)	2ayhA (214)	84	90.2	97.6	1.441
1ajoA (212)	2ayhA (214)	129	97.2	100.0	1.026
1cpmA (214)	2ayhA (214)	59	98.6	100.0	1.001
1cpnA (208)	2ayhA (214)	59	92.5	100.0	1.050
1alqA (259)	3blmA (257)	223	98.8	100.0	0.548
1fw8A (416)	3pgkA (416)	72	72.8	98.1	3.155
1n02A (102)	2ezmA (101)	50	90.1	100.0	1.316
1un2A (186)	1a2jA (188)	100	96.3	100.0	1.156
1bd7A (176)	1blbC (187)	98	85.6	94.9	6.501
1g2bA (62)	1shgA (57)	47	86.0	100.0	1.056
1tucA (61)	1shgA (57)	20	86.0	96.7	1.609
1tudA (60)	1shgA (57)	48	89.5	100.0	0.833
1swfA (116)	1stpA (121)	51	87.9	100.0	1.256
1swgA (112)	1stpA (121)	51	86.6	100.0	0.657
1p5cA (166)	1lw9A (164)	12	98.1	100.0	1.310

[a] Numbered according to the order of residues in the PDB file

[b] Sequence identity between the ***template*** protein and the target CPM (computed by the circularly-permuted structure alignment algorithm CPSARST [25])

[c] These were the structural alignment ratio and RMSD values between the ***model*** and the actual structure of target CPMs. The alignment ratio was defined as the number of aligned residues divided by the target's size

Table 2 Effects of sequence identity on the performance of CirPred

Sequence identity[a] (%)	Alignment ratio[b] (%)	RMSD[b] (Å)	Number of CP pairs
90–100	97.6	1.236	79
80–90	95.7	1.673	39
70–80	94.7	1.504	8
60–70	95.1	3.124	10
50–60	96.7	1.910	31
40–50	98.7	2.002	236
30–40	98.1	2.026	351
20–30	90.9	2.496	161
10–20	74.0	3.904	653

In this experiment, for each pair of circular permutants, one protein was utilized as the template to create the model of the other protein, i.e., the target

[a] Sequence identities between the ***template*** and target proteins

[b] Alignment ratio and RMSD values between the ***model*** and known structure of target proteins

respectively. For instance (see Additional file 4), a RIM2 C_2A-domain (PDB 2bwqA) and a calcium-phospholipid binding domain (PDB 1rlwA) shared only 19.5% identity. The model made based on 2bwqA as the template (CP site: residue 16) achieved a structural alignment ratio of 95.2% and RMSD 2.75 Å when aligned with the known structure of the target CPM, i.e., PDB 1rlwA. In summary, the performance of CirPred would decrease as the sequence identity between template and target lowers, but the quality of the constructed model remains high as long as the identity is not far below 20%.

In addition to sequence identity, we had examined to what extent the performance of CirPred would be influenced by protein size or the location of CP sites. Additional file 5: Tables S1 and S2 demonstrate that protein size exerted little influence on the modeling quality; however, the performance decreased slightly for proteins with CP sites close to the termini (see also Additional file 4 for examples of CirPred modeling for CPMs of different sizes).

Performance of linker design

Connecting the native termini of a protein is a formidable challenge for successful CP engineering [9]. A direct connection may be appropriate when the native termini are close in 3D space; otherwise, a well-designed peptide linker is required. There is still no general method for designing CP linkers in addition to short glycine/serine-rich peptides [9]. In "Methods" section, a linker design protocol for CP is proposed. This subsection provides performance assessments of the proposed protocol.

Evaluation with known circular permutants

A CPDB linker dataset (Additional file 6) containing all non-redundant linkers of CPMs from the CP pair dataset was prepared to evaluate the linker design algorithm of CirPred by performing a strict 500-round independent test described in "Methods" section. Statistical analyses showed that 99.0% of the CPM structures with linkers designed by our algorithm exhibited equivalent or lower potential energy than their original structures. The average sequence similarity between the designed and the original linkers was 68.6% (see Additional file 7 for raw data and Additional file 8: Table S3 for statistics). Since all the original linkers were obtained from naturally occurring CPMs, these data indicated that the proposed algorithm could design linkers with structural stabilities and sequence compositions analogous to linkers that evolved naturally.

To dissect the influence of the distance between native termini on the performance of linker design, we further analyzed the sequence similarities and RMSDs between the designed and original linkers. As the distance between termini increased, the similarity lowered, and RMSD rose, indicating a decline in performance (see Additional file 8: Fig. S3 and Table S3). However, since the output of the CirPred linker design procedure for a given protein was a set of candidate linkers ordered by potential energy, retrieving more candidates helped find better results. Take proteins with a termini distance $\leq$ 10 Å for instance, if the top 5 candidates were retrieved, the average optimal sequence similarity between the designed and original linkers was 83.7%, and the RMSD was 0.32 Å; if the top 10 candidates were retrieved, those values became 90.7% and 0.26 Å, respectively. Before this work, the most accurate linker design method accomplished a 0.50 Å RMSD out of 6 candidates [26]. The CirPred has made a noticeable advance in the linker design for CP engineering.

Evaluation with in silico circular permutants

To perform a thorough assessment of the linker design algorithm, we prepared an in silico synthetic CPM dataset, namely, the Dataset S (Additional file 9), in which every protein shared < 25% identity with any protein either from itself or from the CPDB linker dataset. There were 2141 CPMs in Dataset S, each with a known native protein and a predefined missing linker. After the machine-learning predictor of CirPred's linker design module was trained with the CPDB linker dataset, every CPM in Dataset S was processed by CirPred modeling to redesign its missing linker. The potential energy of CirPred models was compared with that of the actual CPM structures. The results showed that 98.4% of the CPMs with linkers redesigned by CirPred had equivalent or lower potential energy than their original structures, and the average sequence similarity between the redesigned and original linkers was 71.7%. Since Dataset S was highly non-redundant and very different from the CPDB linker dataset (i.e., the training data for machine learning), this independent test demonstrated that the proposed linker design protocol was stable in performance.

Ability to model 3D domain-swapped proteins induced by CP

As we evaluated the CirPred with engineered CP cases, a very interesting situation was recognized for the CPMs of βB2-crystallin with PDB entries 1bd7A and 1blbC. The crystallin 1bd7A was an artificial CPM of 1blbC with residue Glu87 as the CP site [27]. Crystallin 1blbC, before the CP, was a homo-tetramer composed of two dimers, each of which was still a dimer comprising two intertwined subunits with an open conformation. The CP converted the **inter**molecular pairing between subunits into **intra**molecular pairing and thus disentangled the intertwined subunits into two side-by-side monomers with a closed conformation [27]. The conversion between intertwined dimeric "open-form" and disentangled "closed-form" is an example of 3D domain swapping (DS) [23].

DS-related homologs are difficult for sequence or structure alignment methods to identify because of the dramatic conformational difference [28]. The identification of CP is also not straightforward [25]. When CP and DS co-occurred, the situation would become too complicated for conventional modeling systems to construct a model of correct conformation, even if a proper native template was provided. One native βB2-crystallin subunit (182 residues; PDB 1blbC) contains two homologous tandem-repeat domains. Since the CP site 87 (which created crystallin 1bd7A, or the CPM87-crystallin) was situated close to the middle of the subunit, it would be expectable that conventional modeling systems construct a model of CPM87-crystallin that is similar to the native crystallin. Indeed, as Fig. 4 demonstrated, when the CPM87-crystallin sequence was input as the target and the native crystallin (Fig. 4a) as the template, SWISS-MODEL [18] built a model very similar to the native structure (Fig. 4c). However, the truth was complicated. The actual structure of the CPM87-crystallin (PDB 1bd7A; Fig. 4b) had a very different domain orientation from the native crystallin. Thus, superimposition between the model built by SWISS-MODEL and the actual CPM87-crystallin structure showed a low alignment ratio (51.7%) and a large spatial displacement in the unaligned region (Fig. 4d). Contrarily, the CirPred model of CPM87-crystallin showed a very high alignment ratio (99.4%) and a small RMSD (3.61 Å) with the actual CPM structure (Fig. 4e, f). The fact that CirPred could correctly model a 3D domain-swapped CPM implied that it has the potential of detecting the DS phenomenon induced by CP.

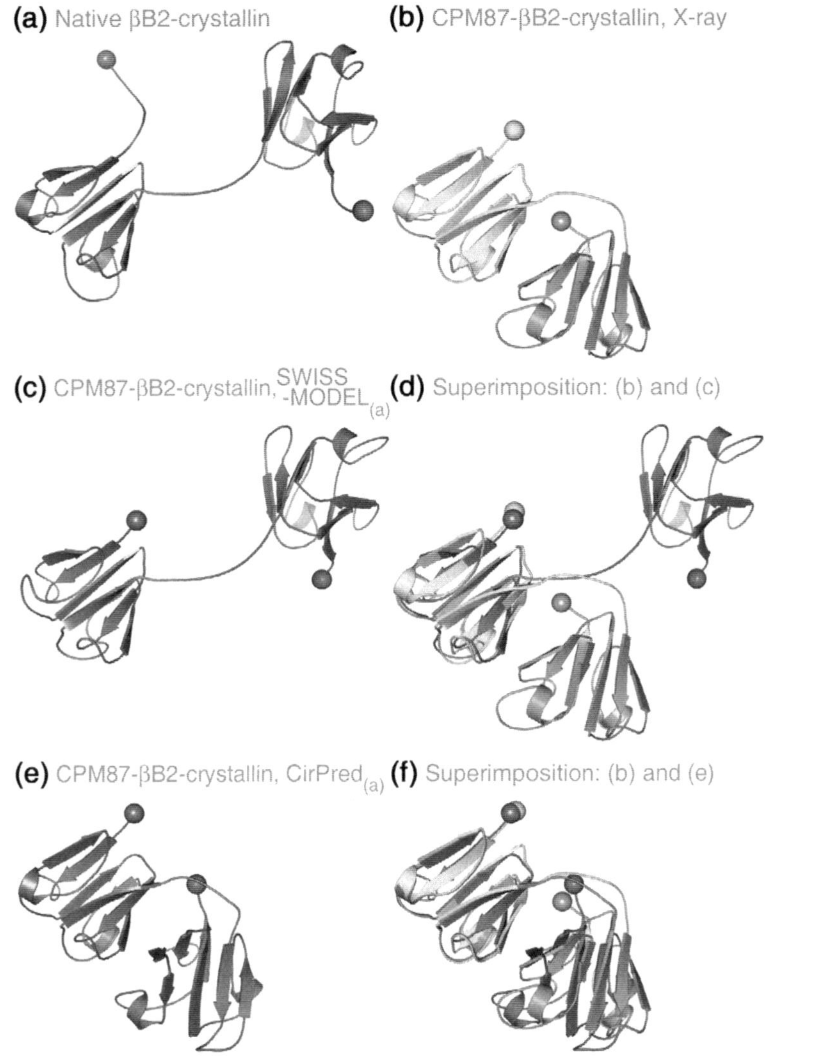

Fig. 4 Modeling of proteins with 3D domain swapping phenomenon induced by CP. The structures of **a** the native βB2-crystallin (PDB 1blbC) and **b** the engineered CPM87 βB2-crystallin (PDB 1bd7A; CP site on the native crystallin: Glu87) have both been solved by X-ray crystallography. This pair of circular permutants is also an example of the 3D domain-swapping phenomenon. Since this crystallin had a two-fold rotational symmetric structure and the CP site was at the middle of the subunit, conventional comparative modeling methods might be able to construct the full model of the CPM by mimicking the template structure, such as **c** the model of CPM87 βB2-crystallin constructed by SWISS-MODEL using **a** as the template. However, the conformation of the models constructed by conventional methods was very different from that of the actual CPM structure, as shown by **d** the superimposition between **b** and **c**. Interestingly, **e** the model of CPM87 βB2-crystallin constructed by CirPred was highly similar to the actual CPM conformation, as shown by **f** the superimposition between **b** and **e**

Web server of CirPred

The CirPred algorithm has been implemented as a web server providing three working modes, (1) structure modeling for a primary CP, (2) structure modeling for a highly modified CPM, and (3) linker design. On the input page, the user will be asked to provide a structure of the native protein and a CP site. If the linker had been designed, using Mode 1 is appropriate. If there are other sequence modifications (substitutions, insertions, and deletions) on the native protein in addition to the CP and the linker, Mode 2 can be a great help. If a linker is necessary and the user needs assistance in designing it, Mode 3 is applicable. CirPred will provide 30 candidate linkers with their CPM models ordered by the DOPE score. The final choice should depend on the user's knowledge of the protein to be engineered. Distributed computation techniques [29] are applied to speed up the execution of CirPred algorithms. If the optional MD simulation step is not enabled, a typical query of any mode requires < 3 min.

Discussion

As a novel comparative modeling method, CirPred is fully capable of modeling circularly permuted proteins, even if the permutant shared a low sequence identity with the native protein. When a polypeptide linker is required to connect the termini before CP, CirPred helps design the linker. Besides, it could be applied to the design of 3D domain-swapped proteins by CP.

The novelty of CirPred was established on three bases. First, a pseudo-CP template was made before modeling. Second, CPDB [22] provided valuable data for establishing the machine learning predictor of the linker design protocol. Third, the CP-site-hinged model refinement procedure (Fig. 1 and "Methods" section) helped overcome the problems met by comparative modeling systems when dealing with circularly permuted proteins (e.g., incorrect domain orientation and expensive time cost). It also enabled CirPred to model domain-swapped CPMs.

On the performance for low identity circular permutants

The quality of sequence alignment between the target and template is crucial to the accuracy of comparative modeling, especially when their sequence identity is low. It has been reported that for cases with target-template identity < 40%, an error of ~ 4 Å would be introduced into the model by just a single-residue misalignment [30]. Using CirPred, even when the identity fell to 20–40%, the RMSDs between the constructed models and the known structures of target permutants were < 2.5 Å, and the proportions of structurally well-aligned residues between them were > 90% (Table 2). CirPred performed well at low identity because it utilized three global sequence alignment methods [21, 31, 32] to find the optimal target-template alignment according to the produced sequence similarity. Among those methods, $(PS)^2$ [21] took the predicted secondary structure of the target and the known secondary structure of the template into consideration and was particularly suitable for making accurate target-template alignments. Notwithstanding its high alignment quality, the current implementation of $(PS)^2$ is dependent on a secondary structure predictor developed two decades ago and a traditional gap penalty scheme. If new secondary structure prediction strategies like [33–35] could be applied, or gaps could be dynamically suppressed within regions of regular secondary structure, buried core, or straight segments, the sequence alignment quality would be significantly improved and make CirPred more accurate for CPMs sharing < 20% sequence identities.

On the accuracy of linker design

We proposed a linker design algorithm to provide bioengineers with energetically-favored choices in addition to glycine-rich linkers. The performance of this algorithm should be owed to its "predict-and-refine" strategy. Even if the accuracy of the initial machine learning prediction of amino acid for each residue position of a linker was only 67.5%, the subsequent probability-guided random modeling and energy screening greatly increased the chance of finding suitable linker sequences. Nevertheless, if the absolute prediction accuracy could be improved, the quality of designed linkers might be enhanced. At least, the algorithm's efficiency could be enhanced by reducing the number of random modeling (see Additional file 10 for advanced discussions).

On 3D domain-swapped circular permutants

DS is a quaternary structural phenomenon defined as proteins exchanging equivalent parts to form oligomers [23]. Understanding DS may help find new treatments for protein deposition diseases [36, 37] and develop bioengineering technology to modify enzyme activities or create auto-assembling biopolymers [38, 39]. Since DS enables homologous proteins to have dramatically different conformations (open *versus* closed) and CP changes the location of N-/C-termini, if they occurred concurrently, obtaining a reliable model by conventional modeling methods would be improbable. The CP-site-hinge model refinement step of CirPred seemed to create a shortcut to overcome the difficulty in modeling DS-related proteins caused by CP. The combination of naturally occurring CP and DS has been reported [9, 40, 41], and the βB2-crystallin we met in this study was an engineered case. These interesting examples implied that it is possible to create "circularly-permuted domain swapping" proteins to accomplish novel functions like molecular switches and auto-assembling biomaterials. CirPred is supposed useful for this combined type of protein engineering (Additional file 10).

Other applications

Previously, we developed a protein viable CP site predictor abbreviated "CPred" [16], the machine learning model of which was constructed when CP (in)viability data were rare. In addition to reconstructing the model with the rapidly increasing data of recent years, integrating the proposed CirPred algorithm into the CPred pipeline may help improve the accuracy of CP site prediction. A preliminary test based on the CP viability data of [42] demonstrated that the energy scores computed according to CirPred models helped calibrate CPred predictions. Details of this preliminary test and discussions about future applications of CirPred in template identification, CP study, and complex structure modeling are also available in Additional file 10.

Conclusions

We proposed a comparative modeling method for circularly permuted proteins to facilitate and broaden the application of CP in protein engineering. This CirPred method could accurately construct the model of a circular permutant at low sequence identity, indicating that it is promising for predicting the structure of a protein engineered by CP even if many other mutations are introduced. CirPred can design polypeptide linkers similar to the linkers of naturally occurring circular permutants. This ability helps ensure the viability of an engineered permutant that requires additional residues to connect the native termini. The model refinement procedure not only accelerated the search for energetically stable conformation for the domains delimited by the CP site but enabled CirPred to identify 3D domain swapping induced by CP. Thus, CirPred may assist in engineering "CP + DS" proteins with different biological properties, fusion sites, and oligomeric states from native proteins. Regular mutagenesis has enabled researchers to manipulate enzymes, antibodies, signal transmitters, cellular structural proteins, etc., for various applications. Circular permutation has brought about many novel engineered proteins that were difficult to create by regular mutations. We believe that many more protein engineering possibilities will be achieved by combining circular permutation with regular mutagenesis. As the first computational method capable of circularly-permuted structure modeling with linker design and co-linear structure modeling, CirPred shall help move forward various fields requiring protein engineering. The CirPred web server is available at http://10.life.nctu.edu.tw/CirPred/ (main server) and http://lo.life.nctu.edu.tw/CirPred/.

Methods

Software

In addition to the algorithms and software developed in our previous works, such as the CPSARST [25], $(PS)^2$ [21], and an integrated machine learning and optimization server [16, 17], several third party software packages were applied in this study, inclusive of the sequence alignment program Stretcher (vEMBOSS:6.6.0.0) [32], the comparative modeling software Modeller (v9.19) [43] and the molecular dynamics simulation package GROMACS (v2016.4) [44]. Many steps of the proposed CirPred method were computationally expensive. The distributed computation technique we developed for the *i*SARST protein structural similarity search server [29] was extensively used to speed up research progress and the CirPred web server. All protein structures shown in this report were rendered using PyMOL [45].

Datasets

The dataset of CP pairs from CPDB

Pairs of circular permutants sharing sequence identities ≥ 10% were downloaded from the CPDB [22]. Since the alignment data of CPMs provided by CPDB were computed by an old version of CPSARST [25], the structural alignment measures (e.g., alignment ratio and RMSD) and CP sites of these CP pairs were updated by the current CPSARST implemented in the *i*SARST server [29]. This dataset of 1568 CP pairs (Additional file 3) was used to perform large-scale evaluations of CirPred.

The CPDB linker dataset

CP-based structural alignments of CP pairs from CPDB were performed by CPSARST to establish this dataset. For each CP pair, as illustrated in Fig. 5, two linkers were obtained based on the algorithm stated below,

1. Let Q and S represent the two proteins of a CP pair.
2. Let x_q represents the first residue of Q aligned with S, and x_s represents the residue of S aligned with x_q. Similarly, let y_q be the last aligned residue of Q and its equivalent residue on S be y_s.
3. The candidate linker for Q was computed as the residues between y_s and x_s on S (excluding y_s and x_s). Let l be the number of residues of this candidate linker.

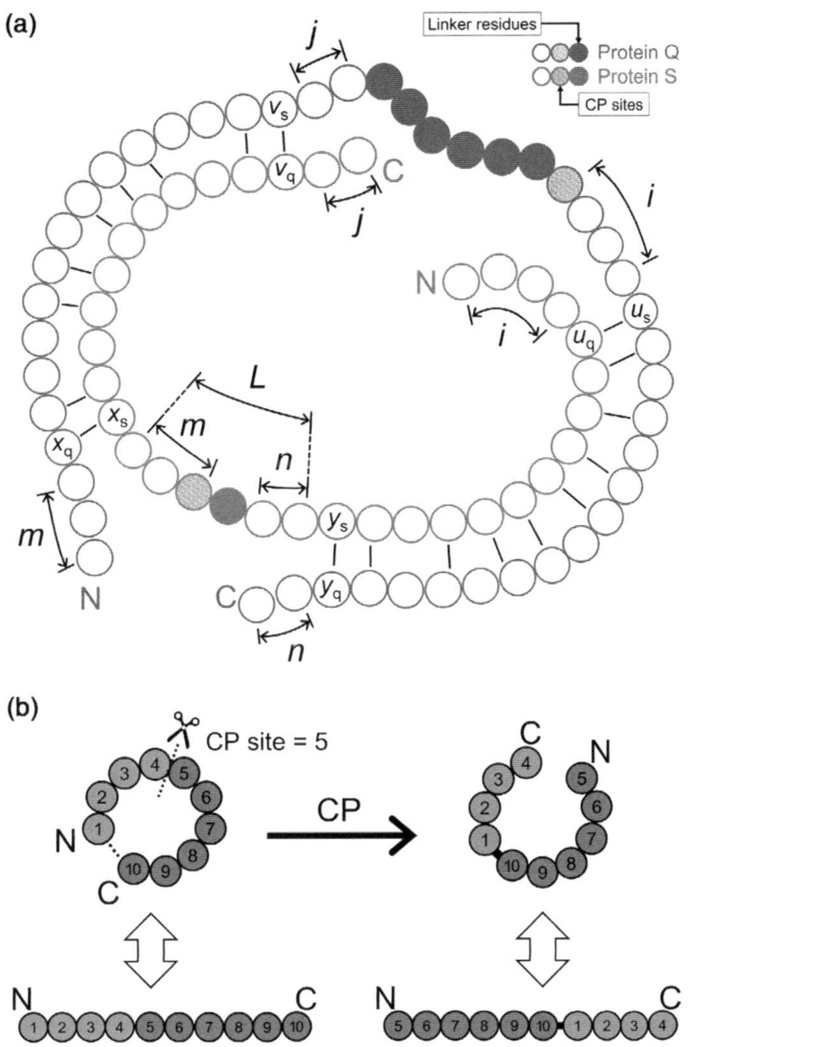

Fig. 5 Determination of termini linkers and, in silico circular permutation. **a** Algorithm for determining the termini linkers for a pair of CPMs. In the implementation, the circularly-permuted structure alignment between CPMs was performed by CPSARST [25]. **b** The in silico circular permutation. The circular permutation is, in essence, a sequence interchange between the N- and C-terminal proportions

4. Let m be the number of unaligned residues on the N-terminus of Q in front of x_q and n be the number of unaligned residues on the C-terminus of Q after y_q.

(i)If $m+n \geq l$, no linker was required to connect the native termini of Q.

(ii) Otherwise, the linker for Q would be refined as the residues between y_s+n and x_s-m on S (excluding y_s+n and x_s-m).

5. Referring to Fig. 5a, let x, y, m, and n be replaced with u, v, i, and j, respectively. Repeat steps 2–4 to find out the linker for protein S.

The CPDB termini linker dataset is available in Additional file 6. If a protein has more than one permutant, it may have multiple linkers in this dataset.

Dataset S, a synthetic termini linker dataset

This dataset consisted of in silico circularly permuted proteins, each with a surface polypeptide fragment removed to serve as the known missing linker. The following steps were taken to establish this dataset.

1. Let the position of a residue be represented by its α-carbon (Cα).
2. Using the 100% sequence identity non-redundant subset of the PDB (snapshot date: Dec. 25th, 2016), calculate the average distance, denoted by d, between two consecutive residues in a protein. The result was $d = 3.36$ (Å).
3. Compute the distance between the N- and C-terminal residues, denoted by D, for each protein in the above PDB subset (73,073 proteins) and filter out any protein with $D > 3d$. Only 3027 proteins remained in the dataset, in which every protein possessed a short distance between termini, i.e., within 3 residues on average.
4. Proteins sharing $\geq 25\%$ sequence identity with any protein in the CPDB termini linker dataset (Additional file 6) were further discarded. The resultant dataset (1802 proteins) was thus suitable, as an independent dataset, for assessing the machine learning predictor trained with the CPDB linker dataset.
5. For each protein remaining in this dataset, a CPM was created as stated below.

   (i) Since CP tends to occur at residues exposed to the solvent [17], each residue's relative solvent accessible surface area (RSA) was computed.
   (ii) Consecutive residues with RSA > 20% were considered an exposed fragment on the surface of the protein. A protein might possess several exposed fragments.
   (iii) Randomly select one exposed fragment, set the CP site at the carboxyl-end of the fragment.
   (iv) Since the native N- and C-termini of the protein were close, they were directly connected in silico. New termini were created at the selected CP site. This step was performed by renumbering residues in the PDB file in a way described in the next subsection.
   (v) Remove the selected exposed fragment from the PDB file.
   (vi) Refine the structure of this CPM by energy minimization using GROMACS [44].

The exposed peptide fragments removed in the above procedure were the known linkers for these in silico synthetic CPMs. A complete list of the PDB entries, CP sites, and linker sequences of this Dataset S is available in Additional file 9.

Generating the pseudo-circularly-permuted template

Because a CPM usually folded into a structure similar to the native structure [1, 9], the native structure would be a good template when modeling a CPM. However, since the termini of the CPM had changed, the native structure should be manipulated first. CP of a protein structure is actually an amino acid sequence rearrangement such that the N- and C-terminal proportions of the sequence were interchanged. As illustrated in Fig. 5b, making a circularly permuted protein in silico was equivalent to cutting the N-terminal proportion and attaching it to the end of the C-terminal proportion. At the sequence level, this manipulation was a simple text string rearrangement. At the structure level with a PDB file, as described below, it was more complicated.

1. To ensure that the PDB file's residues appeared in the same order in the template and model structures, starting with 1, assign serial numbers to the residues from the top of the file.
2. If the CP site was n, move $n-1$ residues from the top of the file to the bottom.
3. Starting with 1, renumber all residues and atoms from the top of the file.
4. In case there were missing or structurally undetermined atoms in the PDB file, the reduce [46] and teLeap [47] programs were utilized to add/restore them.

Comparative structure modeling

The comparative modeling procedure of CirPred consisted of three major steps. First, the target CPM sequence was generated in the way stated above, unless provided by the user. For instance, Mode 2 of the CirPred web server required the user to input the sequence of CPM. Second, the global sequence alignment between the target CPM and template (the pseudo-CP template) was obtained using three methods: the algorithm combining amino acid sequence and secondary structure information that we developed for $(PS)^2$ [21], the Smith-Waterman algorithm [31], and the Stretcher program [32]. Among the alignment results, the one with the largest number of aligned residues was selected. Third, comparative structure modeling was performed using the $(PS)^2$ procedure according to the sequence alignment. In addition to these three pipelined steps, according to the requirements of experiments or web users, the linker design procedure might be integrated into the pipeline (Fig. 1); and, at the end of the pipeline, model refinement and MD simulation procedures might be activated.

Linker design

Overview of the linker design protocol

The linker design algorithm of CirPred utilized several machine learning and structural modeling methods. The basic idea was to coarsely determine the position of residues of a linker at first and then predict for each residue position the amino acid. The whole protocol is outlined as follows,

1. *Determination of the coarse residue positions of a linker.*

(i) Let l denote the number of residues of the linker.
(ii) Randomly make t temporary linkers according to amino acid propensities of known CPM linkers (Additional file 11).
(iii) Before making sequence alignment between the target CPM and the pseudo-CP template, insert each temporary linker into the target CPM.
(iv) Based on the sequence alignment with each temporary linker inserted, generate a coarse model of the target CPM using Modeller, which also computed its DOPE (discrete optimized protein energy) score [43].

(v) Pick up t' from the t coarse models with the lowest energy scores for the next step. In this study, t and t' were set to be 200 and 20, respectively.

2. *Prediction of the amino acid composition of the linker.*

(i) Take one coarse model of the target CPM, for each residue position of the linker, compute the feature values required to perform predictions.

(ii) For each residue position, predict the occurrence probabilities of 20 amino acids by machine learning. See later subsections for algorithms of the feature set and machine learning techniques we applied.

3. *Amino acid sequence assignments of the linker.* According to the occurrence probabilities of amino acids, randomly assign an amino acid to each residue position of the linker. For instance, if the probabilities of valine and leucine at position 1 were 70% and 30%, respectively, the chance that position 1 was assigned with a valine was 70%, while the chance of leucine was 30%. Repeat this step for k times, creating k candidate linker sequences ($k=10$ in this study).
4. *Generation of a candidate linker sequence pool.* Repeat steps 2 and 3 until all coarse models were applied. There would be $t' \times k$ candidate linker sequences generated to form the pool.
5. *Computation of energy score for linkers in the pool.*

   (i) Take one candidate linker, insert it into the target CPM sequence, and make a sequence alignment with the pseudo-CP template.
   (ii) Based on the alignment, generate m models of the target CPM, each with a DOPE score using Modeller [43] ($m=10$ in this study).
   (iii) Select the model with the lowest DOPE to represent the quality of this candidate linker.
   (iv) Repeat (i) to (iii) until all candidate linkers in the pool were processed.
6. *Selection of the final candidate linker(s).* Find the candidate linker(s) with the lowest energy score(s) to be the final designed linker(s). For all experiments of this study, except the ones of Additional file 8, only one best-designed linker was taken. In our web implementation, several candidate linkers with low energy scores are reported to the user.

The feature set for prediction

We speculated that if the position of a residue of interest is known in a protein structure, information (or "features") obtained from its surroundings could be utilized to predict what amino acid the residue is. For a known linker, its residue positions were readily available in the PDB file. For a linker to be designed, residue positions could be coarsely simulated as described above. Features obtained from known linkers were used to train the machine learning kernels for predicting the amino acid composition of a linker to be designed. The feature set used in this study comprised 20 features computed according to this equation,

$$F_A(i) = \sum_{a=1}^{n_A(i,d_r)} \frac{1}{d_{ia}^2} \quad (1)$$

where i denotes the residue of interest, A represents the type of amino acid, $F_A(i)$ means the feature value of amino acid A for i, a stood for a residue of A surrounding i, and d_{ia} is the distance between residues i and a. The $n_A(i, d_r)$ is the number of A residues located within the radius of d_r (which was set as 20 Å in this work) from i. The position of a residue is represented by its Cα.

For a given A, this feature describes how many and how close this type of amino acids appeared near the residue of interest. A high feature value means that many such amino acids are nearby, or the distances between them and the residue of interest are short. Because the sequence of a linker to be designed is unknown, there is no prior knowledge of the amino acids adjacent to each other. Therefore, when computing the feature values, no matter for a known linker or a linker to be designed, five adjacent residues before and five after the residue of interest are discarded.

Establishing the predictor by machine learning methods

For any residue in a known linker, in addition to feature values, the "answer" must be provided for machine learning. The answer in the case of linker design should be the residue's amino acid, and thus there should be 20 candidate answers for learning and prediction. The computation loading of most machine learning algorithms would increase significantly as the number of candidate answers increases. Since 20 answers were beyond our hardware's computation capacity, we classified 20 amino acids into three types (hydrophilic, hydrophobic, and neutral) and reduced the number of candidate answers to 3.

Previously as we studied viable CP sites, we developed an artificial intelligence system that integrated several machine learning, random sampling, and parameter optimization algorithms [16, 17]. This system was applied in this work, and the recruited algorithms included bootstrap sampling, decision tree, and artificial neural network. After obtaining the answers and feature values from a set of known linkers, 250 and 50 bootstrap samples were made to train minor predictors of decision tree and artificial neural network, respectively. The final predictor was then formed by collecting the minor ones, which made predictions by vote. With this procedure, the probabilities of candidate answers for a given case could be estimated as the proportions of votes the answers received.

For preliminarily assessing the performance of the feature set we designed, tenfold cross-validation was performed using the CPDB linker dataset. The average accuracy (rate of correct predictions) was 67.5%. In this preliminary test, for each testing case, only the best-voted candidate answer was considered. However, in the actual application, the probabilities of candidate answers played a prominent role. This procedure constituted just a part of the CirPred linker design protocol (see the previous subsection). A thorough assessment of the complete linker design protocol had been carried out, and the performance is stated in "Results" section.

Tuning the probability estimate of amino acids in a designed linker

Since the 20 amino acids were reduced to 3 classes for machine learning, the predicted amino acid probability estimates were sketchy. Before these estimates could be used in the actual linker design protocol, they should be restored to 20 amino acids.

Linkers of the training dataset had been analyzed, such that for each class, proportions of amino acids were known. After the "class-level" prediction, the occurrence probabilities of 20 amino acids at a given residue were finely estimated using the equation,

$$pe(A) = pe(C) \times p(A|C) \tag{2}$$

where *pe*() and *p*() stand for the probability estimate and proportion, respectively, and *A* and *C* denote amino acid and class. For example, if the probability estimate of the hydrophilic class for a residue is 0.80 and the proportion of aspartic acid in this class is 0.25, the probability estimate of aspartic acid at this position is $0.80 \times 0.25 = 0.20$.

Length estimate of the linker

In our experiments, the length of a linker to be designed could be obtained from the alignment between CPMs. However, before using the CirPred server for linker design, the user might not know how many residues there should be in the linker. Hence, we proposed an algorithm to estimate the length of the linker to be designed. For a protein with a distance of N- and C-termini (represented by Cα atoms) < 20 Å, the length of the linker would be estimated using this equation,

$$l = \text{Round}(21.8 \times \ln(\text{b}) - 52.5);\ \ 1 \geq 0 \tag{3}$$

where l and b denote the number of residues of the linker and the distance of termini, respectively. This equation was established according to CPDB. The 20 Å cutoff was determined based on the fact that the length of known linkers (l) and the distance of the termini they bridged (b) fit the equation within it and that for termini more distant than it, there seemed no rule between l and b (Additional file 12). For proteins with a long distance between the termini, we proposed the following algorithm to estimate the length of the linker,

1. Starting with $l = 20$ (residues), make t temporary linkers of length l based on the amino acid propensities of the CPDB linker dataset (Additional file 11).
2. Add each temporary linker to the target CPM; according to the sequence alignment between the target CPM and the pseudo-CP template, generate a model of the CPM and compute its energy score.
3. Among the t temporary linkers, find the one with the lowest energy score to represent linkers of length l.
4. Increase l by five residues and repeat steps 1–3 until l reaches a given maximum. We empirically suggest the maximum be 1/5 of the size of the target protein.
5. Find the value of l producing the lowest energy score. Scan the length range from $l -$ 4 to $l + 4$ and compute these lengths' energy score by repeating steps 1–3.
6. The length of the linker to be designed is estimated as the length of temporary linkers achieving the lowest energy score.

For evaluating this algorithm, proteins of the CPDB linker dataset with linker length ≥ 20 residues were tested with the DOPE energy score. The average difference in length between the known and designed linkers was 20.3%. The performance of this algorithm was acceptable; it was very time-consuming, however. In the CirPred web server, Eq. (3) is applied by default unless the user changes it.

Evaluation of the linker design protocol by multiple rounds of independent test

Using the CPDB linker dataset (Additional file 6), we performed an independent test of 500 rounds to evaluate the proposed linker design protocol. This procedure was an improved tenfold cross-validation test. It ensured the independence between the training and testing data for each round and reduced the imprecision of evaluation. The training datasets, independent test datasets, and results of each round are available in Additional file 7. The procedure of this test is provided below,

1. Since one protein might have multiple CPMs in CPDB, all CPDB linkers were grouped according to their PDB entries.
2. Repeat the following steps 500 times,
   - (i) Randomly divide the grouped linkers into a training Dataset T and an independent test Dataset I possessing 90% and 10% of the proteins.
   - (ii) For ensuring that Dataset I was highly different from Dataset T, any protein in Dataset I sharing ≥ 15% sequence identity with any protein from Dataset T was discarded.
   - (iii) A machine-learning linker predictor was established using Dataset T as the training data based on the CirPred linker design protocol.
   - (iv) For each case in Dataset I, remove the known linker from the protein, make a pseudo CPM, and then input this pseudo CPM to the CirPred system to redesign the linker using the machine-learning linker predictor established in the previous step. Sequence similarities and the difference of potential energy scores between the designed and known linkers were computed using the BLOSUM45 matrix [48] and GROMACS [44].
3. Statistically analyze the sequence similarity and energy data obtained from the 500 rounds.

Refinement of the generated model

As shown in Fig. 1, a protein is virtually divided by the CP site (equivalent to the middle point of the linker of the native protein), i.e., the hinge, into two proportions. If the proportions have the same size, take the C-terminal proportion to be the small one. An axis r is formed between the hinge and the center of mass of the small proportion. Besides, a plane P is defined by these two points and the center of mass of the large proportion. Then, two kinds of movements are made to the small proportion. First, with the hinge fixed, rotate the small proportion on plane P with a pause per 20 degrees. Second, at each pause, rotate it around axis r with a snapshot per 20 degrees. There are 162 snapshots in total (180°/20° × 360°/20°). Finally, compute the energy score of these snapshots and select the one with the lowest energy to be the refined model of the CPM.

Molecular dynamics simulations

Molecular dynamics simulations were applied as a final optimization of the model. The model was submerged in a box filled with water molecules. When necessary, a suitable amount of Na^+ or Cl^- ions were added into the box to neutralize the system's charge. The neutralized system was first energy minimized before the full MD simulations, in which two rounds of annealing, each with temperature points 298 K, 320 K, and 298 K, were performed. Without applying the CP-site-hinge model refinement described above, the number of steps made in MD was set as 5 million, and the step size was two femtoseconds (total simulation time = 10 ns). If the model refinement was applied, we found that 0.5 million steps were enough to produce results with equivalent quality in terms of the alignment ratio and RMSD between the actual and modeled structures. In all modeling experiments performed in this study, both 5 and 0.5 million time steps were tested, and the reported data were based on the 0.5-million-step results. To reduce server machines' loading, we set the default number of time steps as 0.1 million in the implemented CirPred server, and the user could change the setting.

Abbreviations

CPM: Circular permutant. A circularly permuted protein variant (artificially engineered) or a circularly permuted homolog (engineered or naturally occurring).; CP pair: A pair of proteins related by a circular permutation. Note that in a CP pair, one protein is the CPM of the other.

Supplementary Information

The online version contains supplementary material available at https://doi.org/10.1186/s12859-021-04403-1.

Additional file 1: Data S1. Modeling results of SWISS-MODEL, RaptorX, and Robetta for viable circular permutants of the DHFR.

Additional file 2: Fig. S1. Co-linear modeling quality of the CirPred and several state-of-the-art modeling methods.

Additional file 3: Data S2. The CP pair dataset.

Additional file 4: Fig. S2. Models constructed by CirPred for circular permutants with high identity, low identity, or large sizes.

Additional file 5: Tables S1, S2. Performance of CirPred for proteins of various sizes and various CP site positions.

Additional file 6: Data S3. The CPDB linker dataset. A full list of termini linkers of all circular permutants from the CP pair dataset (Additional file 3).

Additional file 7: Data S4. The training datasets, independent datasets, and results of the 500-round independent test for the linker design protocol of CirPred.

Additional file 8: Fig. S3, Table S3. Performance of the linker design algorithm of CirPred for proteins with various termini distances.

Additional file 9: Data S5. Dataset S. A dataset of 2,141 in silico synthetic circular permutants, each with a known native protein and a predefined missing linker. Every protein in this dataset shares <25% sequence identity with any other protein either from itself or the CPDB linker dataset (Additional file 6).

Additional file 10: Discussion. Advanced discussions on the utilization, limitation, and future developments of the CirPred method.

Additional file 11: Table S4. Amino acid propensities of the CPDB linker dataset.

Additional file 12: Fig. S4. Relation between the length of linkers and the distance of the termini they bridged.

Acknowledgements

We thank Chia-Hua Lo, Hao-Wen Ke, Yu-Hsiu Huang, and Chia-Tzu Ho, students of WCL, for their help in data analyses and preparation of the supplementary information files. Chia-Hua Lo is also acknowledged for proofreading the manuscript, so is Prof. Eric Hwang at the Department of Biological Science and Technology, National Yang Ming Chiao Tung University, Taiwan.

About this supplement
This article has been published as part of BMC Bioinformatics Volume 22 Supplement 10 2021: Selected articles from the 19th Asia Pacific Bioinformatics Conference (APBC 2021): bioinformatics. The full contents of the supplement are available at https://bmcbioinformatics.biomedcentral.com/articles/supplements/volume-22-supplement-10.

Authors' contributions
TRC prepared software, performed experiments and data analyses, established the web server, and helped draft the manuscript. YCL performed experiments, participated in data analyses, and helped the web implementation of algorithms. YWH performed experiments and data analyses. CCC helped design algorithms and draft the manuscript. WCL conceived of the study, designed algorithms, performed experiments, analyzed results, and was the major contributor in writing the manuscript. All authors read and approved the final manuscript.

Funding
This work was funded by the Ministry of Science (MOST), Taiwan, with grant numbers NSC 101-2311-B-009-006-MY2 to WCL and MOST-108-2221-E-110-061-MY2 to CCC. The funder had no role in the design of the study, the collection, analysis, and interpretation of data, or in writing the manuscript. The publication costs of this paper were covered by WCL with personal savings and CCC with MOST grant number MOST-108-2221-E-110-061-MY2.

Availability of data and materials
All data generated or analyzed during this study are included in this published article and its supplementary information files.

Declarations

Ethics approval and consent to participate
Not applicable.

Consent for publication
Not applicable.

Competing interests
The authors declare that they have no competing interests.

Author details
[1]Institute of Bioinformatics and Systems Biology, National Chiao Tung University, Hsinchu, Taiwan. [2]Institute of Bioinformatics and Systems Biology, National Yang Ming Chiao Tung University, Hsinchu, Taiwan. [3]Department of Biological Science and Technology, National Chiao Tung University, Hsinchu, Taiwan. [4]Department of Biological Science and Technology, National Yang Ming Chiao Tung University, Hsinchu, Taiwan. [5]Institute of Medical Science and Technology, National Sun Yat-sen University, Kaohsiung, Taiwan. [6]The Center for Bioinformatics Research, National Yang Ming Chiao Tung University, Hsinchu, Taiwan.

Received: 12 September 2021 Accepted: 24 September 2021
Published online: 12 October 2021

References

1. Lindqvist Y, Schneider G. Circular permutations of natural protein sequences: structural evidence. Curr Opin Struct Biol. 1997;7(3):422–7.
2. Vogel C, Morea V. Duplication, divergence and formation of novel protein topologies. BioEssays. 2006;28(10):973–8.
3. Todd AE, Orengo CA, Thornton JM. Plasticity of enzyme active sites. Trends Biochem Sci. 2002;27(8):419–26.
4. Gebhard LG, Risso VA, Santos J, Ferreyra RG, Noguera ME, Ermacora MR. Mapping the distribution of conformational information throughout a protein sequence. J Mol Biol. 2006;358(1):280–8.
5. Nakamura T, Iwakura M. Circular permutation analysis as a method for distinction of functional elements in the M20 loop of *Escherichia coli* dihydrofolate reductase. J Biol Chem. 1999;274(27):19041–7.
6. Schwartz TU, Walczak R, Blobel G. Circular permutation as a tool to reduce surface entropy triggers crystallization of the signal recognition particle receptor beta subunit. Protein Sci. 2004;13(10):2814–8.
7. Arnold FH. Fancy footwork in the sequence space shuffle. Nat Biotechnol. 2006;24(3):328–30.
8. Qian Z, Lutz S. Improving the catalytic activity of *Candida antarctica* lipase B by circular permutation. J Am Chem Soc. 2005;127(39):13466–7.
9. Yu Y, Lutz S. Circular permutation: a different way to engineer enzyme structure and function. Trends Biotechnol. 2011;29(1):18–25.
10. Kojima M, Ayabe K, Ueda H. Importance of terminal residues on circularly permutated *Escherichia coli* alkaline phosphatase with high specific activity. J Biosci Bioeng. 2005;100(2):197–202.
11. Ostermeier M. Engineering allosteric protein switches by domain insertion. Protein Eng Des Sel. 2005;18(8):359–64.
12. Baird GS, Zacharias DA, Tsien RY. Circular permutation and receptor insertion within green fluorescent proteins. Proc Natl Acad Sci U S A. 1999;96(20):11241–6.

13. Lee YT, Su TH, Lo WC, Lyu PC, Sue SC. Circular permutation prediction reveals a viable backbone disconnection for split proteins: an approach in identifying a new functional split intein. PLoS ONE. 2012;7(8):e43820.
14. Lee YZ, Lo WC, Sue SC. Computational prediction of new intein split sites. Methods Mol Biol. 2017;1495:259–68.
15. Iwakura M, Nakamura T, Yamane C, Maki K. Systematic circular permutation of an entire protein reveals essential folding elements. Nat Struct Biol. 2000;7(7):580–5.
16. Lo WC, Wang LF, Liu YY, Dai T, Hwang JK, Lyu PC. CPred: a web server for predicting viable circular permutations in proteins. Nucleic Acids Res. 2012;40(Web Server issue):W232–237.
17. Lo WC, Dai T, Liu YY, Wang LF, Hwang JK, Lyu PC. Deciphering the preference and predicting the viability of circular permutations in proteins. PLoS ONE. 2012;7(2):e31791.
18. Arnold K, Bordoli L, Kopp J, Schwede T. The SWISS-MODEL workspace: a web-based environment for protein structure homology modelling. Bioinformatics. 2006;22(2):195–201.
19. Kallberg M, Margaryan G, Wang S, Ma J, Xu J. RaptorX server: a resource for template-based protein structure modeling. Methods Mol Biol. 2014;1137:17–27.
20. Park H, Kim DE, Ovchinnikov S, Baker D, DiMaio F. Automatic structure prediction of oligomeric assemblies using Robetta in CASP12. Proteins. 2018;86(Suppl 1):283–91.
21. Huang TT, Hwang JK, Chen CH, Chu CS, Lee CW, Chen CC. (PS)2: protein structure prediction server version 3.0. Nucleic Acids Res. 2015;43(W1):W338–342.
22. Lo WC, Lee CC, Lee CY, Lyu PC. CPDB: a database of circular permutation in proteins. Nucleic Acids Res. 2009;37(Database issue):D328–332.
23. Bennett MJ, Schlunegger MP, Eisenberg D. 3D domain swapping: a mechanism for oligomer assembly. Protein Sci. 1995;4(12):2455–68.
24. Hahn M, Piotukh K, Borriss R, Heinemann U. Native-like in vivo folding of a circularly permuted jellyroll protein shown by crystal structure analysis. Proc Natl Acad Sci U S A. 1994;91(22):10417–21.
25. Lo WC, Lyu PC. CPSARST: an efficient circular permutation search tool applied to the detection of novel protein structural relationships. Genome Biol. 2008;9(1):R11.
26. Correia BE, Holmes MA, Huang PS, Strong RK, Schief WR. High-resolution structure prediction of a circular permutation loop. Protein Sci. 2011;20(11):1929–34.
27. Wright G, Basak AK, Wieligmann K, Mayr EM, Slingsby C. Circular permutation of betaB2-crystallin changes the hierarchy of domain assembly. Protein Sci. 1998;7(6):1280–5.
28. Chu CH, Lo WC, Wang HW, Hsu YC, Hwang JK, Lyu PC, Pai TW, Tang CY. Detection and alignment of 3D domain swapping proteins using angle-distance image-based secondary structural matching techniques. PLoS ONE. 2010;5(10):e13361.
29. Lo WC, Lee CY, Lee CC, Lyu PC. *i*SARST: an integrated SARST web server for rapid protein structural similarity searches. Nucleic Acids Res. 2009;37(Web Server issue):W545–551.
30. Fiser A. Template-based protein structure modeling. Methods Mol Biol. 2010;673:73–94.
31. Smith TF, Waterman MS. Identification of common molecular subsequences. J Mol Biol. 1981;147(1):195–7.
32. McWilliam H, Li W, Uludag M, Squizzato S, Park YM, Buso N, Cowley AP, Lopez R. Analysis tool web services from the EMBL-EBI. Nucleic Acids Res. 2013;41(Web Server issue):W597–600.
33. Juan SH, Chen TR, Lo WC. A simple strategy to enhance the speed of protein secondary structure prediction without sacrificing accuracy. PLoS ONE. 2020;15(6):e0235153.
34. Chen TR, Juan SH, Huang YW, Lin YC, Lo WC. A secondary structure-based position-specific scoring matrix applied to the improvement in protein secondary structure prediction. PLoS ONE. 2021;16(7):e0255076.
35. Chen TR, Lo CH, Juan SH, Lo WC. The influence of dataset homology and a rigorous evaluation strategy on protein secondary structure prediction. PLoS ONE. 2021;16(7):e0254555.
36. Jaskolski M. 3D domain swapping, protein oligomerization, and amyloid formation. Acta Biochim Pol. 2001;48(4):807–27.
37. Knaus KJ, Morillas M, Swietnicki W, Malone M, Surewicz WK, Yee VC. Crystal structure of the human prion protein reveals a mechanism for oligomerization. Nat Struct Biol. 2001;8(9):770–4.
38. Ogihara NL, Ghirlanda G, Bryson JW, Gingery M, DeGrado WF, Eisenberg D. Design of three-dimensional domain-swapped dimers and fibrous oligomers. Proc Natl Acad Sci U S A. 2001;98(4):1404–9.
39. Nagarkar RP, Hule RA, Pochan DJ, Schneider JP. Domain swapping in materials design. Biopolymers. 2010;94(1):141–55.
40. Ostermeier M, Benkovic SJ. Evolution of protein function by domain swapping. Adv Protein Chem. 2000;55:29–77.
41. Lee MC, Scanlon MJ, Craik DJ, Anderson MA. A novel two-chain proteinase inhibitor generated by circularization of a multidomain precursor protein. Nat Struct Biol. 1999;6(6):526–30.
42. Chuang YC, Hu IC, Lyu PC, Hsu SD. Untying a protein knot by circular permutation. J Mol Biol. 2019;431(4):857–63.
43. Shen MY, Sali A. Statistical potential for assessment and prediction of protein structures. Protein Sci. 2006;15(11):2507–24.
44. Pronk S, Pall S, Schulz R, Larsson P, Bjelkmar P, Apostolov R, Shirts MR, Smith JC, Kasson PM, van der Spoel D, et al. GROMACS 4.5: a high-throughput and highly parallel open source molecular simulation toolkit. Bioinformatics. 2013;29(7):845–54.
45. The PyMOL Molecular Graphics System. In., v2.0.6 edn: Schrödinger, LLC.
46. Word JM, Lovell SC, Richardson JS, Richardson DC. Asparagine and glutamine: using hydrogen atom contacts in the choice of side-chain amide orientation. J Mol Biol. 1999;285(4):1735–47.
47. Case DA, Cheatham TE 3rd, Darden T, Gohlke H, Luo R, Merz KM Jr, Onufriev A, Simmerling C, Wang B, Woods RJ. The Amber biomolecular simulation programs. J Comput Chem. 2005;26(16):1668–88.
48. Henikoff S, Henikoff JG. Amino acid substitution matrices from protein blocks. Proc Natl Acad Sci U S A. 1992;89(22):10915–9.

Publisher's Note

Liao et al. BMC Bioinformatics (2021) 22:515
https://doi.org/10.1186/s12859-021-04404-0

BMC Bioinformatics

RESEARCH Open Access

Predicting subcellular location of protein with evolution information and sequence-based deep learning

Zhijun Liao[1,2†], Gaofeng Pan[2†], Chao Sun[2] and Jijun Tang[2,3*]

From The 19th Asia Pacific Bioinformatics Conference (APBC 2021) Tainan, Taiwan. 3-5 February 2021

*Correspondence: jtang@cse.sc.edu
†Zhijun Liao and Gaofeng Pan contributed equally to this work.
[3] College of Electrical and Power Engineering, Taiyuan University of Technology, No. 79 Yinze West Street, Taiyuan 030024, SX, China
Full list of author information is available at the end of the article

Abstract

Background: Protein subcellular localization prediction plays an important role in biology research. Since traditional methods are laborious and time-consuming, many machine learning-based prediction methods have been proposed. However, most of the proposed methods ignore the evolution information of proteins. In order to improve the prediction accuracy, we present a deep learning-based method to predict protein subcellular locations.

Results: Our method utilizes not only amino acid compositions sequence but also evolution matrices of proteins. Our method uses a bidirectional long short-term memory network that processes the entire protein sequence and a convolutional neural network that extracts features from protein sequences. The position specific scoring matrix is used as a supplement to protein sequences. Our method was trained and tested on two benchmark datasets. The experiment results show that our method yields accurate results on the two datasets with an average precision of 0.7901, ranking loss of 0.0758 and coverage of 1.2848.

Conclusion: The experiment results show that our method outperforms five methods currently available. According to those experiments, we can see that our method is an acceptable alternative to predict protein subcellular location.

Keywords: Subcellular prediction, Protein sequence, Evolution information, Deep learning, Multiple label classification

Background

Subcellular locations of a protein is crucial in understanding its function and physicochemical characteristics; computational methods are necessary in the research of protein analysis since traditional protein subcellular localization methods are laborious and time-consuming [1]. There are a group of membrane-bound organelles in the eukaryotic cells, such as Nucleus, Centriole, Ribosome, vesicle, Mitochondria, Golgi apparatus and so on [2]. Since such compartments perform various functions decided by protein complexes, ascertain the location of a protein in a cell can direct protein function discovery [3]. Traditional protein subcellular localization methods, such as immunofluorescence techniques and expression system fused with green-fluorescent protein, recognize the subcellular location of proteins from the vision of protein layout which is formed with fluorescent protein and fluorescence microscope [4]. However, those traditional methods are laborious and time-consuming, which become limitations to widely utilize [5]. In order to mitigate those limitations, a lot of computational methods were developed in the past few decades [6].

One major branch of subcellular localization prediction method is to use features extracted from protein sequence and a classifier that separates those features into different patterns to predict the protein subcellular location in cells. The method SubLoc, developed by Hua et al. in 2000, uses basic sequence vector with a dimension of 20 as there are 20 natural amino acids, and Support Vector Machine (SVM) [7] to classify proteins [8]. Each unit of this method's input vector stands for one amino acid. SVM constructs an optimal hyperplane in the high dimensional feature space to separate samples into positive group and negative group. By using SVM as sample classifier, the protein subcellular localization prediction problem can be solved by finding a suitable kernel function to define the inner product in the feature space.

In order to improve the subcellular localization prediction accuracy, many feature extraction methods based on protein sequence were defined, and many classifiers inherited from SVM were generated. For example, the method published by Shen et al. in 2018 takes multi-kernel SVM as a classifier, and applies features extracted from evolution and physicochemical information other than sequence vectors as the input of classifier [9]. Extra features, such as PsePSSM, PsePP and so on, expand the feature vector of the model to thousands of dimensions, and extend the ability of protein representation that also includes the correlation between amino acid sites that are not adjacent to each other. At the same time, multi-kernel SVM enables the classifier to combine multiple kernel functions that are defined for different feature spaces into one kernel function with suitable weights [10, 11]. To improve the proficiency, Ding et al. applied Fuzzy Support Vector Machine (FSVM) instead of multi-kernel SVM in the classification task. Fuzzy support vector machine was developed by Lin et al. based on standard SVM with an extra value for each sample to denote its membership value, with a goal of adjusting the weights of outliers in the trained classification model [12, 13]. Ding et al. applied KNR (Kernelized Neighborhood Representation) function, a kernel function proposed by them to identify protein crystallization that generates the membership values from input samples for FSVM [14], to fully represent the spatial distribution of feature space and control the interference of outliers effectively [15].

Features of proteins can be extracted not only from raw protein sequences but also from correlations between amino acid sites indirectly. The method proposed by He et al., Imbalanced Multi-Model Multi-Label Learning with Gaussian Process (IMMMLGP), is one of the methods that uses sequence information and Gaussian Process Prior to predict subcellular localization of protein [16]. The authors of this method applied a gaussian process model to effectively exploit the correlations among multiple locations that are important in improving the prediction accuracy of multi-localized proteins [17, 18]. Furthermore, this method is robust on imbalanced data sets because optimal linear combinations of various feature extraction technologies were included. Currently, a Java library named MULAN, which was proposed by Tsoumakas et al., offers programmatic application interface for multi-label learning [19]. Wei et al. selected 11 simple multi-label classifiers from MULAN library to fuse a mean ensemble classifier after each single classifier was trained and clustered [20]. Overall, the general procedure of building up a subcellular localization prediction method is to first extract features from protein sequence comprehensively; then find a suitable classifier algorithm according to the feature space and adjust the classifier to fit the problem; lastly analyze the performance of the classifier after training and testing on some benchmark datasets.

The prediction method mGOASVM utilizes BLAST [21] search to find out the accession numbers of a given protein's homologs, then it searches the Gene Ontology annotation database for the GO terms of the given protein. The GO terms are processed into feature vectors, then those features are used to train a SVM classifier which can predict the subcellular locations of protein accurately. In addition to the achievement of mGOASVM, the prediction method mLASSO-Hum takes one step further. It uses a one-vs-rest LASSO-based classifier, instead of a SVM classifier to improve the interpretability of the classification results and the robustness of classification model [22]. Since the GO terms have high dimensional features, classifiers using this kind of information tend to overfit the train dataset. The LASSO-based classifier identified important features from all the GO terms, by which the dimension of the feature vector was significantly reduced. With these advantages introduced by informative GO terms and functional classifiers, the method mLASSO-Hum outperforms several other subcellular localization methods on prediction accuracy. Furthermore, its result has strong interpretability which provides insight into the relationship between proteins and subcellular locations [23].

Protein sequence information and its GO annotations are widely used in subcellular location prediction methods. For example, the method Hum-mPLoc 2.0, proposed by Shen et al. in 2009 established on Hum-mPLoc, is a synthesized prediction method that incorporates both protein sequence and GO terms [24, 25]. The method Hum-mPLoc takes the GO terms into account but is still restricted by the shortage of accession numbers and evolution information, so the performance of this predictor is not as satisfactory on some specific problem sets. For this reason, Shen et al. made improvements on this method by adding domain and evolution information extracted from protein sequence to the feature list, which boosted the performance of the classifier on the problem sets [26]. The method iLoc-Hum is another method that uses GO terms besides protein sequence [27]. Prediction methods, such as pLoc_bal-mAnimal [28] and pLoc-mGneg [29], apply over-sampling algorithm and under-sampling algorithm on imbalanced datasets to balance the training dataset, and train the prediction model with balanced samples in order to overcome the limitation of machine learning algorithm.

Deep neural networks extract features from input data automatically and predict the output of the model based on those features, which can predict the subcellular locations of a protein sequence without feature extraction methods. Deep learning has become more and more popular in the research field of artificial intelligence in the past few years [30]. There are a lot of deep neural network models that focus on accurate prediction of protein subcellular localization. The method DeepLoc, proposed by Armenteros et al. in

2017, takes recurrent neural network to processes the entire protein sequences, and it also applies attention mechanism which identifies the important protein regions for the subcellular localization in the neural network model [31]. With the support of the recurrent neural network, feature extraction methods are omitted from the construction of prediction method. And the attention mechanism improves the accuracy of prediction.

Before training the neural networks, protein sequences can be reformulated to generate an effective representation of the biological sequence samples. Sequence based subcellular localization prediction methods, such as pLoc_Deep-mHum [32], pLoc_Deep-mEuk [33], pLoc_Deep-mvirus [34] and so on, pre-process the protein sequences with Pseudo Amino Acid Composition algorithm [35], then train a convolutional neural network or a recurrent neural network with the new generated property vectors. The neural network can learn the patterns in proteins more effectively from the reformulated properties than from the original protein sequences. The method DeepPSL, developed by Wei et al., uses an adaptive skip dipeptide composition to calculate the fraction of adjacent residues in the protein sequences, then uses a stacked auto-encoder to encode those fraction properties into subcellular location labels [36]. Protein physicochemial properties are also used by DeepPSL as a sequence representation; those properties enlarge the feature space and improve the prediction accuracy [37].

Deep learning models can predict protein subcellular locations from images. The method ImPLoc, proposed by Long et al., predicts protein subcellular locations from IHC (immunohistochemistry) images [38]. In this method, deep convolutional neural network extracts image features, self-attention encoder aggregates the extracted feature vectors, fully connected network predicts the location labels. Similar to ImPLoc, the method DeepYeast also applies deep convolutional neural network in the subcellular localization prediction from high throughput microscopy images [39]. Image preprocessing algorithms, such as Gamma correction, Morphological Closing and so on, are widely used to improve the quality of images. The method proposed by Masurkar et al. takes this strategy to revise the result predicted from microscopy images [40]. Deep learning was also applied in the prediction of messenger RNA subcellular location or long non-coding RNA subcellular location. Those RNA subcellular localization prediction methods include IncLocator [41], RNATracker [42], DeepLncRNA [43] and so on.

In this paper, we introduce a deep learning-based protein subcellular localization prediction method which uses bidirectional long short-term memory network as an encoder and takes evolution information into account. Experiment results shows that our method can predict protein locations accurately.

Method

Protein sequence encoding

Amino acids construct various proteins that can fulfill the basic functions in cells. Each kind of amino acid has a unique side chain that determines its property and action. The amino acid sequence represents the primary structure of a protein. The composition of amino acids decides the structure and function of proteins, so the amino acid sequence becomes the major research target to analyze proteins. Given a protein sequence P which consists of n amino acids, it can be defined as:

$$p = [a_0, a_1, a_2, a_3, \ldots, a_n]$$

and a_i represents the amino acid at ith position in the protein.

One of the most widely used encoding algorithm in computational biology methods is one-hot encoding, which converts nucleotide sequence or amino acid sequence into a list of binary vectors [44, 45]. Each amino acid will be encoded into a binary vector with only one specific position being 1 and other positions being 0s based on its type. The one-hot encoding of protein p is:

$$P = \begin{bmatrix} p_{0,0}, p_{0,1}, p_{0,2}, \ldots, p_{0,22} \\ p_{1,0}, p_{1,1}, p_{1,2}, \ldots, p_{1,22} \\ \vdots \\ p_{n,0}, p_{n,1}, p_{n,2}, \ldots, p_{n,22} \end{bmatrix}$$

in which $p_{i,j}$ is 0 or 1, and $\sum_{j=0}^{22} p_{i,j} = 1$. The encoded matrix can be used to train neural network.

Position specific scoring matrix

Position Specific Scoring Matrix (PSSM), introduced by Gary Stormo et al. in 1982, is the consensus score and evolution information of protein sequences [46, 47]. The score in the matrix is the substitution probability of amino acids at specific positions in a protein sequence. High score means more frequent substitution in an alignment, which low value means less frequency.

PSSM can be gnerated by BLASTing the protein sequence with Position Specific Iterated BLAST (PSI-BLAST) tool on the server of National Institutes of Health, of which the URL is: https://blast.ncbi.nlm.nih.gov/Blast.cgi?PROGRAM=blastp.

The PSSM of a protein sequence p is:

$$S = \begin{bmatrix} s_{0,0}, s_{0,1}, s_{0,2}, \ldots, s_{0,42} \\ s_{1,0}, s_{1,1}, s_{1,2}, \ldots, s_{1,42} \\ \vdots \\ s_{n,0}, s_{n,1}, s_{n,2}, \ldots, s_{n,42} \end{bmatrix}$$

in which $s_{i,j}$ is the substitution frequency from ith amino acid in the protein sequence to amino acid j. Positive values indicate the substitution occurs more frequently, while negative values indicate less frequent substitution. A convolutional neural network extracts features from the generated PSSM. Those features represent the consensus properties of protein functions.

The protein sequences used in the experiment of this project had been pre-processed and the corresponding PSSM was uploaded into github repository (https://github.com/Paeans/subcellular).

Bidirectional long short-term memory network

Long short-term memory (LSTM) network is a special kind of recurrent neural network (RNN), which is widely used in the field of deep learning to process time series data [48]. The LSTM network has four interactive neural network layers which combine the previous states and current input. The architecture of LSTM avoid the long term dependency problem, and the output formulate the entire sequence of input data [49]. Bidirectional LSTM (BLSTM) contains two LSTM networks, of which one LSTM network process protein sequence from beginning to end while another one process the sequence from end to beginning [50]. Since the LSTM network can memory the states at different locations of a sequence, it was widely used to encode sequence samples [51].

With a one hot encoding matrix of a protein P, the LSTM network can encode the matrix into a state matrix which includes the functions expressed by different protein sites. For an input P_t, the state and output of the model node are defined as:

$$\begin{aligned}
f_t &= \sigma(W_f \times [h_{t_1}, P_t] + b_f) \\
i_t &= \sigma(W_i \times [h_{t-1}, P_t] + b_i) \\
\tilde{C}_t &= tanh(W_C \times [h_{t-1}, P_t] + b_C) \\
C_t &= f_t \times C_{t-1} + i_t \times \tilde{C}_t \\
h_t &= \sigma(W_h \times [h_{t-1}, P_t] + b_h) \times tanh(C_t)
\end{aligned}$$

and W_f, W_i, W_C and W_h are the weight matrices applied on input matrix, b_f, b_i, b_C and b_h are the bias matrices, f_t is the forgetting coefficient, i_t is the remembering coefficient, C_t is the current state and h_t is the output.

The Bidirectional LSTM network processes the protein sequence from two directions. The raw vector of amino acids is translated into another numerical vector which includes position information and relation information between different function sites.

Convolutional neural network

Convolutional neural network (ConvNet) utilizes regularized filters and activation nodes to discover patterns from input matrices; it had been widely used in image processing, natural language processing, financial trends analysis, and so on [52–55]. ConvNet also has a lot of applications in the research fields of computational biology, such as non-coding gene function prediction [56], methylation sites classification on whole genome sequence [57], protein-protein interaction analysis [58], and low quality Hi-C data denoising [59]. ConvNet can extract features from raw input automatically, and the important features can be filtered out with kernel weights.

The convolutional layers are the core building blocks of ConvNet; those layers consist of a group of kernels that can be adjusted according to the loss values of prediction during the backward pass [60]. The kernels in ConvNet are small size matrices. The convolution function between kernel matrix ω and input volume X is:

$$C_{i,j} = \sum_{m,n=0}^{|\omega|} \omega_{m,n} X_{i+m,j+n}$$

and $C_{i,j}$ is the convolution value of kernel ω on region $X_{[i:i+m,j:j+n]}$. With backpropagation algorithm, the values in kernel ω can be optimized according to the value $\frac{\partial L}{\partial \omega}$ which is the partial derivative of error L with respect to ω [30]. Figure 1 shows the architecture of our method.

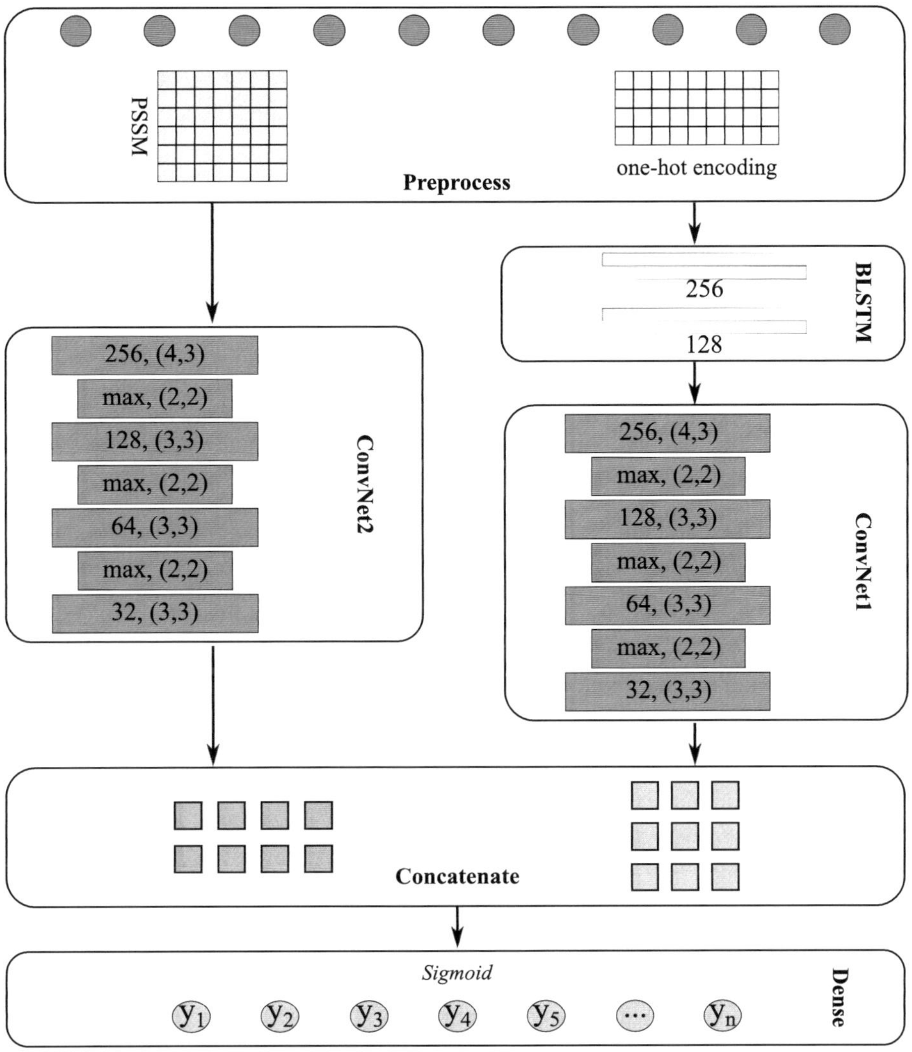

Fig. 1 The workflow of our method. Our method composes 5 components which are sequence pre-process layer, bidirectional LSTM encoder, 2 convolutional neural network and prediction layer. **a** Sequence pre-process layer processes protein sequences into one-hot encoding matrix and position-specific scoring matrix. One protein sequence is converted into two matrix which are the inputs of following deep learning neural networks. **b** Bidirectional LSTM encoder takes an one-hot encoding matrix as input and processes the items in the matrix sequentially. This encoder includes two LSTM layers. One layer processes the matrix from beginning to end while another layer processes the matrix backwards from end to beginning. After two direction encoding, the original one-hot encoding matrix is encoded to 256 values. **c** Convolutional neural network. The convolutional neural network is used to extract features from pssm matrix and encoded matrix. Two identical neural networks are used in our method. One network learns pssm matrix and another one learns encoded matrix. In the network, 4 convolution layers are included to filter out main features and 3 maxpooling layers are inserted among the convolution layers to choose outstanding features. The number of kernels in the 4 convolution layers are 256, 128, 64 and 32 respectively, and the kernel size of each layer is 4×3, 3×3, 3×3 and 3×3 correspondingly. d). Prediction layer produces the possibility of each subcellular location. At the beginning of this layer, the outputs from previous two convolutional neural networks are concatenated together, then the concatenated matrix is flattened into one dimensional array. A fully connected network integrate those features together and Sigmoid function computes the corresponding possibility at each location. Based on the output possibility, the subcellular locations of a protein can be decided

Activation function

Activation function transforms the weighted input into an activation output of the neural network node. This function simulates the stimulation happened between biological neurons [61]. In this method, rectified linear activation function (ReLU) is applied in the convolutional layer nodes and sigmoid function is used to generate output labels. The ReLU in the node is defined as:

$$r(x) = max(0, x)$$

which means the node can only be stimulated by signals that are strong enough [62]. It is a linear function when the input is greater than 0, but it also introduces non-linearity property into the neural network model at point 0 [63].

Sigmoid function translates the input into a value in the range 0 to 1. The formula of this function is:

$$S(x) = \frac{e^x}{e^x + 1}$$

with x as the input of node. When the input signal is strong, the output possibility value is close to 1, otherwise the output value is close to 0. On multi-label classification problems, the sigmoid function is used in the output layer of the neural network to generate the possibility for each output label [64].

Loss function

Loss function maps prediction values onto real values representing the distance between model outputs and targets. The optimization algorithms minimize the loss function during training process. Protein subcellular localization prediction method needs to generate multiple binary labels for a protein; in order to train the model in our method effectively, binary cross entropy loss function is applied to compute the cross-entropy loss between predicted labels and true labels [65].

$L2$ regularization prevents model overfitting by imposing a cost on the loss function [66]. The $L2$ penalty $||\omega||_2^2$ is the squared magnitude of kernel in ConvNet, and this penalty can regulate the weights applied on input signals [67]. The loss function with $L2$ regularization is:

$$L = -\frac{1}{n}\sum_{i=1}^{n}\left(y_i \times log\tilde{y_i} + (1 - y_i) \times log(1 - \tilde{y_i})\right) + \lambda||\omega||_2^2$$

in which $\tilde{y_i}$ is the prediction on ith label while y_i is label i with value 0 or 1, n is the size of labels, ω is the kernel of ConvNet while λ is the coefficient of regularization penalty.

Results

Benchmark datasets

Our method was tested on two benchmark datasets, $D3106$ and $D4802$. Benchmark dataset $D3106$, generated by Shen et al. in 2009, includes 3106 protein sequences which are located at 14 subcellular locations [26]. The number of protein sequences located at

Table.1 Number of protein sequences in benchmark dataset *D*3106 created by Shen et al.

Subcellular location	Number of protein sequences
Nucleus	1021
Cytoplasm	817
Extracellular	385
Mitochondrion	364
Plasma membrane	354
Endoplasmic reticulum	229
Golgi apparatus	161
Cytoskeleton	79
Centriole	77
Lysosome	77
Peroxisome	47
Endosome	24
Microsome	24
Synapse	22
Total	3681

The dataset *D*3106 covers 14 subcellular which are listed at the first column of this table. And the numbers of proteins located at each subcellular location are listed at the second column. There are 3106 protein sequences in this dataset, and the total number of subcellular locations is 3681 since many certain sequences can be found in multiple locations. The sequences distribute at those 14 locations unevenly. 32.9% sequences are located at Nucleus and 26.3% sequences are at Cytoplasm, while less than 1% sequences are located at Synapse. This dataset is unbalanced. 3681 positive cases take only 8.47% of all 3106 × 14 cases

each subcellular location in dataset *D*3106 is listed in Table 1 and the detailed information about correlations of subcellular locations is listed in Additional file 2. Benchmark dataset *D*4802, generated by Wei et al. in 2016, contains 4802 protein sequences located at 33 subcellular locations [20]. The number of proteins at each subcellular location in dataset *D*4802 is listed in Table 2 and the correlations between each subcellular location are listed in Additional file 3. The protein sequence distribution in Tables 1 and 2 shows that most of the proteins are located in Nucleus and Cytoplasm, and only a small number of proteins are in Peroxisome and Synapse; the samples in the datasets are unbalanced.

The correlations between each subcellular location of dataset D3106 which are listed in Additional file 2 respectively show that there are 480 proteins in D3106 which are located in two different subcellular, while 43 proteins exist at three different locations and 3 proteins are in four cell constituents at the same time. Dataset D4802 has 1354 two-subcellular samples, and the number of those protein samples is in Additional file 3. From the table in Additional file 3, we can see that 401 protein sequences exist at Nucleus and Cytoplasm at the same time and 268 protein sequences exist at both Nucleus and Nucleolus, between which high correlations are exhibited.

Evaluation metrics

To evaluate the accuracy of prediction results, ranking loss, coverage and average precision are commonly used in multi-label deep learning [68, 69]. Those functions are defined as:

Table.2 Number of protein sequences in benchmark dataset 4802 created by Wei et al.

Subcellular location	Number of protein sequences
Nucleus	1720
Cytoplasm	1050
Plasma membrane	836
Extracellular	487
Mitochondria	407
Endosome	342
Golgi apparatus	272
Nucleolus	268
Lysosomes	125
Endoplasmic reticulum	120
Cytoskeleton	89
Centrosome	81
Peroxisome	67
Early endosomes	52
Nuclear envelope	47
Cytoplasmic vesicles	46
Basolateral plasma membrane	29
Synaptic vesicles	28
Microtubule	26
Apical plasma membrane	16
Late endosomes	16
Golgi trans face	11
Secretory granule	10
Tight junction	9
Golgi cis cisterna	7
Medial-golgi	7
Melanosome	6
Secretory vesicles	5
Cellular component	4
ERGIC	4
Inner mitochondrial membrane	4
Transport vesicle	4
Golgi trans cisterna	3
Total	6198

In this dataset, 4802 protein sequences are identified in 33 subcellular locations. The first column is the name of the subcellular covered by this dataset, and the second column is the number of proteins located at each subcellular location. The total number of subcellular locations is 6198 since each sequence can be found in multiple subcellular locations. The sequences distribute at those 33 locations unevenly. 35.8% of sequence samples are located at Nucleus and 21.9% sequences are located in Cytoplasm, while only 3 sequences are identified in Golgi Trans Cisterna. The number of positive cases in this dataset is 6198, and the positive case rate is 3.9% (6198/(4802 $\times$ 33)) ranking loss

$$RL = \frac{1}{n}\sum_{i=0}^{n-1}\frac{|\{(k,l): \tilde{y_{ik}} \leq \tilde{y_{il}}, y_{ik} = 1, y_{il} = 0\}|}{||y_i||_0(n - ||y_i||_0)}$$

coverage

$$Cov = \frac{1}{n}\sum_{i=0}^{n-1}\max_{j:y_{ij}=1}|\{k : \tilde{y_{ik}} \geq \tilde{y_{ij}}\}|$$

average precision

$$AP = \frac{1}{n}\sum_{i=1}^{n-1}\frac{1}{||y_i||_0}\sum_{j:y_{ij}=1}\frac{|\{k : y_{ik} = 1, \tilde{y_{ik}} \geq \tilde{y_{ij}}\}|}{|k : \tilde{y_{ik}} \geq \tilde{y_{ij}}|}$$

in those functions, n is the number of samples, y_{ij} is the ground truth label of sample i at label j, and $\tilde{y}$ is the output score matrix of all the samples. $|X|$ is the number of elements in the set X while $||X||_0$ is the number of nonzero elements in the set. Those three functions are defined based on the label pairs that are incorrectly ordered. For example, ranking loss function computes the proportion of true labels which get lower score than false labels, coverage function computes the average rank of lowest prediction values for true labels, and average precision is the average fraction of true labels among all the labels which are higher-ranked than the lowest true label rank. Those three functions will be used to evaluate the performance of our method on the test samples.

The F1 score (F_1), Matthews Correlation Coefficient (MCC) and Area under the ROC curve (AUC) are defined to measure the quality of binary classification [70]. The balanced F1 score in machine learning can measure the quality of binary classifications on unbalanced datasets. It is a weighted average of the precision and recall with values range from 0 to 1. The function of F1 score is:

$$F_1 = 2 \cdot \frac{PPV \cdot TPR}{PPV + TPR}$$

with precision $PPV = tp/(tp + fp)$ and recall $TPR = tp/(tp + fn)$ in which tp, tn, fp and fn are true positive, true negative, false positive and false negative, respectively.

The MCC is a balanced measure of the quality of binary classifications even when the test dataset is an unbalanced dataset [71]. The value of MCC is between -1 an inverse prediction and 1 a perfect prediction. The MCC function for binary classification is:

$$MCC = \frac{tp \times tn - fp \times fn}{\sqrt{(tp + fp)(tp + fn)(tn + fp)(tn + fn)}}$$

and tp, tn, fp and fn are the true positive, true negative, false positive and false negative, respectively.

The AUC computes the area under the receiver operating characteristic (ROC) curve, which is created with true positive rate and false positive rate at various threshold [72]. It reflects the probability that a classifier generates higher scores on true instances than on false instances. We will compute the F1 score, MCC and AUC on each subcellular label to evaluate the performance of our method.

Model performance

In this experiment, we test four different models with benchmark datasets D3106 and D4802 and compare the prediction accuracy of those models with five existing subcellular localization methods. The four models tested in this experiment are model BLSTM which uses 2 bidirectional LSTM layers to encode the protein sequences into subcellular location labels, model BLSTM plus ConvNet1 which takes 4 convolution layers and 3 maxpooling layers following 2 bidirectional LSTM layers, model ConvNet2 which utilizes PSSM of protein sequences and a 7 layer convolutional neural network to predict protein subcellular locations, and model BLSTM plus 2 ConvNet which uses BLSTM, ConvNet1 and ConvNet2. At first, we evaluate the performance of each model on single subcellular location prediction, then evaluate the accuracy of multi-subcellular locations prediction. The detailed testing results can be found in additional file document.

The performance of the four models on single label prediction is listed in Table 3. On dataset D3106, the AUC of model BLSTM reaches to 0.9242 on subcellular Lysosome, but the worst AUC is 0.8315 on subcellular Peroxisome. The average AUC of this model on dataset D3106 is 0.8841. The model BLSTM performs better on dataset D3106 than on dataset D4802, while the best AUC on D4802 is 0.9121 which is worse than the best result on D3106 and the average AUC on D4802 is 0.1178 less than the average result on D3106. The values of MCC and *F*1 score also show that model BLSTM works better on dataset D3106.

Full list of AUC values of each subcellular location prediction on dataset *D*3106 is in Additional file 4 and statistical values are listed in Additional file 1: Table S1. The statistical values in Additional file 1: Table S1 show that the model BLSTM plus 2 ConvNet has best performance on 10 of 14 subcellular location predictions, while the model BLSTM plus ConvNet1 performs best on other 4 predictions. Predictions on 8 subcellular locations have average AUC values greater than 0.9 and the standard deviations of those predictions are less than 0.03 on AUC values. On dataset *D*4802, all AUC values of prediction are in Additional file 5 and statistical values are listed in Additional file 1: Table S2. Additional file 1: Table S2 shows that the model BLSTM

Table 3 The F_1 score, MCC and AUC of subcellular location prediction generated by model BLSTM, BLSTM + ConvNet1, ConvNet2 and BLSTM + ConvNet1 + ConvNet2

	D3106			*D4802*		
	F_1	*MCC*	*AUC*	F_1	*MCC*	*AUC*
BLSTM	0.7473	0.6001	0.9242	0.7419	0.5705	0.9121
BLSTM + ConvNet1	0.7775	**0.6419**	0.9255	0.7801	0.6284	0.9327
ConvNet2	0.6475	0.4819	0.8785	0.6696	0.4259	0.9297
BLSTM + ConvNet1 + ConvNet2	**0.7843**	0.6410	**0.9458**	**0.7842**	**0.6411**	**0.9434**

The four models were tested on datasets *D*3106 and *D*4802. On dataset *D*3106, the highest F_1 score and AUC are achieved by the model BLSTM + ConvNet1 + ConvNet2, while the model BLSTM + ConvNet1 has the highest MCC. On dataset *D*4802, the model BLSTM + ConvNet1 + ConvNet2 was the best among the four models plus ConvNet1 is the best model to prediction subcellular locations based on samples in dataset *D*4802 and the model BLSTM plus 2 ConvNet can not out perform it. Model BLSTM plus ConvNet1 gets highest AUC on 30 subcellular location predictions and most of those 30 predictions have AUC value greater than 0.9. However, the standard deviation of predictions by model BLSTM plus 2 ConvNet is lower than the value of predictions by model BLSTM plus ConvNet1. The standard deviation reduction shows that the extra ConvNet2 makes the model more stable.

The model BLSTM plus ConvNet1 achieves similar performance on dataset D3106 as the model BLSTM with best AUC 0.9255, MCC 0.6419 and *F*1 score 0.7775. However, the best AUC of model BLSTM plus ConvNet1 on D4802, 0.9327 is better than the achievement on D3106, which is different from model BLSTM. The *F*1 score of model BLSTM plus ConvNet1 on D4802 is 0.7801, and the MCC is 0.6284. Overall, the model BLSTM plus ConvNet1 has better performance than model BLSTM. So, we can see that the ConvNet1 takes effects during the learning process, since the convolution layers can extract features from the encoded vectors. The performance of our method on single subcellular location prediction shows that BLSTM component in our method is the most important part to make accurate prediction on protein subcellular localization. In our method, BLSTM component encodes protein sequence into feature matrices. So the key factor of subcellular localization model is encoding method. This encoding method can also be applied to other biological computational methods, especially methods which take protein sequences as input.

Model ConvNet2 with PSSM as inputs is out of expectations. On dataset D3106, model ConvNet2 only gets 0.8785 of AUC, 0.4819 of MCC and 0.6475 of *F*1 score. On dataset *D*4802, the best MCC of this model is .4259 which is worse than the performance of this model on dataset *D*3106. When comparing the testing results of this model and BLSTM model on datasets *D*3106 and *D*4802, we can see that BLSTM model performs better than this model. The reason of bad performance is that the PSSM can represent the evolution information of proteins but can not replace the sequence information in subcellular location prediction. With bidirectional LSTM, protein sequences can be translated into matrix form which is rich of function information.

When appending ConvNet2 to the model BLSTM plus ConvNet, the accuracy of single subcellular location prediction has no big difference. The model BLSTM plus 2 ConvNet gets best AUC on most of the 14 subcellular locations in dataset *D*3106 according to the results listed in Additional file 1: Table S1. However, when using t test to evaluate the difference between models with and without ConvNet2, we can see that the difference is not significant to prove that ConvNet2 makes the model better on single subcellular prediction ($p_{value} > 0.5$). The t test results are listed in Additional file 1: Table S3. Testing results show similar conclusion with dataset *D*4802. The AUC of each subcellular location prediction has no significant changes when ConvNet2 was appended to the model BLSTM plus ConvNet. So, ConvNet2 has no effect to single subcellular prediction. The statistical values in Additional file 1: Table S3 show that ConvNet1 makes great improvement to the accuracy of single subcellular location prediction on both dataset *D*3106 and dataset *D*4802.

The best AUC and average AUC of the joined model on dataset D3106 are 0.9458 and 0.9046, respectively. According to the AUC values, the joined model has great improvements on protein subcellular localization prediction. The *F*1 score and MCC of this model, 0.7843 and 0.6410 respectively, are also higher than the values of another three models. We plot the ROC curves of predictions in Figs. 2 and 3. All the ROC curves of fivefold cross validation are plotted in Additional file 1: Figures S1 and S2. The plots in supplement figures are similar to the corresponding ones in Figs. 2 and 3 since each model has similar performance on each fold data. Those plots show that model BLSTM plus ConvNet1 is better than the joined model since some predictions created by joined model are not good enough. The reason of accuracy deduction is that the ConvNet2 network introduces extra parameters into the joint model and those parameters affect the training process and model optimization.

When evaluating the performance of the four models as multi-label classification models, we can get similar conclusion from the results of ranking loss, coverage and average precision. The results of experiment are listed in Table 4. The model ConvNet2 has the worst performance on both dataset D3106 and dataset D4802. And the best one is the joined model with average precision of 0.7901, ranking loss of 0.0758 and coverage of 1.2848 on dataset D3106. The prediction accuracy of the joined model on dataset D4802 is not as good as the accuracy on dataset D3106, but the model can still get average precision of 0.7414 and ranking loss of 0.0637. We use fivefold cross validation to test our method comprehensively and plot the box chart of accuracy in Figure 4. Additional file 1: Table S4 contains ranking loss, coverage and average precision of fivefold cross validation on dataset *D*3106 and *D*4802. All the four models were tested on dataset *D*3106 and *D*4802 and the corresponding validation results are in Additional file 1: Table S4.

Discussion

The experiment results listed in Additional file 1: Table S4 show that model BLSTM plus 2 ConvNet has highest accuracy on dataset *D*3106 and model BLSTM plus ConvNet1 is the best with dataset *D*4802. According to average precision and coverage results, model BLSTM plus 2 ConvNet performs better on dataset *D*3106 than on dataset *D*4802, but the ranking loss value of it on *D*4802 is lower than the value on *D*4802. The possible reason of bad performance on dataset *D*4802 is that the unbalanced samples takes higher proportion in *D*4802 which makes model over-fitting under same training procedure. Samples which have high correlation between subcellular locations are suitable to our

Table 4 The average precision, ranking loss and coverage of model BLSTM, BLSTM + ConvNet1, ConvNet2 and BLSTM + ConvNet1 + ConvNet2

	D3106			*D4802*		
	RL	*Cov*	*AP*	*RL*	*Cov*	*AP*
BLSTM	0.0967	1.5895	0.7523	0.0820	3.3916	0.6901
BLSTM + ConvNet1	0.0778	1.3113	0.7876	**0.0603**	**2.9225**	**0.7453**
ConvNet2	0.1294	2.0113	0.6430	0.0673	3.2868	0.6214
BLSTM + ConvNet1 + ConvNet2	**0.0758**	**1.2848**	**0.7901**	0.0637	3.0528	0.7414

On dataset D3106, the BLSTM + ConvNet1 + ConvNet2 has the best performance with lowest ranking loss, coverage and highest average precision which are 0.0758, 1.2848 and 0.7901, respectively. However, the model BLSTM + ConvNet1 + ConvNet2 is not as good as model BLSTM + ConvNet1 when tested on dataset *D*4802. The best values are marked out with bold text

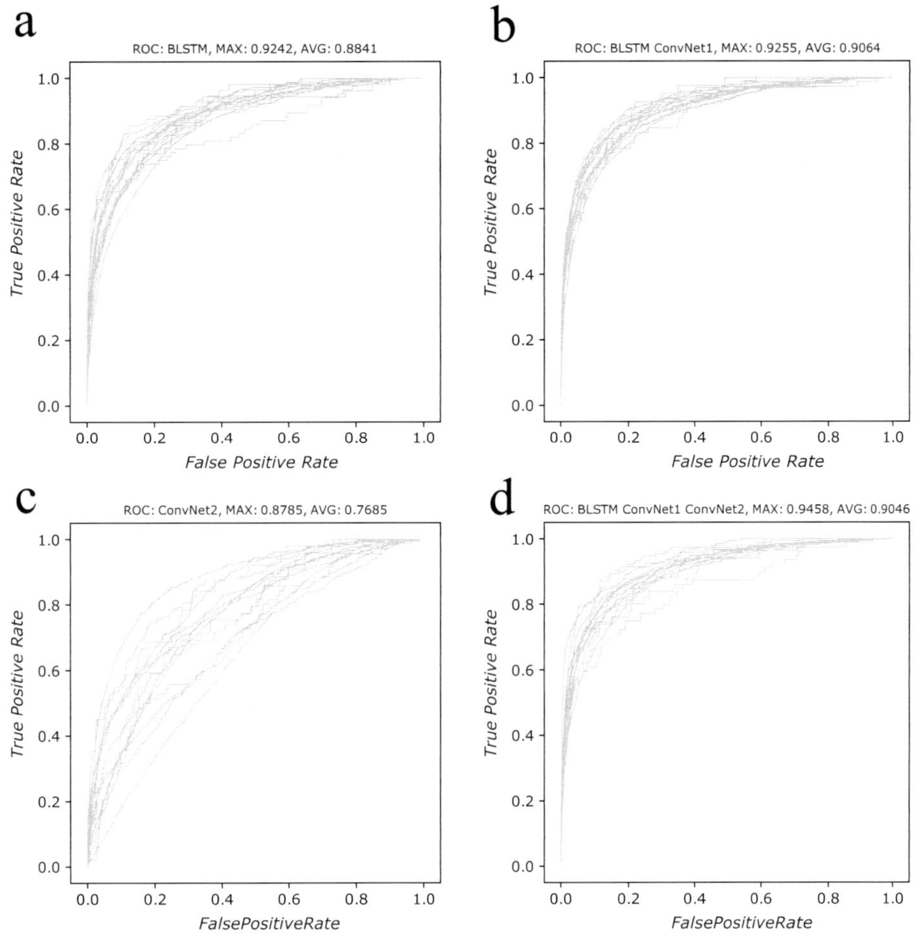

Fig. 2 ROC curves (Receiver operating characteristic curve) of model BLSTM, BLSTM + ConvNet1, ConvNet2 and BLSTM + ConvNet1 + ConvNet2, on benchmark dataset *D*3106. Those models were tested with fivefold cross validation. This figure plots the result of one fold validation, and all five fold testing results can be found at Additional file 1: Table S3 and Additional file 1: Figure S1. **a** 14 curves of model BLSTM. The best one with AUC value 0.9242 is achieved on location prediction of subcellular Lysosome, while prediction of subcellular Peroxisome is the worst one with AUC value 0.8480. The average AUC value of this model is 0.8841. The plot shows that the curves of predictions on 13 subcellular locations, except subcellular Peroxisome, are centered together. **b** Validation of model BLSTM + ConvNet1. With an extra convolutional neural network to extract features from encoded protein sequences, the AUC on subcellular Lysosome and Peroxisome reached to 0.9255 and 0.9134. The average AUC value of this model is 0.9064. All 14 curves are convergent together. This model is the most robust model among all the four models. **c** ROC curves of model ConvNet2. Different from curves in (**b**), those 14 curves are divergent. The max AUC value is 0.8785 and the min AUC value is 0.7058. Since it only extracts features from evolution information, this model performance isn't as good as the other three models. **d** With combination of sequence information and evolution information, model BLSTM + ConvNet1 + ConvNet2 improves the best AUC value to 0.9458. However, predictions of several subcellular locations are effected by the parameters in Convnet2, so the curves of those predictions are divergent from center

method. Since our method considers relations between each subcellular location as a feature, high correlation improves the prediction accuracy of subcellular location, of which the training samples are insufficient. Our method takes advantage of this relation and the accuracy of prediction has significant improvement.

The t test statistic of average precision, ranking loss and coverage is in Additional file 1: Table S5. In order to make t test statistic on prediction result, the prediction task was taken out on sub-group testing samples multiple times. The detailed information about average precision, ranking loss and coverage on dataset *D*3106 and *D*4802 is listed in Additional files 6 and 7 respectively. The data in Additional file 1: Table S5 shows that ConvNet2 has nothing to do with average precision when the method was tested on dataset *D*3106 as the change between average precision values is insignificant ($p_{value} > 0.5$). The differences of ranking loss and coverage are significant. With ConvNet2, our method gets better ranking loss and coverage on dataset *D*3106 with significant decrease ($p_{value} < 0.5$). However, on dataset *D*4802, ConvNet2 makes the situation worse. The reason caused this situation could be the limitation of our method on datasets which has less multi-cellular samples. From Additional file 1: Figure S3 we can see that dataset *D*4802 has more 2-cellular samples than dataset *D*3106 while 3-cellular samples and 4-cellular samples in *D*4802 are less than the ones in *D*3106. The percentages of 3-cellular samples and 4-cellular samples in dataset *D*3106 are 8.17% and 0.57% respectively, while the percentages of those kind of samples are 2.96% and 0.22% in dataset *D*4802. Since *D*4802 covers total of 33 subcellular locations and *D*3106 covers only 14 subcellular locations, the cellular position density in *D*4802 is lower than the one in *D*3106. Then ConvNet2 reduces ranking loss and coverage when the dataset contains more multi-cellular samples and the density of cellular coverage is higher. However, this conclusion needs more evaluation of those models on different datasets, which could be done in future researches.

We compared our method with five currently available methods which had been tested on the benchmark datasets D3106 and D4802. The evaluation results are listed in Table 5. The results in this table show that our method performs better on protein subcellular localization prediction than the other five methods. Our method gets average precision 0.7901 which is 0.08 higher than the average precision of method FSVM-KNR on dataset D3106, and both the ranking loss and coverage are also better than method FSVM-KNR. Only the coverage of our method on D4802 is worse than the coverage of FSVM-KNR. The t test statistic results of average precision are listed in Additional file 1: Table S6. The results in this table show that our method has higher average precision

Table 5 The average precision, ranking loss and coverage of IMMMLGP, Hum-mPloc, mGOF-loc, MKSVM, FSVM-KNR and out method when they are tested on datasets *D*3106 and *D*4802

	D3106			*D4802*		
	RL	*Cov*	*AP*	*RL*	*Cov*	*AP*
IMMMLGP	0.4190	4.3030	0.5810	0.2436	4.9772	0.5725
Hum-mPloc	0.4906	5.3170	0.5790	0.3145	5.6830	0.5644
mGOF-loc	–	–	–	**0.0606**	3.0227	0.6482
MKSVM	0.1085	1.7193	0.7065	0.0662	2.9753	0.6889
FSVM-KNR	0.1071	1.7025	0.7108	0.0971	**2.6339**	0.6916
Our Method	**0.0758**	**1.2848**	**0.7901**	0.0637	3.0528	**0.7414**

Our method has great improvements on subcellular localization prediction than five currently available methods when tested on dataset *D*3106. The average precision of our method is 0.7901 which is 0.08 greater than the average precision of FSVM-KNR. The ranking loss and coverage of our method are lower than the values of other five methods. On dataset *D*4802, our method did not get the lowest ranking loss and coverage. However, the average precision of our method on dataset *D*4802 is the highest among those six methods with 0.7414. The best values are listed out with bold text

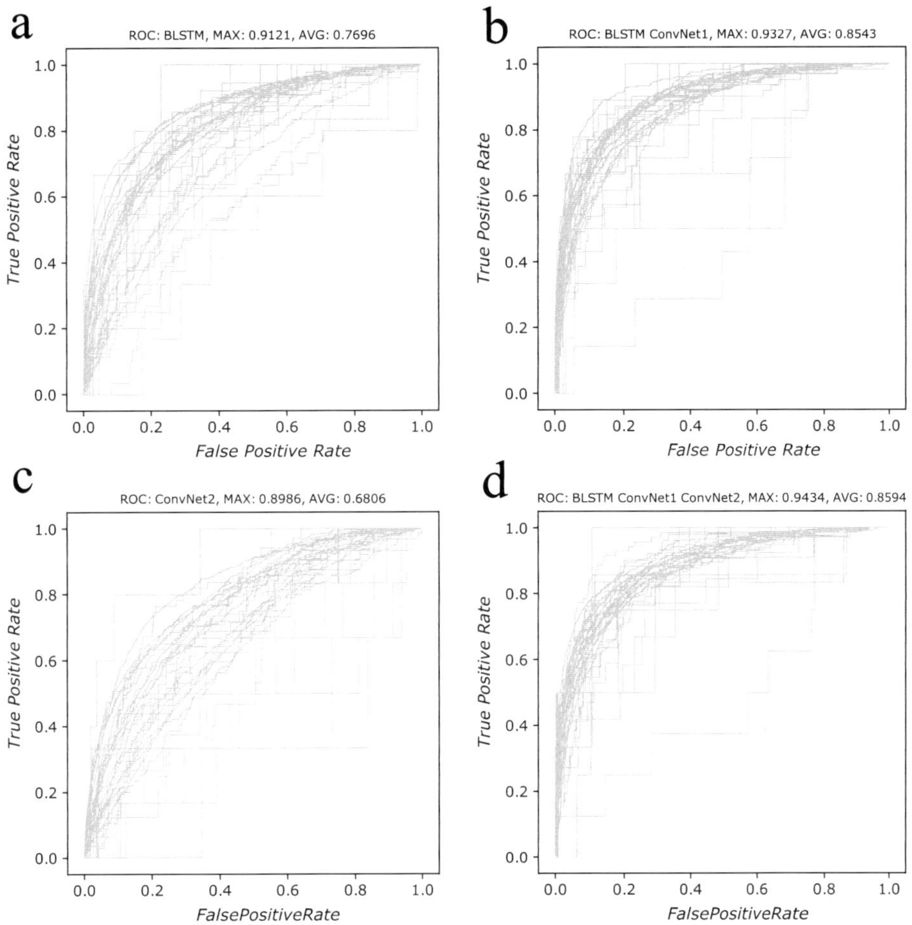

Fig. 3 ROC curves (Receiver operating characteristic curve) of model BLSTM, BLSTM + ConvNet1, ConvNet2 and BLSTM + ConvNet1 + ConvNet2, on benchmark dataset *D*4802. This figure plots one fold validation result. Detailed information about the fivefold cross validation results can be found at Additional file 1: Table S4 and Figure S2. Those plots shows that those four models perform worse on dataset *D*4802 than on dataset *D*3106. **a** 33 curves of model BLSTM. The best AUC is 0.9121, however the worst one is 0.7072. The average AUC value of this model is 0.7696. The plot shows that a small group of curves are centered together, while other curves are divergent from one another. **b** 33 curves of model BLSTM + ConvNet1. Two curves around the diagonal show that the accuracy of prediction on two subcellular positions is bad. The average AUC value of this model is 0.8543 which is better than the average value of model BLSTM. **c** ROC curves of model ConvNet2. This plot is similar to the (**c**) in Figure2 but with several curves under diagonal. So the prediction on some subcellular locations isn't better than random guess. The max AUC value is 0.8986 and the min value is 0.5080. The average AUC value is 0.6806. **d** In this plot, there is still one curve that is under diagonal. However, the best AUC value reaches to 0.9434 and the average AUC value reaches to 08594. In general this model performs better than the other three models

than the other four prediction methods on dataset *D*3106 and *D*4802. The improvements on average precision by our method are significant.

Additional file 1: Table S7 is the t test statistic results of ranking loss. On dataset *D*3106, our method is better than the other four methods with great significance. Model BLSTM plus ConvNet1 is also better than the other four methods. On dataset *D*4802, our method has no big advantage than other four methods even it performs better than methods IMMMLGP, Hum-mPloc and FSVM-KNR. Our method only got similar ranking loss values with method MKSVM since the difference between them is insignificant ($p_{value} > 0.5$).

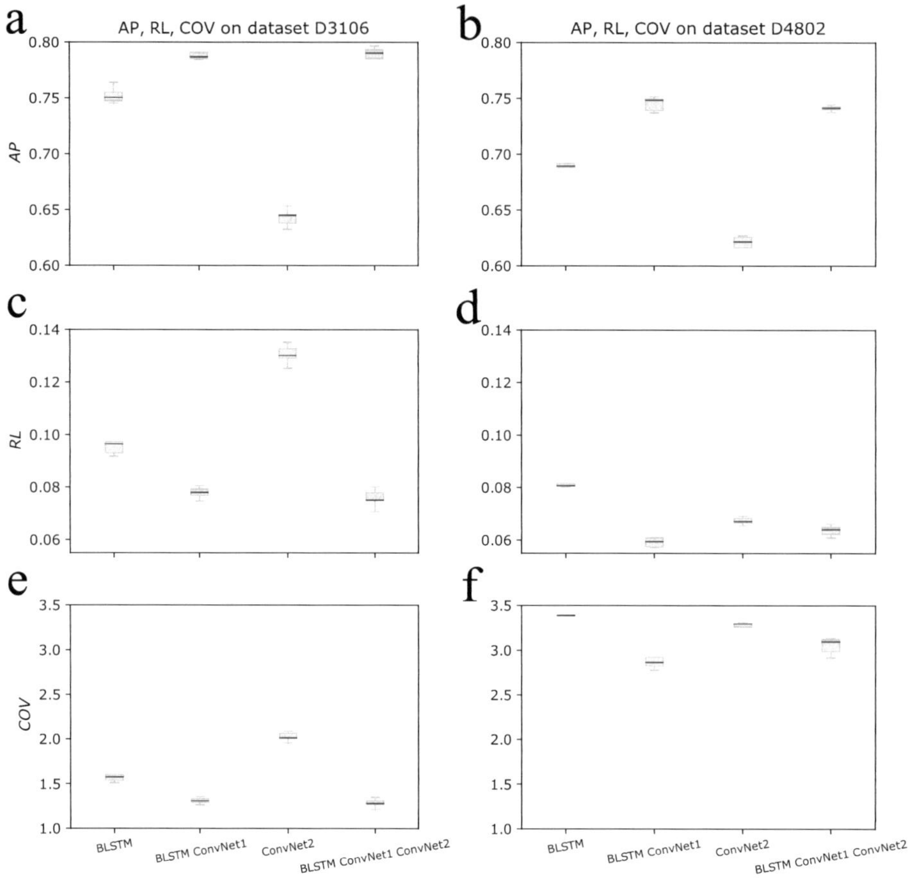

Fig. 4 Box chart of prediction accuracy on benchmark datasets *D*3106 and *D*4802. The three plots in left column are box charts of average precision (AP), ranking loss (RL), coverage (COV) on dataset *D*3106, and the plots in right column are box charts of corresponding results on dataset *D*4802. Model BLSTM + ConvNet1 + ConvNet2 is the best one among the four models with highest average precision and smallest divergence on dataset *D*4802. However, the ranking loss and coverage of this model is greater than the ones of model BLSTM + ConvNet1. On dataset *D*3106, model BLSTM + ConvNet1 + ConvNet2 is better than the other three models in average precision, ranking loss and coverage. In average precision and coverage, those four models perform better on dataset *D*3106 than on dataset *D*4802. The ranking loss values of those four models on dataset *D*4802 are lower than the values on dataset *D*3106. **a** This plot shows the average precisions of model BLSTM, model BLSTM + ConvNet1, model ConvNet2 and model BLSTM + ConvNet1 + ConvNet2 when they are tested on dataset *D*3106 with five-fold cross validation. Model BLSTM + ConvNet1 + ConvNet2 has the greatest average precision than the other three models, while model BLSTM + ConvNet1 has stable performance. **b** The average precisions of those four models when they are tested on dataset **D**4802. Model BLSTM + ConvNet1 has the best performance on this dataset, however model BLSTM + ConvNet1 + ConvNet2 is more stable on this dataset. **c** The ranking loss values of those four models when they are tested on dataset *D*3106. **d** The ranking loss of them with dataset *D*4802. **e** The coverage results of predictions by model BLSTM, model BLSTM + ConvNet1, model ConvNet2 and model BLSTM + ConvNet1 + ConvNet2 on dataset *D*3106. **f** The coverage results of predictions by those four models on dataset *D*4802

When compared with method mGOF-loc, our method is worse than mGOF-loc. The ranking loss of our method increases to 0.0637 with a difference of 0.0031 from ranking loss of mGOF-loc and t test shows that this difference is significant ($p_{value} < 0.05$). The t test statistic results of coverage, which are listed in Additional file 1: Table S8, deliver the same conclusion as the statistic results of ranking loss. Our method outperforms methods IMMMLGP, Hum-mPloc, MKSVM and FSVM-KNR on dataset *D*3106 with significant decrease in coverage values ($p_{value} < 0.05$). However our method is only better than methods IMMMLGP, Hum-mPloc on dataset *D*4802. It has similar coverage as methods MKSVM and mGOF-loc with insignificant differences ($p_{value} > 0.05$), but it is worse than method FSVM-KNR by an increase of 0.4 on coverage. So, from those results, we can see that our method performs well on *D*3106 while it does not have good performance on *D*4802, which means it has limitations on some kinds of datasets, such as datasets have few multi-cellular samples and datasets have low cellular density. Previous research shows that almost half of human proteins localize to multiple subcellular locations [73], so our method has potential to perform well at real biological problems.

Conclusion

The experiment results in the previous section show that bidirectional LSTM and ConvNet are able to accurately predict protein subcellular localization with information of protein sequences and evolution matrices. By comparing our method with five currently available methods, it is obvious that our method outperforms those five. Therefore, the method we proposed is excellent in protein subcellular localization prediction. With the prediction results of our method, the locations and functions of proteins can be refined to remove bias or errors existed in traditional methods. Also, predictions at different stage can find out the changes of protein locations, which can be used to make biological pathway analysis. At the same time, we also realize that our method performs worse on some datasets due to the composition of training samples. For example, the performance on dataset *D*4802 is not as good as the performance on dataset *D*3106 because the unbalance property of samples and the high correlations between subcellular locations.

Furthermore, our experiments show that evolution information alone is insufficient to classify the proteins into groups, and amino acid composition of proteins is crucial in the prediction of protein subcellular localization. However, it is remain reasonable to apply evolution information in biological prediction problems as an additional information, since our method which applies evolution information achieves small improvements on prediction accuracy of multi-cellular localization, especially when a dataset has a large proportion of multi-cellular samples. Besides PSSM, there are also other forms of evolution information such as mutation rate, mutation bias, phylogenetic tree, and so on. Those kinds of information are also helpful to biological predictions.

Sequence information encoding is very important to the success of biological prediction methods. The performance of model BLSTM shows that protein sequence encoding method is essential in deep learning-based protein subcellular localization prediction methods. There is an increasing number of deep learning models that are being proposed to process sequence data, which are beneficial to protein classification. However, a fine encoding method is pivotal in translating raw protein sequences into learn-able matrices in order to utilize the proposed models in bioinformatics research.

Our method can not only predict subcellular locations of proteins, but also can predict other properties of proteins based on protein sequence and evolution information. For example, our method can predict protein functions besides locations. Novel application such as protein structure prediction can be developed based on the architecture of our model. However, this needs further work to make it work.

Abbreviations
AP: Average precision; AUC: Area under the ROC curve; BLAST: Basic local alignment search tool; BLSTM: Bidirectional long short-term memory network; ConvNet: Convolutional neural network; Cov: Coverage; FSVM: Fuzzy support vector machine; GO: Gene ontology; GP: Gaussian process; KNR: Kernelized neighborhood representation; LASSO: Least absolute shrinkage and selection operator; lnc-RNA: Long non-coding RNA; LSTM: Long short-term memory network; MCC: Matthews correlation coefficient; PPV: Positive predictive value; PsePP: Pseudo-physicochemical properties; PsePSSM: Pseudo-position specific scoring matrix; PSI-BLAST: Position specific iterated BLAST; PSSM: Position specific scoring matrix; ReLU: Rectified linear activation function; RNN: Recurrent neural network; RL: Ranking loss; ROC: Receiver operating characteristic curve; SVM: Support vector machine; TPR: True positive rate.

Supplementary Information
The online version contains supplementary material available at https://doi.org/10.1186/s12859-021-04404-0.

Additional file 1. This file contains detailed information about the performance of this method when it is trained and tested on dataset D3106 and D4802.

Additional file 2. This file contains additional information about dataset D3106, correlation between each pair of subcellular locations in dataset D3106.

Additional file 3. This file contains additional information about dataset D4802, correlation between each pair of subcellular locations in this dataset.

Additional file 4. This file lists the AUC values of this method on dataset D3106.

Additional file 5. This file lists the AUC values of this method on dataset D4802.

Additional file 6. The average precision and ranking loss of this mothod when tested on dataset D3106.

Additional file 7. The average precision and ranking loss of this method when tested on dataset D4802.

Acknowledgements
This work is supported by a grant from National Key R&D Program of China (2020YFA0908400), the Shenzhen KQTD Project (KQTD20200820113106007), the National Natural Science Foundation of China (No. 62072107), and the Natural Science Foundation of Fujian Province of China (No. 2020J01610)

About this supplement
This article has been published as part of BMC Bioinformatics Volume 22 Supplement 10 2021: Selected articles from the 19th Asia Pacific Bioinformatics Conference (APBC 2021): bioinformatics. The full contents of the supplement are available at https://bmcbioinformatics.biomedcentral.com/articles/supplements/volume-22-supplement-10.

Authors' contributions
ZL and JT conceived of the presented idea. ZL analyzed the data. GP developed the theory. GP and CS performed the computations and wrote the manuscript. JT supervised the research. All authors read and approved the final manuscript.

Funding
National Key R&D Program of China (2020YFA0908400) Shenzhen KQTD Project (KQTD20200820113106007) National Natural Science Foundation of China (No. 62072107) Natural Science Foundation of Fujian Province of China (No. 2020J01610)

Availability of data and materials
The datasets used and analysed during the current study are available in the figshare repository, https://doi.org/10.6084/m9.figshare.14378630.v1. The python code described in this manuscript are available in the github reposiroty, https://github.com/Paeans/subcellular.

Declarations

Ethics approval and consent to participate
Not applicable.

Consent to publish
Not applicable.

Competing interests
The authors declare that they have no competing interests.

Author details
[1]Department of Biochemistry and Molecular Biology, School of Basic Medical Sciences, Fujian Medical University, 1 Xuefu North Road, University Town, Fuzhou 350122, FJ, China. [2]Department of Computer Science and Engineering, University of South Carolina, 550 Assembly St, Columbia, SC 29208, USA. [3]College of Electrical and Power Engineering, Taiyuan University of Technology, No. 79 Yinze West Street, Taiyuan 030024, SX, China.

Received: 20 September 2021 Accepted: 24 September 2021
Published online: 22 October 2021

References

1. Gardy JL, Brinkman FS. Methods for predicting bacterial protein subcellular localization. Nat Rev Microbiol. 2006;4(10):741–51.
2. Karp G. Cell and molecular biology: concepts and experiments. Hoboken: Wiley; 2009.
3. Nakai K, Kanehisa M. A knowledge base for predicting protein localization sites in eukaryotic cells. Genomics. 1992;14(4):897–911.
4. Tsien RY. The green fluorescent protein. Annu Rev Biochem. 1998;67(1):509–44.
5. Rey S, Gardy JL, Brinkman FS. Assessing the precision of high-throughput computational and laboratory approaches for the genome-wide identification of protein subcellular localization in bacteria. BMC Genomics. 2005;6(1):162.
6. Shen Y, Ding Y, Tang J, Zou Q, Guo F. Critical evaluation of web-based prediction tools for human protein subcellular localization. Brief Bioinform. 2020;21(5):1628–40.
7. Cortes C, Vapnik V. Support-vector networks. Mach Learn. 1995;20(3):273–97.
8. Li H, Tian S, Li Y, Fang Q, Tan R, Pan Y, Huang C, Xu Y, Gao X. Modern deep learning in bioinformatics. J Mol Cell Biol. 2020;12:823–7.
9. Shen Y, Tang J, Guo F. Identification of protein subcellular localization via integrating evolutionary and physicochemical information into Chou's general Pseaac. J Theor Biol. 2019;462:230–9.
10. Wu Q, Ying Y, Zhou D-X. Multi-kernel regularized classifiers. J Complex. 2007;23(1):108–34.
11. Chou K-C. Advances in predicting subcellular localization of multi-label proteins and its implication for developing multi-target drugs. Curr Med Chem. 2019;26(26):4918–43.
12. Lin C-F, Wang S-D. Fuzzy support vector machines. IEEE Trans Neural Netw. 2002;13(2):464–71.
13. Abe S, Inoue T. Fuzzy support vector machines for multiclass problems. In: ESANN; 2002.
14. Ding Y, Tang J, Guo F. Protein crystallization identification via fuzzy model on linear neighborhood representation. IEEE/ACM Trans Comput Biol Bioinform. 2019. https://doi.org/10.1109/TCBB.2019.2954826.
15. Ding Y, Tang J, Guo F. Human protein subcellular localization identification via fuzzy model on kernelized neighborhood representation. Appl Soft Comput. 2020;96:106596.
16. He J, Gu H, Liu W. Imbalanced multi-modal multi-label learning for subcellular localization prediction of human proteins with both single and multiple sites. PLoS ONE. 2012;7(6):37155.
17. Lawrence ND. Gaussian process latent variable models for visualisation of high dimensional data. In: Advances in neural information processing systems; 2004. p. 329–336.
18. Casale FP, Dalca A, Saglietti L, Listgarten J, Fusi N. Gaussian process prior variational autoencoders. In: Advances in neural information processing systems; 2018. pp. 10369–80.
19. Tsoumakas G, Spyromitros-Xioufis E, Vilcek J, Vlahavas I. Mulan: a java library for multi-label learning. J Mach Learn Res. 2011;12:2411–4.
20. Wei L, Liao M, Gao X, Wang J, Lin W. mgof-loc: a novel ensemble learning method for human protein subcellular localization prediction. Neurocomputing. 2016;217:73–82.
21. Altschul SF, Gish W, Miller W, Myers EW, Lipman DJ. Basic local alignment search tool. J Mol Biol. 1990;215(3):403–10.
22. Hans C. Bayesian lasso regression. Biometrika. 2009;96(4):835–45.
23. Wan S, Mak M-W, Kung S-Y. mlasso-hum: a lasso-based interpretable human-protein subcellular localization predictor. J Theor Biol. 2015;382:223–34.
24. Shen H-B, Chou K-C. Hum-mploc: an ensemble classifier for large-scale human protein subcellular location prediction by incorporating samples with multiple sites. Biochem Biophys Res Commun. 2007;355(4):1006–11.
25. Shen H-B, Chou K-C. A top-down approach to enhance the power of predicting human protein subcellular localization: Hum-mploc 2.0. Anal Biochem. 2009;394(2):269–74.
26. Zhou H, Yang Y, Shen H-B. Hum-mploc 3.0: prediction enhancement of human protein subcellular localization through modeling the hidden correlations of gene ontology and functional domain features. Bioinformatics. 2017;33(6):843–53.
27. Chou K-C, Wu Z-C, Xiao X. iloc-hum: using the accumulation-label scale to predict subcellular locations of human proteins with both single and multiple sites. Mol BioSyst. 2012;8(2):629–41.
28. Cheng X, Lin W-Z, Xiao X, Chou K-C. ploc_bal-manimal: predict subcellular localization of animal proteins by balancing training dataset and pseaac. Bioinformatics. 2019;35(3):398–406.
29. Cheng X, Xiao X, Chou K-C. ploc-mgneg: predict subcellular localization of gram-negative bacterial proteins by deep gene ontology learning via general pseaac. Genomics. 2018;110(4):231–9.
30. LeCun Y, Bengio Y, Hinton G. Deep learning. Nature. 2015;521(7553):436–44.
31. Almagro Armenteros JJ, Sønderby CK, Sønderby SK, Nielsen H, Winther O. Deeploc: prediction of protein subcellular localization using deep learning. Bioinformatics. 2017;33(21):3387–95.
32. Shao Y-T, Liu X-X, Lu Z, Chou K-C. ploc_deep-mhum: predict subcellular localization of human proteins by deep learning. Nat Sci. 2020;12(7):526–51.
33. Shao Y, Chou K-C, et al. ploc_deep-meuk: predict subcellular localization of eukaryotic proteins by deep learning. Nat Sci. 2020;12(06):400.
34. Shao Y, Chou K-C. ploc_deep-mvirus: a CNN model for predicting subcellular localization of virus proteins by deep learning. Nat Sci. 2020;12(6):388–99.
35. Chou K-C. Prediction of protein cellular attributes using pseudo-amino acid composition. Proteins Struct Funct Bioinform. 2001;43(3):246–55.
36. Wei L, Ding Y, Su R, Tang J, Zou Q. Prediction of human protein subcellular localization using deep learning. J Parallel Distrib Comput. 2018;117:212–7.
37. Lin C, Zou Y, Qin J, Liu X, Jiang Y, Ke C, Zou Q. Hierarchical classification of protein folds using a novel ensemble classifier. PLoS ONE. 2013;8(2):56499.
38. Long W, Yang Y, Shen H-B. Imploc: a multi-instance deep learning model for the prediction of protein subcellular localization based on immunohistochemistry images. Bioinformatics. 2020;36(7):2244–50.
39. Pärnamaa T, Parts L. Accurate classification of protein subcellular localization from high-throughput microscopy images using deep learning. G3 Genes Genomes Genet. 2017;7(5):1385–92.

40. Masurkar SR, Rege PP. Human protein subcellular localization using convolutional neural network as feature extractor. In: 2019 10th international conference on computing, communication and networking technologies (ICCCNT). IEEE; 2019. pp. 1–7.
41. Cao Z, Pan X, Yang Y, Huang Y, Shen H-B. The lnclocator: a subcellular localization predictor for long non-coding RNAs based on a stacked ensemble classifier. Bioinformatics. 2018;34(13):2185–94.
42. Yan Z, Lécuyer E, Blanchette M. Prediction of mRNA subcellular localization using deep recurrent neural networks. Bioinformatics. 2019;35(14):333–42.
43. Gudenas BL, Wang L. Prediction of lncRNA subcellular localization with deep learning from sequence features. Sci Rep. 2018;8(1):1–10.
44. Okada S, Ohzeki M, Taguchi S. Efficient partition of integer optimization problems with one-hot encoding. Sci Rep. 2019;9(1):1–12.
45. Li J, Si Y, Xu T, Jiang S. Deep convolutional neural network based ECG classification system using information fusion and one-hot encoding techniques. Math Problems Eng. 2018;7354081;1–10.
46. Stormo GD, Schneider TD, Gold L, Ehrenfeucht A. Use of the 'perceptron' algorithm to distinguish translational initiation sites in *E. coli*. Nucl Acids Res. 1982;10(9):2997–3011.
47. Stormo GD. DNA binding sites: representation and discovery. Bioinformatics. 2000;16(1):16–23.
48. Hochreiter S, Schmidhuber J. Long short-term memory. Neural Comput. 1997;9(8):1735–80.
49. Cho K, Van Merriënboer B, Gulcehre C, Bahdanau D, Bougares F, Schwenk H, Bengio Y. Learning phrase representations using RNN encoder–decoder for statistical machine translation. arXiv preprint arXiv:1406.1078. 2014.
50. Schuster M, Paliwal KK. Bidirectional recurrent neural networks. IEEE Trans Signal Process. 1997;45(11):2673–81.
51. Graves A, Jaitly N, Mohamed A. Hybrid speech recognition with deep bidirectional LSTM. In: 2013 IEEE workshop on automatic speech recognition and understanding. IEEE; 2013; pp. 273–8.
52. Collobert R, Weston J. A unified architecture for natural language processing: deep neural networks with multitask learning. In: Proceedings of the 25th international conference on machine learning, 2008. p. 160–7.
53. Krizhevsky A, Sutskever I, Hinton GE. Imagenet classification with deep convolutional neural networks. In: Advances in neural information processing systems; 2012. p. 1097–105.
54. Xu L, Ren JS, Liu C, Jia J. Deep convolutional neural network for image deconvolution. In: Advances in neural information processing systems; 2014. p. 1790–8.
55. Lin X, Zhao C, Pan W. Towards accurate binary convolutional neural network. In: Advances in neural information processing systems; 2017. p. 345–353.
56. Zhou J, Troyanskaya OG. Predicting effects of noncoding variants with deep learning-based sequence model. Nat Methods. 2015;12(10):931–4.
57. Angermueller C, Lee HJ, Reik W, Stegle O. Deepcpg: accurate prediction of single-cell DNA methylation states using deep learning. Genome Biol. 2017;18(1):1–13.
58. Sun T, Zhou B, Lai L, Pei J. Sequence-based prediction of protein protein interaction using a deep-learning algorithm. BMC Bioinform. 2017;18(1):1–8.
59. Zhang Y, An L, Xu J, Zhang B, Zheng WJ, Hu M, Tang J, Yue F. Enhancing hi-c data resolution with deep convolutional neural network hicplus. Nat Commun. 2018;9(1):1–9.
60. Géron A. Hands-on machine learning with Scikit-learn, keras, and tensorflow: concepts, tools, and techniques to build intelligent systems. California: O'Reilly Media; 2019.
61. Hodgkin AL, Huxley AF. A quantitative description of membrane current and its application to conduction and excitation in nerve. J Physiol. 1952;117(4):500.
62. Hahnloser RH, Sarpeshkar R, Mahowald MA, Douglas RJ, Seung HS. Digital selection and analogue amplification coexist in a cortex-inspired silicon circuit. Nature. 2000;405(6789):947–51.
63. LeCun YA, Bottou L, Orr GB, Müller K-R. Efficient backprop. In: Neural networks: tricks of the trade. Berlin: Springer; 2012. p. 9–48.
64. Han J, Moraga C. The influence of the sigmoid function parameters on the speed of backpropagation learning. In: International workshop on artificial neural networks. Springer; 1995. p. 195–201.
65. Murphy KP. Machine learning: a probabilistic perspective. Cambridge, MA: MIT Press; 2012.
66. Bishop CM. Pattern recognition and machine learning. New York: Springer; 2006.
67. Zaremba W, Sutskever I, Vinyals O. Recurrent neural network regularization. arXiv preprint arXiv:1409.2329. 2014.
68. Tsoumakas G, Katakis I, Vlahavas I. Mining multi-label data. In: Data mining and knowledge discovery handbook. Berlin: Springer; 2009. p. 667–85.
69. Zhang M-L, Zhou Z-H. A review on multi-label learning algorithms. IEEE Trans Knowl Data Eng. 2013;26(8):1819–37.
70. Powers DM. Evaluation: from precision, recall and f-measure to ROC, informedness, markedness and correlation; 2011.
71. Baldi P, Brunak S, Chauvin Y, Andersen CA, Nielsen H. Assessing the accuracy of prediction algorithms for classification: an overview. Bioinformatics. 2000;16(5):412–24.
72. Fawcett T. An introduction to ROC analysis. Pattern Recogn Lett. 2006;27(8):861–74.
73. Thul PJ, Åkesson L, Wiking M, Mahdessian D, Geladaki A, Blal HA, Alm T, Asplund A, Björk L, Breckels LM, et al. A subcellular map of the human proteome. Science. 2017;356:6340.

Publisher's Note

Wu *et al. BMC Bioinformatics* (2021) 22:613
https://doi.org/10.1186/s12859-022-04561-w

BMC Bioinformatics

METHODOLOGY

Open Access

GRAde: a long-read sequencing approach to efficiently identifying the *CYP11B1/CYP11B2* chimeric form in patients with glucocorticoid-remediable aldosteronism

Yu-Ching Wu[1], Chia-I Chen[1], Peng-Ying Chen[2], Chun-Hung Kuo[1], Yi-Hsuan Hung[1], Kang-Yung Peng[2], Vin-Cent Wu[2], Jyy-Jih Tsai-Wu[1*], Chia-Lang Hsu[1,3,4*] and TAIPAI group[5]

From The 19th Asia Pacific Bioinformatics Conference (APBC 2021) Tainan, Taiwan. 3-5 February 2021

*Correspondence:
jjtsai@ntuh.gov.tw;
chialanghsu@ntuh.gov.tw
[1] Department of Medical Research, National Taiwan University Hospital, Taipei, Taiwan
Full list of author information is available at the end of the article

Abstract

Background: Glucocorticoid-remediable aldosteronism (GRA) is a form of heritable hypertension caused by a chimeric fusion resulting from unequal crossing over between 11β-hydroxylase (*CYP11B1*) and aldosterone synthase (*CYP11B2*), which are two genes with similar sequences. Different crossover patterns of the *CYP11B1* and *CYP11B2* chimeric genes may be associated with a variety of clinical presentations. It is therefore necessary to develop an efficient approach for identifying the differences between the hybrid genes of a patient with GRA.

Results: We developed a long-read analysis pipeline named GRAde (GRA deciphering), which utilizes the nonidentical bases in the *CYP11B1* and *CYP11B2* genomic sequences to identify and visualize the chimeric form. We sequenced the polymerase chain reaction (PCR) products of the *CYP11B1/CYP11B2* chimeric gene from 36 patients with GRA using the Nanopore MinION device and analyzed the sequences using GRAde. Crossover events were identified for 30 out of the 36 samples. The crossover sites appeared in the region exhibiting high sequence similarity between *CYP11B1* and *CYP11B2*, and 53.3% of the cases were identified as having a gene conversion in intron 2. More importantly, there were six cases for whom the PCR products indicated a chimeric gene, but the GRAde results revealed no crossover pattern. The crossover regions were further verified by Sanger sequencing analysis.

Conclusions: PCR-based target enrichment followed by long-read sequencing is an efficient and precise approach to dissecting complex genomic regions, such as those involved in GRA mutations, which could be directly applied to clinical diagnosis. The scripts of GRAde are available at https://github.com/hsu-binfo/GRAde.

Keywords: Glucocorticoid-remediable aldosteronism, Chimeric genes, Long-read sequencing

Background

Primary aldosteronism (PA) is the most common and curable form of secondary arterial hypertension. Most diagnosed cases of PA are mainly caused by either aldosterone overproduction by both adrenal glands or unilateral aldosterone-producing adenomas (APA) [1]. However, about 5% of cases are inherited forms of familial hyperaldosteronism (FH) [2]. According to current understanding, there are three well-established forms of FH (FH-I–III), and some germline mutations associated with PA, such as CACN1H and CACN1D, have been identified [3]. However, data from genetic analyses reveal a more complex situation, and more heritable forms of PA may still be undiscovered [2].

Familial hyperaldosteronism type I (FH1), also called glucocorticoid-remediable aldosteronism (GRA), is transmitted as an autosomal-dominant disorder and accounts for 0.5–1.0% of PA cases [4]. GRA is caused by a chimeric gene resulting from a nonhomologous crossing-over event on chromosome 8q24.3, between the 11β-hydroxylase (*CYP11B1*) and aldosterone synthase (*CYP11B2*) genes. This chimeric enzyme contains the promoter of *CYP11B1* at the 5′ end and the coding sequences from *CYP11B2* at the 3′ end. It can therefore synthesize aldosterone under adrenocorticotropic hormone (ACTH) control (Fig. 1A).

Although GRA is a genetic disease, the clinical and biochemical characteristics of patients are highly variable, and even patients from the same familial type may present different symptoms [5, 6]. In addition, different crossover patterns of the chimeric *CYP11B1*/*CYP11B2* gene within a familial type have been described [7]. These findings

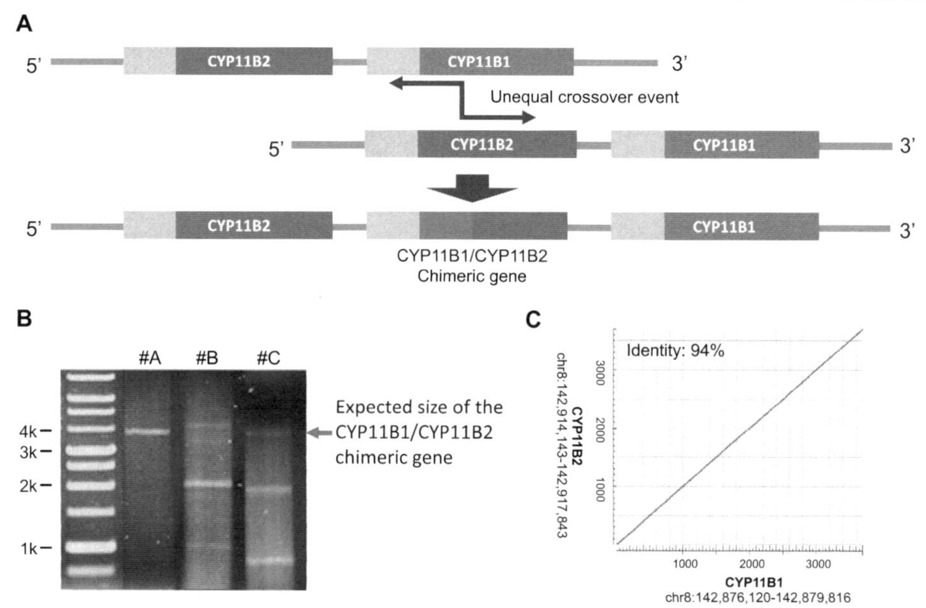

Fig. 1 The challenge of *CYP11B1/CYP11B2* chimeric form identification. **A** Scheme of the *CYP11B1/CYP11B2* chimeric gene. **B** Long-range PCR shows the PCR products of the *CYP11B1/CYP11B2* chimeric gene and unexpected fragments. Amplification of a chimeric gene is expected to produce a 3.9 kb product. Case A has a clear single band of the expected size of the PCR product, but cases **B** and **C** have multiple bands with weak signals of the expected size of the main product. **C** Dot plot revealing the high degree of similarity between the *CYP11B1* and *CYP11B2* genomic sequences. The plot was generated by blast2seq using default parameters

suggest that variability in clinical presentations might be related to heterogeneity in the hereditary factor—in particular, in the crossover pattern of the hybrid gene. Currently, the genetic diagnosis of GRA is made using Southern blotting or long-range polymerase chain reaction (PCR) techniques [8–12]. However, these methods are unable to precisely determine the crossover pattern. Therefore, it is crucial to develop a more precise method for identifying the specific type of hybrid gene that is carried by a patient.

The detection of pathogenic gene fusion in inherited diseases and oncology is particularly useful. In general, fusion events cause a loss or gain of function in one of the fused partners. However, unlike with oncogenic fusion genes, it is relatively difficult to detect gene conversion in genes with highly similar sequences. New detection strategies for this kind of gene fusion are urgently required to facilitate diagnostic and therapeutic decisions.

Since the introduction of next-generation sequencing (NGS), tremendous progress has been achieved in all fields of biology. The declining costs and growing availability of NGS have made it the method of choice for genetic analysis and related applications. Nevertheless, NGS suffers from a limited availability of data on several aspects, such as repetitive elements, polymorphic regions, camouflaged genes, and large structural variations (SVs), which prevents full extraction of information associated with the genome [13, 14]. However, recently developed single-molecule sequencing techniques such as single-molecule real-time sequencing (SMRT) and nanopore sequencing provide access to larger variations, because the read lengths are typically several thousands of bases [15, 16]. These techniques provide the opportunity to investigate or diagnose diseases caused by pathogenic structural variations and gene conversion, such as GRA. In this study, we used Oxford nanopore technology (ONT) to sequence the PCR products of chimeric *CYP11B1*/*CYP11B2* genes and developed a long-read analysis pipeline that can efficiently identify and visualize the chimeric forms.

Results

The challenge of GRA chimeric form identification

Previous studies have suggested that a variety of *CYP11B1*/*CYP11B2* chimeric forms are associated with different clinical presentations [5, 6]. Long-range PCR is one of the standard approaches to GRA diagnosis. However, in a few of GRA cases, multiple bands with weak signals for the expected PCR products are obtained (Fig. 1B), making diagnosis difficult. To avoid misdiagnosis of patients with GRA, it is crucial to identify the exact crossover site. The current method for identifying crossover sites is multiplex PCR with Sanger sequencing [12]. However, the genomic sequences of *CYP11B1* and *CYP11B2* are about 94% identical in the main crossover region, making primer design for the enrichment of regions that are specific to only one of the two genes challenging (Fig. 1C). In addition, this procedure is time-consuming and laborious. To address this issue, we proposed an alternative strategy that combines long-range PCR with nanopore sequencing to confirm the chimeric gene and identify the crossover site, and we developed an analysis pipeline to decipher these long-read sequences.

Overview of GRAde

The analysis workflow of GRAde is illustrated in Fig. 2A. The input for GRAde was FASTQ files, which were generated from the long sequencing reads of PCR products derived from GRA samples. This pipeline consists of two components: quality control for the sequencing reads, and the procedure for determining the crossover regions of the *CYP11B1/CYP11B2* chimeric form. High-quality reads were obtained by correcting them using a nonhybrid approach, Canu [17], and then mapping them to the human reference genome. Only the reads that aligned with the loci of *CYP11B1* and *CYP11B2* were considered for further analysis. To accurately dissect each qualifying read, GRAde used the Smith–Waterman algorithm to align reads to the *CYP11B1* and *CYP11B2* genomic sequences, respectively, and then analyzed the alignment results to identify the crossover region.

Due to the high degree of sequence similarity between the *CYP11B1* and *CYP11B2* genes, we first identified the "discriminating" and "ambiguous" bases by comparing them to the *CYP11B1* (chr8:142876120–142879816) and *CYP11B2* (chr8:142914143–142917843) sequences (Fig. 2B). We then used the discriminating bases to distinguish the sequences from the genomic source (Fig. 2C). There were 180 and 188 discriminating bases for *CYP11B1* and *CYP11B2*, respectively (Additional file 1). The median of the positional distribution of the neighbor discriminating bases was 11 bp, with a range of 1–221 bp. Ideally, the ambiguous bases should perfectly match both genes, but there are typically some mismatches due to errors in the PCR or sequencing processes and genetic polymorphisms (Fig. 2C). Mismatches of ambiguous bases may affect the interpretation,

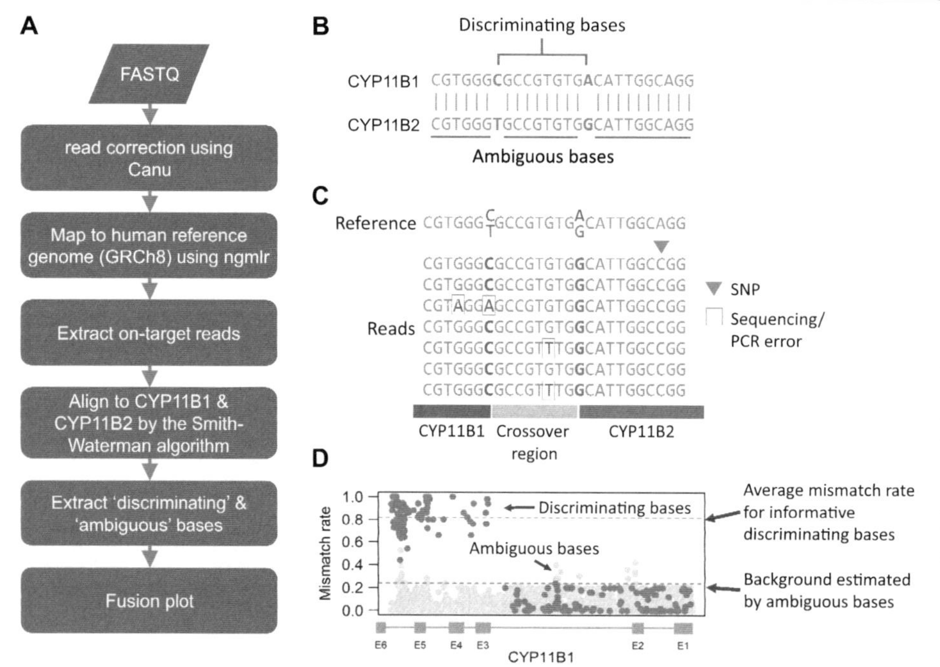

Fig. 2 Overview of GRAde. **A** Analysis workflow of GRAde. **B** The identification of discriminating and ambiguous bases in the *CYP11B1* and *CYP11B2* genomic sequences. **C** The concept of using discriminating and ambiguous bases to recognize the *CYP11B1* and *CYP11B2* genes. **D** An example of a fusion plot based on the *CYP11B1* genomic sequence. Red and grey points represent the discriminating and ambiguous bases

so the ambiguous bases were discarded if polymorphisms were reported at these positions in the single-nucleotide polymorphism (SNP) database (dbSNP). Based on the alignment results of all reads mapped to *CYP11B1* and *CYP11B2*, we calculated the mismatch rates of each discriminating and ambiguous base and visualized them as a fusion plot (Fig. 2D). Because of the relatively high error rate of nanopore sequencing, if the bases aligned to discriminating sites did not exactly match either *CYP11B1* or *CYP11B2*, we considered them to be sequencing or PCR errors and did not include them in the fusion plot. This fusion plot provides the intuitive and interpretable chimeric form for each sample.

Variety of *CYP11B1/CYP11B2* chimeric forms in GRA samples

We collected 36 samples from patients who were diagnosed with GRA based on clinical practice guidelines, which was confirmed by using the long-range PCR technique to reveal the chimeric genes. The chimeric genes were amplified using the long-range PCR technique and then subjected to nanopore sequencing. We also amplified *CYP11B2* genes from six other patients with PA as negative controls. The results of the GRAde analysis of the 36 GRA samples are summarized in Table 1, and the fusion forms are shown in Fig. 3A and Additional file 2. Sixteen of the 36 cases had fusion sites located at intron 2, and the crossover region of the other cases was within the ranges of exon 3–intron 3 (seven cases), exon 4–intron 4 (five cases) and exon 5–intron 5 (two cases). There were no fusion patterns in the fusion plots for

Table 1 GRAde analysis results for GRA samples

Sample no.	Crossover region	Sample no.	Crossover region
#1	E2-I2	#19	E2-I2
#2	E3-I3	#20	E2-I2
#3	E2-I2	#21	E4-I4
#4	E3-I3	#22	No fusion observed
#5	E2-I2	#23	No fusion observed
#6	E3-I3	#24	No fusion observed
#7	E2-I2	#25	E2-I2
#8	E3-I3	#26	E3-I3
#9	E3-I3	#27	E4-I4
#10	E3-I3	#28	E2-I2
#11	E2-I2	#29	No fusion observed
#12	E5-I5	#30	E2-I2
#13	No fusion observed	#31	E5-I5
#14	No fusion observed	#32	E4-I4
#15	E2-I2	#33	E2-I2
#16	E2-I2	#34	E2-I2
#17	E2-I2	#35	E2-I2
#18	E4-I4	#36	E4-I4

E2-I2: located in exon 2 and intron 2

E3-I3: located in exon 3 and intron 3

E4-I4: located in exon 4 and intron 4

E5-I5: located in exon 5 and intron 5

No fusion observed: no clear fusion pattern was observed in the GRAde fusion plot

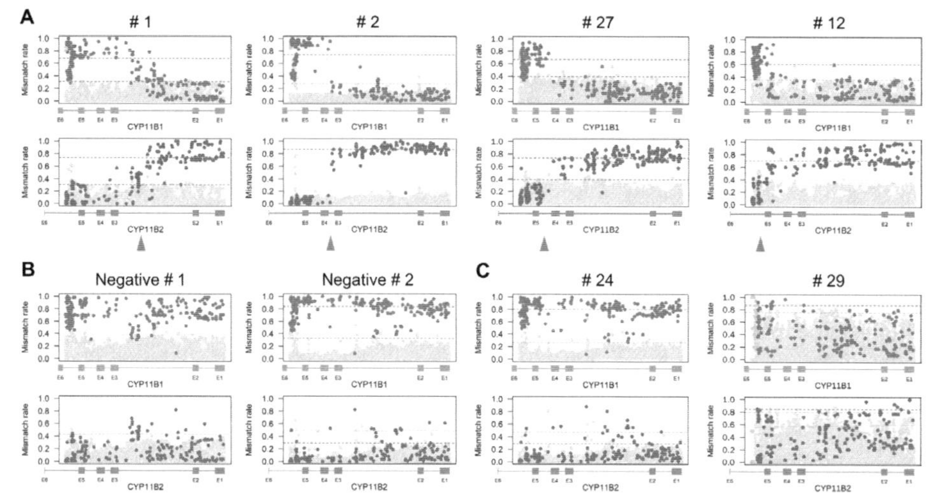

Fig. 3 A variety of cross-over regions could be identified using GRAde. Fusion plots for **A** representative fusion forms, **B** negative controls (normal *CYP11B2* gene), and **C** false-positive cases that were diagnosed with GRA based on PCR but without any fusion observable in the sequences. The pink triangle denotes the possible cross-over region

any of the negative control samples of normal *CYP11B2* genes (Fig. 3B). Notably, there were also six GRA cases for whom no fusion patterns were apparent (Fig. 3C and Additional file 2), suggesting that their diagnoses might be based on false-positive test results.

Validating the crossover site via multiplex PCR with Sanger sequencing

To demonstrate the validity of our approach and GRAde, we selected a case for whom the PCR produced multiple products and the signal for the chimeric gene was relatively weak (Fig. 4A). After sequencing the PCR products using the nanopore technology and analyzing the reads using GRAde, a clear fusion pattern was revealed in the fusion plot, although the background noise was high at the 3′ end of the chimeric gene (Fig. 4B). Using the fusion plot, we designed primers to amplify the identified crossover region and sequenced the amplified products via Sanger sequencing. Nucleotide sequence analysis showed that the gene-conversion site was in the middle of intron 2, which was consistent with the GRAde result.

Runtime and robustness of GRAde

We evaluated the runtime and robustness of GRAde for use as a diagnostic tool. We generated testing sets that were randomly sampled from case No. 34, with a varying number of reads (100, 200, 500, 1000, 2000, or 3000). As shown in Additional file 3, analysis of input with 3000 reads could be completed within six minutes, and the most time-consuming step was performing the Canu algorithm for hybridization-based sequence correction. We were able to achieve stable results using the sample with 200 reads, so we consider that to be the minimum number of reads required for analysis (Additional file 3).

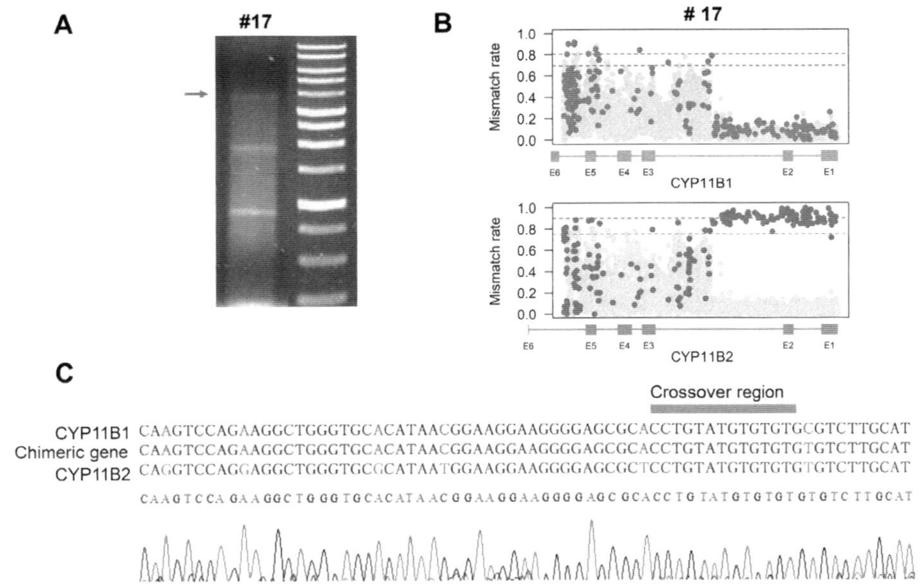

Fig. 4 A case in which the cross-over region was validated by Sanger sequence analysis. **A** Long-range PCR revealed PCR products of the *CYP11B1/CYP11B2* chimeric gene in sample #17. Amplification with a chimeric gene is expected to produce a 3.9 kb product. **B** The fusion plot of sample #17 reveals a possible cross-over region located in intron 2, with high background noise. **C** Sequencing of the chimeric PCR product demonstrated that the crossover site was located in intron 2. The nucleotides that differed between *CYP11B1*, *CYP11B2*, and the chimeric gene are highlighted in different colors

Discussion

The emerging long-read sequencing technologies offer improvements in the characterization of genetic variation and regions that are difficult to assess using short reads. These techniques have been used to investigate genetic disorders with previously known or strongly suspected disease loci [18] and are considered a diagnostic tool. The general strategy of long-read–based diagnosis is to enrich the locus of causality and then perform long-read sequencing. Long-range PCR is a technique currently used for the detection of GRA [8–12], so we chose it as the enrichment method. However, GRA is a special case because the locus of causality is formed by two genes with a high degree of sequence similarity (Fig. 1C). This results in challenging primer design and poor PCR products for evaluation, due to issues like the presence of multiple bands and weak signals for the expected chimeric products (Fig. 1B). Also, the presence of the expected band in the PCR product does not necessarily mean it is a chimeric gene, which can lead to incorrect diagnoses based on false-positive test results (Fig. 3C). Therefore, long-range PCR in addition to long-read sequencing is a more efficient method of chimeric gene detection than traditional Sanger sequencing and may reduce the false-positive rate of traditional PCR testing.

However, PCR-based enrichment still has several limitations. PCR amplification can be influenced by improper PCR conditions [19, 20], thus producing false negatives or incorporating PCR errors. Analysis of sequencing reads from PCR products with multiple bands may also be hampered by a high level of background noise and an unclear fusion pattern (Fig. 3A). To avoid this, the development of alternative target-enrichment methods for GRA chimeric genes is required, such as capture-based [21, 22] and CRISPR-based enrichment methods [23, 24].

In our GRA cohort, there were six patients who had PCR products for sequencing in which we could identify no fusion pattern. The possible reasons for this include poor integrity and low purity of the genomic DNA, which could have resulted from the state of storage and the process of DNA extraction. Other possible factors include poor specificity of the primers, inappropriate DNA input, and nucleic-acid contamination. It is difficult to design extremely specific primers for the crossover region of chimeric genes, so this pipeline can assist in excluding nonspecific PCR products and distinguishing the correct signal.

In addition to the *CYP11B1*/*CYP11B2* chimeric gene, there are some SNPs at the *CYP11B1* and *CYP11B2* loci that are also associated with hypertension [25]. Theoretically, these single nucleotide variants (SNV) can be detected in the current long-read sequence data. To provide precise diagnosis and treatment of GRA, however, it is critical to report all possible variants in genetic testing. Although GRAde was developed for the identification of chimeric forms, therefore, we will incorporate the function of variant-calling to identify SNVs and small insertions and deletions (indels).

Although GRAde is specifically designed for the identification of crossover sites between *CYP11B1* and *CYP11B2* in GRA patients, the analysis strategy used in GRAde could also be applied to other diseases that are caused by dysfunctional proteins due to unequal crossover events. For example, chronic granulomatous disease (CGD) is caused by the chimeric form of *NCF1* (neutrophil cytosolic factor 1) and its pseudogenes *NCF1B* (neutrophil cytosolic factor 1B pseudogene) and *NCF1C* (neutrophil cytosolic factor 1C pseudogene) [26]. Because these two pseudogenes are on either side of *NCF1* and have 99% sequence identity to *NCF1*, distinguishing *NCF1* from its pseudogenes in CGD patients relies on a set of SNPs [27] and an analysis similar in concept to that of GRAde. In addition to CGD, several other diseases are caused by the chimeric products of the crossover between a gene and its pseudogenes, such as congenital adrenal hyperplasia, caused by the chimeric genes *CYP21A1P*/*CYP21A2* [28], and Gaucher disease, caused by a fusion gene formed from *GBA* (glucosylceramidase beta) and its pseudogene *GBAP1* (glucosylceramidase beta pseudogene 1) [29]. GRAde could easily be modified and used for crossover-site identification in other diseases. Besides the detection of specific fusion genes, our strategy could be also applied to genome-wide detection of this kind of gene fusion by systematically identifying discriminating bases in homologous genes.

Conclusions

In this study, we proposed the strategy of combining long-range PCR with long-read sequencing techniques to identify gene conversions, such as the one that causes GRA. This approach is not only more efficient than general multiplex PCR followed by Sanger sequencing, but also reduces the false-positive rate for PCR-based genetic testing. This analysis procedure could be applied to the diagnosis of other diseases caused by unequal crossover between two genes with highly similar sequences.

Methods

Patients

This study was approved by the institutional review board of the National Taiwan University Hospital, Taipei, Taiwan (No. 200611031R) (ClinicalTrials.gov number NCT00746070). All participants provided written informed consent before inclusion in the study. The Taiwan Primary Aldosteronism Investigation (TAIPAI) group enrolled possible PA patients who first had their aldosterone-to-renin ratio (ARR) screened for PA detection and were then followed-up. Screening, confirmation, and subtype identification of the PA were performed in hypertensive patients according to the standard TAIPAI protocol and aldosteronism consensus [30–33]. Fulfillment of the following three conditions confirmed a diagnosis of PA: (1) autonomous excess aldosterone production evidenced by an ARR > 35; (2) a TAIPAI score [34] of > 60%; (3) seated post-saline loading plasma aldosterone concentration (PAC) > 16 ng/dL [35], or PAC/plasma renin activity (PRA) > 35 (ng/dL)/(ng/mL/h) in a post-captopril/losartan test [30].

Sample preparation

For the detection of the chimeric gene, PCR was performed using the method described by MacConnachie et al. [12] with some modifications. We used the following primer sets to amplify the normal *CYP11B2* gene and the chimeric *CYP11B1/CYP11B2* gene with PfuUltra II Fusion HS DNA Polymerase (Agilent): forward: 5′CAGGTCCAGAGCCAG TTCTCCCAT/reverse: 5′ACCCTCCTTCTCCTTGTACACCCA forward: 5′CAGTTC TCCCATGACGTGATCCCT/reverse: 5′ACCCTCCTTCTCCTTGTACACCCA.

The touchdown PCR process was as follows: 95 °C for 2 min; 38 cycles of denaturation at 95 °C for 1 min; annealing at 70–61 °C for 1 min; and extension at 72 °C for 5 min and 72 °C for 3 min. The annealing temperature began at 70 °C and was lowered by 1 °C every two cycles until it reached 61 °C; this annealing temperature was maintained until the end of the cycling process. The PCR amplicons were evaluated in a 0.8% agarose gel, cleaned up with the 0.45X Agencourt AMPure XP beads (Beckman Coulter), and quantified using a Qubit fluorometer (Life Technologies).

Sample barcoding

For simultaneous detection of multiple samples, we used the ligation method to tag the amplified sample with the native barcoding adaptor (Oxford Nanopore Technologies), which allows up to 24 different libraries (barcodes 1–24) to be combined and loaded onto a single flow cell at the same time. The amplified fragment end was repaired and dA-tailed using the End Repair/dA-tailing Module (KAPA Roche). The end-repaired product was purified using 1X Agencourt AMPure XP beads. Next, a unique dT-tailed barcode adaptor was ligated on the dA-tailed template using ligation Master Mix (KAPA Roche). The barcoded samples were then purified with 0.45X Agencourt AMPure XP beads. The quality and quantity of each sample were evaluated using Nanodrop and a Qubit fluorometer, respectively. The barcoded samples were equally pooled for the sequencing library preparation.

Library preparation and sequencing

For the construction of the sequencing library, we used the KAPA hyper prep kit (Roche). First, the amplified and barcoded samples were pooled together, and end repair and dA-tailing were performed using the End Repair/dA-tailing Module. The end-repaired product was purified using 1X Agencourt AMPure XP beads. Next, adapter ligation and tethering were carried out with sequencing adapter (Oxford Nanopore Technologies) and ligation Master Mix. The sequencing-adapter–ligated DNA library was then purified with 0.5X Agencourt AMPure XP beads, Adapter Bead binding buffer (Oxford Nanopore Technologies) was added, and the samples were eluted in the Elution Buffer (Oxford Nanopore Technologies).

Before sequencing, the sequencing-adapter–ligated DNA library was mixed with Library Loading beads, and SpotON Flow Cells (R9.4) (FLO-MIN106D) were primed with running buffer. The samples were run on a MinION sequencing device for approximately 24 h, and the sequencing runs were operated by the MinKNOW software. Base-calling from the electrical data generated by the sequencer was performed using Guppy (v3.0.3) and the resulting raw sequence data were demultiplexed using qcat (v1.1.0).

Implementation of GRAde

Reads with lengths of 3000–5000 bp were considered for downstream analysis. Reads were corrected using a nonhybrid-based approach Canu (v1.4) [17] with the default parameters altered only as follows: genomeSize = 5 k, overlapper = mhap, utgReAlign = true, and stopOnReadQuality = false. Corrected reads were compared to the human reference genome (GRCh38) using ngmlr (v0.2.7) using default parameters [36]. The reads that aligned on the loci of *CYP11B1* and *CYP11B2* were defined as on-target reads. Each read was aligned to the conserved regions of *CYP11B1* and *CYP11B2* using the Smith–Waterman algorithm, implemented by the SSW library [37]. The discriminating and ambiguous bases were extracted from the alignment between *CYP11B1* and *CYP11B2* by the Smith–Waterman algorithm. The mismatch rate of each discriminating and ambiguous base was calculated by parsing the alignment results of reads aligned to *CYP11B1* and *CYP11B2*. The background noise was defined as the 99th percentile of the mismatch rate of all ambiguous bases, which is represented by the lower dashed line in the fusion plot. The foreground, represented by the upper dashed line in the fusion plot, was the average of the mismatch rates of the discriminating bases that had mismatch rates higher than the background. In addition, the mismatch rates of the discriminating bases were fitted using a sigmoid function to identify the crossover site, using the drm function implemented in the drc R package. All scripts were implemented in Python (v2.7) and R (v3.5.1).

Abbreviations

ACTH: Adrenocorticotropic hormone; APA: Aldosterone-producing adenomas; ARR: Aldosterone-to-renin ratio; CGD: Chronic granulomatous disease; CRISPR: Clustered regularly interspaced short palindromic repeat; FH: Familial hyperaldosteronism; GRA: Glucocorticoid-remediable aldosteronism; Indel: Insertion and deletion; NGS: Next-generation sequencing; ONT: Oxford Nanopore Technology; PA: Primary aldosteronism; PAC: Plasma aldosterone concentration; PCR: Polymerase chain reaction; PRA: Plasma renin activity; SMRT: Single-molecule real-time sequencing; SNP: Single-nucleotide polymorphism; SNV: Single nucleotide variant; SV: Structural variation; TAIPAI: The Taiwan Primary Aldosteronism Investigation.

Supplementary Information

The online version contains supplementary material available at https://doi.org/10.1186/s12859-022-04561-w.

Additional file 1: Nucleotides that differed between *CYP11B1* and *CYP11B2*.

Additional file 2: Fusion plots for all GRA patients.

Additional file 3: Runtime and robustness of GRAde.

Acknowledgements

The authors thank the National Center for High-performance Computing for computer time and facilities. We also thank the Second Core Lab and Sequencing Core, Department of Medical Research, National Taiwan University Hospital for providing laboratory facilities. We would like to thank Uni-edit (www.uni-edit.net) for editing and proofreading this manuscript.

About this supplement

This article has been published as part of BMC Bioinformatics Volume 22 Supplement 10 2021: Selected articles from the 19th Asia Pacific Bioinformatics Conference (APBC 2021): bioinformatics. The full contents of the supplement are available at https://bmcbioinformatics.biomedcentral.com/articles/supplements/volume-22-supplement-10.

Authors' contributions

YCW, JJTW, and CLH designed the study. YCW, CIC, PYC, CHK, and KYP performed the experiments. YHH and CLH implemented the tool and performed the analysis. VCW provided sample materials and clinical information. YCW, KYP, VCW, JJTW, and CLH interpreted the data. YCW and CLH drafted the manuscript. All authors read, revised, and approved the manuscript.

Funding

This work was supported by the Ministry of Science and Technology, Taiwan (MOST 107–2314-B-002-254-MY3, MOST110-2221-E-002-129-MY3 to CLH) and National Taiwan University Hospital, Taipei, Taiwan (108-N4206 to YCW; 109-N4491 to CIC; 110-S4909 and 110-O01 to CLH). The funders did not play any role in the design of the study, the collection, analysis, and interpretation of data, or in writing of the manuscript. Publication costs are funded by National Taiwan University Hospital (110-S4909 and 110-O01).

Availability of data and materials

The scripts used for the analyses have been deposited at GitHub and are available at https://github.com/hsu-binfo/GRAde.

Declarations

Ethics approval and consent to participate

This study was conducted in accordance with the mandates of the Helsinki Declaration and the guidelines of the ethics committee of National Taiwan University Hospital. All participants provided written informed consent before inclusion in the study.

Consent for publication

Not applicable.

Competing interests

The authors declare that they have no competing interests.

Author details

[1]Department of Medical Research, National Taiwan University Hospital, Taipei, Taiwan. [2]Department of Internal Medicine, National Taiwan University Hospital, Taipei, Taiwan. [3]Graduate Institute of Oncology, National Taiwan University College of Medicine, Taipei, Taiwan. [4]Graduate Institute of Medical Genomics and Proteomics, National Taiwan University College of Medicine, Taipei, Taiwan. [5]TAIPAI, Taiwan Primary Aldosteronism Investigator Group and TSA, Taiwan Society of Aldosteronism, Taipei, Taiwan.

Received: 9 December 2021 Accepted: 3 January 2022
Published online: 10 January 2022

References

1. Farrugia FA, Zavras N, Martikos G, Tzanetis P, Charalampopoulos A, Misiakos EP, Sotiropoulos D, Koliakos N. A short review of primary aldosteronism in a question and answer fashion. Endocr Regul. 2018;52(1):27–40.
2. Perez-Rivas LG, Williams TA, Reincke M. Inherited forms of primary hyperaldosteronism: new genes, new phenotypes and proposition of a new classification. Exp Clin Endocrinol Diabetes. 2019;127(2–03):93–9.
3. Zennaro MC, Boulkroun S, Fernandes-Rosa F. An update on novel mechanisms of primary aldosteronism. J Endocrinol. 2015;224(2):R63-77.

4. Mulatero P, Tizzani D, Viola A, Bertello C, Monticone S, Mengozzi G, Schiavone D, Williams TA, Einaudi S, La Grotta A, et al. Prevalence and characteristics of familial hyperaldosteronism: the PATOGEN study (Primary Aldosteronism in TOrino-GENetic forms). Hypertension. 2011;58(5):797–803.
5. Aglony M, Martinez-Aguayo A, Carvajal CA, Campino C, Garcia H, Bancalari R, Bolte L, Avalos C, Loureiro C, Trejo P, et al. Frequency of familial hyperaldosteronism type 1 in a hypertensive pediatric population: clinical and biochemical presentation. Hypertension. 2011;57(6):1117–21.
6. Lin YF, Peng KY, Chang CH, Hu YH, Wu VC, Chueh JS, Wu KD. Adrenalectomy completely cured hypertension in patients with familial hyperaldosteronism type I who had somatic KCNJ5 mutation. J Clin Endocrinol Metab. 2019;104(11):5462–6.
7. Carvajal CA, Campino C, Martinez-Aguayo A, Tichauer JE, Bancalari R, Valdivia C, Trejo P, Aglony M, Baudrand R, Lagos CF, et al. A new presentation of the chimeric CYP11B1/CYP11B2 gene with low prevalence of primary aldosteronism and atypical gene segregation pattern. Hypertension. 2012;59(1):85–91.
8. Malagon-Rogers M. Non-glucocorticoid-remediable aldosteronism in an infant with low-renin hypertension. Pediatr Nephrol. 2004;19(2):235–6.
9. Vonend O, Altenhenne C, Büchner NJ, Dekomien G, Maser-Gluth C, Weiner SM, Sellin L, Hofebauer S, Epplen JT, Rump LC. A German family with glucocorticoid-remediable aldosteronism. Nephrol Dial Transplant Off Publ Eur Dial Transpl Assoc Eur Ren Assoc. 2007;22(4):1123–30.
10. Fallo F, Pilon C, Williams TA, Sonino N, Morra Di Cella S, Veglio F, De Iasio R, Montanari P, Mulatero P. Coexistence of different phenotypes in a family with glucocorticoid-remediable aldosteronism. J Hum Hypertens. 2004;18(1):47–51.
11. Fardella CE, Pinto M, Mosso L, Gómez-Sánchez C, Jalil J, Montero J. Genetic study of patients with dexamethasone-suppressible aldosteronism without the chimeric CYP11B1/CYP11B2 gene. J Clin Endocrinol Metab. 2001;86(10):4805–7.
12. MacConnachie AA, Kelly KF, McNamara A, Loughlin S, Gates LJ, Inglis GC, Jamieson A, Connell JM, Haites NE. Rapid diagnosis and identification of cross-over sites in patients with glucocorticoid remediable aldosteronism. J Clin Endocrinol Metab. 1998;83(12):4328–31.
13. Treangen TJ, Salzberg SL. Repetitive DNA and next-generation sequencing: computational challenges and solutions. Nat Rev Genet. 2011;13(1):36–46.
14. Ebbert MTW, Jensen TD, Jansen-West K, Sens JP, Reddy JS, Ridge PG, Kauwe JSK, Belzil V, Pregent L, Carrasquillo MM, et al. Systematic analysis of dark and camouflaged genes reveals disease-relevant genes hiding in plain sight. Genome Biol. 2019;20(1):97.
15. Huddleston J, Chaisson MJP, Steinberg KM, Warren W, Hoekzema K, Gordon D, Graves-Lindsay TA, Munson KM, Kronenberg ZN, Vives L, et al. Discovery and genotyping of structural variation from long-read haploid genome sequence data. Genome Res. 2017;27(5):677–85.
16. Norris AL, Workman RE, Fan Y, Eshleman JR, Timp W. Nanopore sequencing detects structural variants in cancer. Cancer Biol Ther. 2016;17(3):246–53.
17. Koren S, Walenz BP, Berlin K, Miller JR, Bergman NH, Phillippy AM. Canu: scalable and accurate long-read assembly via adaptive k-mer weighting and repeat separation. Genome Res. 2017;27(5):722–36.
18. Mantere T, Kersten S, Hoischen A. Long-read sequencing emerging in medical genetics. Front Genet. 2019;10:426.
19. Grunenwald H. Optimization of polymerase chain reactions. In: Bartlett JMS, Stirling D, editors. PCR protocols. Totowa: Humana Press; 2003. p. 89–99.
20. Lorenz TC. Polymerase chain reaction: basic protocol plus troubleshooting and optimization strategies. J Vis Exp. 2012;63:e3998.
21. Peng Z, Paudel D, Wang L, Luo Z, You Q, Wang J. Methods for target enrichment sequencing via probe capture in legumes. Methods Mol Biol. 2020;2107:199–231.
22. Eckert S, Chan J, Houniet D, Breuer J, Speight G. Enrichment by hybridisation of long DNA fragments for Nanopore sequencing. Microbial Genomics. 2016;2:e000087.
23. Kang S-H, Lee W, An J-H, Lee J-H, Kim Y-H, Kim H, Oh Y, Park Y-H, Jin YB, Jun B-H, et al. Prediction-based highly sensitive CRISPR off-target validation using target-specific DNA enrichment. Nat Commun. 2020;11(1):3596.
24. Gilpatrick T, Lee I, Graham JE, Raimondeau E, Bowen R, Heron A, Downs B, Sukumar S, Sedlazeck FJ, Timp W. Targeted nanopore sequencing with Cas9-guided adapter ligation. Nat Biotechnol. 2020;38(4):433–8.
25. Alvarez-Madrazo S, Mackenzie SM, Davies E, Fraser R, Lee WK, Brown M, Caulfield MJ, Dominiczak AF, Farrall M, Lathrop M, et al. Common polymorphisms in the CYP11B1 and CYP11B2 genes: evidence for a digenic influence on hypertension. Hypertension. 2013;61(1):232–9.
26. Hayrapetyan A, Dencher PC, van Leeuwen K, de Boer M, Roos D. Different unequal cross-over events between NCF1 and its pseudogenes in autosomal p47(phox)-deficient chronic granulomatous disease. Biochim Biophys Acta. 2013;1832(10):1662–72.
27. Chanock SJ, Roesler J, Zhan S, Hopkins P, Lee P, Barrett DT, Christensen BL, Curnutte JT, Gorlach A. Genomic structure of the human p47-phox (NCF1) gene. Blood Cells Mol Dis. 2000;26(1):37–46.
28. Coeli FB, Soardi FC, Bernardi RD, de Araujo M, Paulino LC, Lau IF, Petroli RJ, de Lemos-Marini SH, Baptista MT, Guerra-Junior G, et al. Novel deletion alleles carrying CYP21A1P/A2 chimeric genes in Brazilian patients with 21-hydroxylase deficiency. BMC Med Genet. 2010;11:104.
29. Cormand B, Diaz A, Grinberg D, Chabas A, Vilageliu L. A new gene-pseudogene fusion allele due to a recombination in intron 2 of the glucocerebrosidase gene causes Gaucher disease. Blood Cells Mol Dis. 2000;26(5):409–16.
30. Wu VC, Hu YH, Er LK, Yen RF, Chang CH, Chang YL, Lu CC, Chang CC, Lin JH, Lin YH, et al. Case detection and diagnosis of primary aldosteronism: the consensus of Taiwan Society of Aldosteronism. J Formos Med Assoc. 2017;116(12):993–1005.
31. Wu VC, Chang HW, Liu KL, Lin YH, Chueh SC, Lin WC, Ho YL, Huang JW, Chiang CK, Yang SY, et al. Primary aldosteronism: diagnostic accuracy of the losartan and captopril tests. Am J Hypertens. 2009;22(8):821–7.
32. Kuo CC, Wu VC, Huang KH, Wang SM, Chang CC, Lu CC, Yang WS, Tsai CW, Lai CF, Lee TY, et al. Verification and evaluation of aldosteronism demographics in the Taiwan Primary Aldosteronism Investigation Group (TAIPAI Group). J Renin Angiotensin Aldosterone Syst. 2011;12(3):348–57.

33. Wu VC, Kuo CC, Wang SM, Liu KL, Huang KH, Lin YH, Chu TS, Chang HW, Lin CY, Tsai CT, et al. Primary aldosteronism: changes in cystatin C-based kidney filtration, proteinuria, and renal duplex indices with treatment. J Hypertens. 2011;29(9):1778–86.
34. Wu VC, Yang SY, Lin JW, Cheng BW, Kuo CC, Tsai CT, Chu TS, Huang KH, Wang SM, Lin YH, et al. Kidney impairment in primary aldosteronism. Clin Chim Acta. 2011;412(15–16):1319–25.
35. Wu CH, Wu VC, Yang YW, Lin YH, Yang SY, Lin PC, Chang CC, Tsai YC, Wang SM, T group. Plasma aldosterone after seated saline infusion test outperforms captopril test at predicting clinical outcomes after adrenalectomy for primary aldosteronism. Am J Hypertens. 2019;32(11):1066–74.
36. Sedlazeck FJ, Rescheneder P, Smolka M, Fang H, Nattestad M, von Haeseler A, Schatz MC. Accurate detection of complex structural variations using single-molecule sequencing. Nat Methods. 2018;15(6):461–8.
37. Zhao M, Lee WP, Garrison EP, Marth GT. SSW library: an SIMD Smith–Waterman C/C++ library for use in genomic applications. PLoS ONE. 2013;8(12):e82138.

Publisher's Note

Huang *et al. BMC Bioinformatics* (2021) 22:624
https://doi.org/10.1186/s12859-022-04682-2

BMC Bioinformatics

RESEARCH Open Access

CoMI: consensus mutual information for tissue-specific gene signatures

Sing-Han Huang[1,4], Yu-Shu Lo[1], Yong-Chun Luo[1], Yi-Hsuan Chuang[1], Jung-Yu Lee[1] and Jinn-Moon Yang[1,2,3]*

From The 19th Asia Pacific Bioinformatics Conference (APBC 2021) Tainan, Taiwan. 3-5 February 2021

*Correspondence:
moon@faculty.nctu.edu.tw
[1] Institute of Bioinformatics and Systems Biology, National Yang Ming Chiao Tung University, Hsinchu 300193, Taiwan
Full list of author information is available at the end of the article

Abstract

Background: The gene signatures have been considered as a promising early diagnosis and prognostic analysis to identify disease subtypes and to determine subsequent treatments. Tissue-specific gene signatures of a specific disease are an emergency requirement for precision medicine to improve the accuracy and reduce the side effects. Currently, many approaches have been proposed for identifying gene signatures for diagnosis and prognostic. However, they often lack of tissue-specific gene signatures.

Results: Here, we propose a new method, consensus mutual information (CoMI) for analyzing omics data and discovering gene signatures. CoMI can identify differentially expressed genes in multiple cancer omics data for reflecting both cancer-related and tissue-specific signatures, such as *Cell growth and death* in multiple cancers, *Xenobiotics biodegradation and metabolism* in LIHC, and *Nervous system* in GBM. Our method identified 50-gene signatures effectively distinguishing the GBM patients into high- and low-risk groups (log-rank $p = 0.006$) for diagnosis and prognosis.

Conclusions: Our results demonstrate that CoMI can identify significant and consistent gene signatures with tissue-specific properties and can predict clinical outcomes for interested diseases. We believe that CoMI is useful for analyzing omics data and discovering gene signatures of diseases.

Keywords: Tissue-specific gene signature, Prognostic gene signature, Omics data

Background

Genome-wide gene expression profiling has been used to identify genetic signatures that could be associated with the outcome of cancer patients [1]. Several different gene signatures have been developed, and many of these approaches have been shown to better define the prognosis of cancer patients as compared with conventional clinical and pathological characteristics of the tumors [1]. Some studies began with genome-wide gene expression profiling from microarray datasets or next-generation sequencing (NGS). For example, the PAM50 gene signature can use to classify breast tumors into one of these four subtypes and to predict clinical outcomes [2, 3]. Moreover, the gene signature identified from specific tissues is a promising avenue to maximize efficacy in target tissues while minimizing the safety risks of affecting unrelated tissues [4].

Many methods have been proposed to identify the gene signatures between normal and disease states [5]. The simple and common way is to use the T-test, which is a statistics-based method. According to the distributions between two states, the T-test evaluates the probability (p-value) in each gene. The significant difference of gene expression is often defined if the probability is less than 0.05 or 0.01. Another method is calculating the fold change (FC) of gene expression, which is not a statistical test and has no associated values for indicating the level of confidence in the genes as differentially expressed or not [6]. Therefore, some methods were developed, such as Significance Analysis of Microarrays (SAM) and Cancer Outlier Profile Analysis (COPA), to choose the biomarkers in high confidence [7–9]. The SAM approach proposed false discovery rate (FDR) to reduce genes showing significantly different expression by chance (i.e., false positive) [7, 10]. The COPA, based on the hypothesis of tumor heterogeneity, identified candidate genes overexpressing in subsets of samples by using median and median absolute deviation of gene expression profiles [8]. In another study, Parker et al. determined 50 genes (i.e., PAM50) for classified four breast subtypes with clinical means by using Predictive Analysis of Microarray (PAM) algorithm [2, 11], but this approach relied on sample labels that was difficult to propose new subtype of diseases. Recently, Gentles et al. applied CIBERSORT computational methods and PRECOG tools to identify cancer prognostic biomarkers and therapeutic targets [12], however, they only focused on the genes that related to clinical outcomes, but the ones might not be suitable for diagnosis due to the non-significant gene expression changes between the sample subgroups [13].

In this study, we hypothesize that the biological processes of tissues/organs are dysregulated during tumorigenesis, and the perturbed genes (i.e., tissue-specific genes) are often involved in corresponding functions of tissues [14–16]. To address these issues, we propose consensus mutual information (CoMI) to analyze omics data and to identify gene signatures. We utilized mutual information (MI) and gene expression distance as the basis to find the significantly and consistently expressed genes and gene signatures between normal and disease states. For multiple cancer omics data, our identified gene signatures could reflect cancer-related signatures and have tissue-specific properties (mean odds ratio = 2.89). Based on our previously developed global omics data analysis method [17], our CoMI identified gene signatures not only involved in common cancer-related progress, such as *Cell growth and death*, but also reflect tissue unique functions of *Xenobiotics biodegradation and metabolism* in LIHC and *Nervous system* in GBM. For clinical prognosis, a 50-gene signature identified by CoMI could distinguish the GBM patients into high- and low-risk groups based on gene expression patterns, and could predict clinical outcomes at 12-month survival (log-rank $p = 0.006$). We believe that our method and results are useful for analyzing omics data, discovering gene signatures with tissue-specific properties, and predicting clinical outcomes of diseases.

Results

CoMI for identifying gene signatures in multiple cancers

To identify consistent patterns of gene expression for gene signatures development in multiple cancers, we utilized CoMI to analyze genome-wide gene expression profiles in

LIHC, GBM, BLCA, BRCA, and COAD. To evaluate the cancer associations of selected genes from CoMI in multiple cancer datasets, we collected 1675 cancer-related genes derived from HPA database [18]. The results show that the gene signatures selected from the different ranking cut-off of CoMI have a high probability to be cancer-related genes compared with T-test, FC, and Significance Analysis of Microarrays (SAM) (Fig. 1). For example, the top-ranked 200 significant genes were 190, 182, 181, and 126 genes for CoMI, T-test, FC, and SAM, respectively.

To investigate the statistical meanings of CoMI, we calculated Spearman's correlation coefficient (ρ) between CoMI scores and FC values, SAM scores and p-values of T-test on all of 20,531 genes in NGS profiles of TCGA, respectively (Additional file 1: Fig. S1). For these methods, the average ρ values of these five cancer types were 0.81 (FC), 0.95 (T-test), and 0.94 (SAM). These results suggest that CoMI has statistical meanings and is able to identify significantly and consistently expressed genes and disease-related gene signatures from omics data.

Tissue-specific properties of gene signatures

To investigate the biological meanings of gene signatures identified by CoMI in different cancers, we collected tissue-specific genes from HPA with protein annotation of tissue specificity in liver, brain, urinary bladder, breast, or colon. Here, we used the odds ratio to assess whether gene signatures have tissue-specific properties or not in five cancers (Fig. 2 and Additional file 1: Fig. S2). In the comparison of CoMI and T-test, the average odds ratios of gene signatures selected from eight kinds of top-ranked thresholds were 2.89, 5.40, 1.90, 1.74, and 2.99 in LIHC, GBM, BLCA, BRCA, and COAD, respectively. For example, the top-ranked 200 genes selected by CoMI and T-test in GBM were 89 and 23 with annotated brain specificity, respectively. The odds were 0.8 (89/111) and 0.13 (23/177), then, the odds ratio

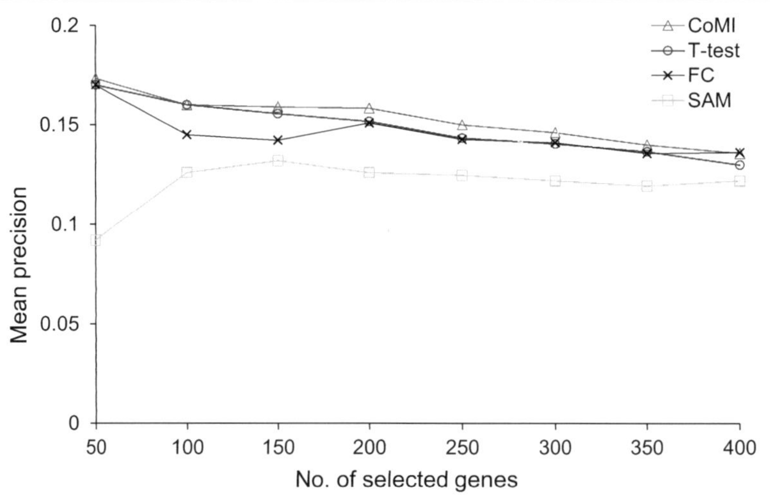

Fig. 1 The prediction accuracies between CoMI, T-test, FC, and SAM in cancer-related genes. The mean precision of cancer-related genes recorded in HPA was selected by CoMI (red line), T-test (blue line), FC (black line), and SAM (yellow line). The predicting precision was the average of genes selected from different cut-offs in LIHC, GBM, BLCA, BRCA, and COAD

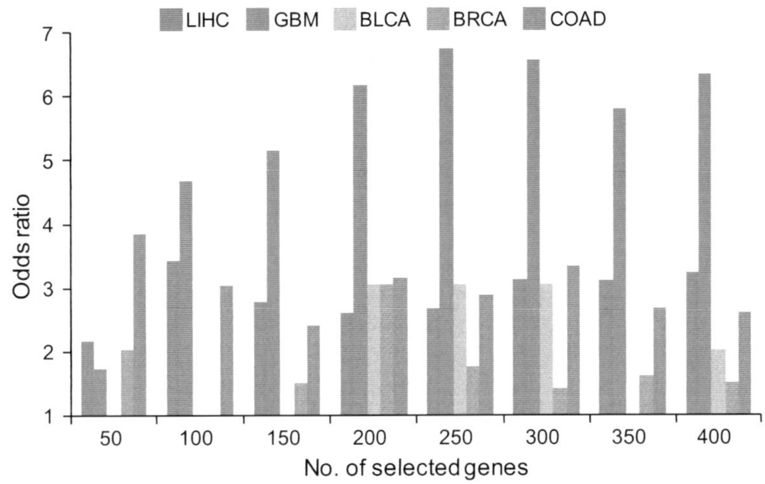

Fig. 2 The odds ratios of tissue-specific genes selected by CoMI and T-test in different cancer types. The results of tissue-specific genes identified by CoMI and T-test in five cancer types, including LIHC (blue), GBM (orange), BLCA (yellow), BRCA (purple), and COAD (green)

was calculated as 6.15 (0.8/0.13). The results were similar when compared to SAM (Additional file 1: Fig. S2). Besides, we observed that CoMI has a lower ranking on average of descending order than the ones of T-test and SAM in all of the tissue-specific genes in each cancer (Additional file 1: Figs. S3 and S4). These results demonstrate that gene signatures identified by CoMI are more like to have tissue-specific properties.

Furthermore, we used our previously developed global omics data analysis method, Hierarchical System Biology Model (HiSBiM), to investigate the involved pathways as well as biological subsystems and systems of gene signatures [17]. We observed that biological subsystems of the 200-gene signatures identified by CoMI and T-test were involved in common cancer-related pathways, such as *Cell growth and death* and *Replication and repair* in five cancer types (Fig. 3). In particular, the gene signature of LIHC identified by CoMI was enriched in *Xenobiotics biodegradation and metabolism* (meta-z score = 3.72) and *Lipid metabolism* (meta-z score = 2.74), which could reflect unique functions to liver tissue. In GBM, we observed that neurotransmission-related functions were highly enriched, such as *Nervous system* (meta-z score = 7.49) and *Signaling molecules and interaction* (meta-z score = 2.46). In addition, the *Digestive system* (meta-z score = 7.03) was highly regulated in COAD.

We also evaluated the tissue-specific properties of CoMI and COPA [9] based on HPA database and HiSBiM analysis. The results show that the genes identified by CoMI are related to both common cancer-related pathways and tissue-specific properties, such as *Endocrine system* (meta-z score = 5.40) in BRCA (Additional file 1: Fig. S5). Additionally, CoMI outperformed COPA and our results indicated that CoMI not only can identify tissue-specific gene signatures in different cancers, but also can reflect corresponding biological pathways and functions unique to those tissues.

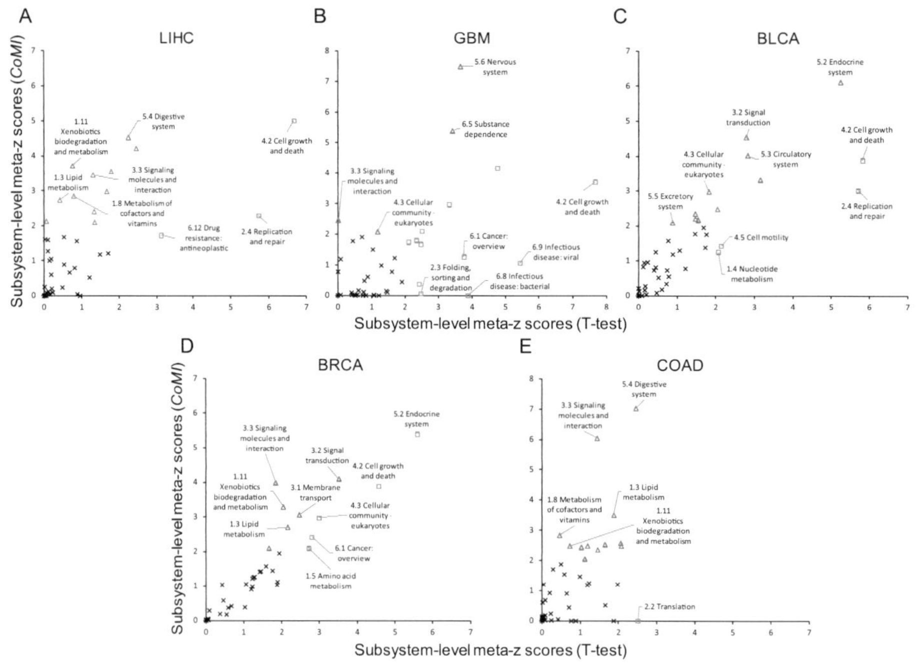

Fig. 3 The genes identified from different cancers reflected tissue-specific pathways. The subsystem-level meta-z scores of the top-ranked 200 genes identified by CoMI and T-test in **A** LIHC, **B** GBM, **C** BLCA, **D** BRCA, and **E** COAD. The red triangle was denoted if meta-z scores > 2 and CoMI > T-test, the green square was represented if meta-z scores > 2 and T-test > CoMI

Clinical prognosis of gene signatures

The grade IV astrocytomas (i.e., GBM) are an aggressive class of brain cancer, it is difficult to treat and has a poor median 12-month overall survival, and the urgent need to develop a prognostic gene signature [19]. We selected a 50-gene signature with significant and consistent expressions by using our CoMI from the gene expression profile in GBM, as well as, five of those genes were brain tissue specificity (Table 1). According to the gene expression patterns, we found that these 50 genes could cluster 156 tumor samples into four groups (Fig. 4A). According to the 12-month overall survival analysis, the patients in GBM-C1 group ($n = 15$) have a significantly lower survival probability (30%) than GBM-C3 group (62%; log-rank $p = 0.006$; Fig. 4B). Moreover, we found that 70% of CoMI top-ranked 10 genes can distinguish patients into high- and low-risk groups (log-rank $p < 0.01$), such as PBK (log-rank $p = 0.0058$) and CCNB2 (log-rank $p = 0.009$), however, only five of 10 genes with log-rank $p < 0.01$ in T-test (Additional file 1: Fig. S6). These results indicated that our identified 50-gene signature could predict clinical outcomes and provide the available clues for developing the new therapeutic strategies in GBM.

Discussion

Genome-wide gene expression profiling has been used to identify genetic signatures that could be associated with the outcome of cancer patients [1]. Some studies using genome-wide gene expression profiling for developing gene signatures, and many of

Table 1 The 50-gene signature identified by CoMI in GBM

Genes	CoMI	*p*-value[a]	Log2 (FC)[b]	Tissue specificity[c]	GO BP	GO CC
RRM2	2.80	9.23E−33	7.48	–	Deoxyribonucleotide biosynthetic process [GO:0009263]; DNA replication [GO:0006260]; negative regulation of G0 to G1 transition [GO:0070317]	Cytosol [GO:0005829]; ribonucleoside-diphosphate reductase complex [GO:0005971]
UBE2C	2.71	4.17E−29	7.69	–	Anaphase-promoting complex-dependent catabolic process [GO:0031145]; cell division [GO:0051301]; exit from mitosis [GO:0010458]	Anaphase-promoting complex [GO:0005680]; cytosol [GO:0005829]; nucleoplasm [GO:0005654]
PBK	2.50	2.07E−27	7.30	–	Cellular response to UV [GO:0034644]; mitotic cell cycle [GO:0000278]; negative regulation of inflammatory response [GO:0050728]	Nucleus [GO:0005634]
CCNB2	2.38	5.82E−27	6.89	–	Cell division [GO:0051301]; G2/M transition of mitotic cell cycle [GO:0000086]; in utero embryonic development [GO:0001701]	Centrosome [GO:0005813]; cyclin-dependent protein kinase holoenzyme complex [GO:0000307]; cytoplasm [GO:0005737]
KIF20A	2.38	5.82E−27	6.72	–	Microtubule-based movement [GO:0007018]; microtubule bundle formation [GO:0001578]; midbody abscission [GO:0061952]	Cleavage furrow [GO:0032154]; Golgi apparatus [GO:0005794]; intercellular bridge [GO:0045171]
MYBL2	2.30	5.50E−25	7.77	–	Cellular response to leukemia inhibitory factor [GO:1990830]; mitotic cell cycle [GO:0000278]; mitotic spindle assembly [GO:0090307]	Myb complex [GO:0031523]; nucleoplasm [GO:0005654]; nucleus [GO:0005634]
NDC80	2.29	1.11E−27	6.70	–	Attachment of mitotic spindle microtubules to kinetochore [GO:0051315]; attachment of spindle microtubules to kinetochore [GO:0008608]; cell division [GO:0051301]	Centrosome [GO:0005813]; chromosome, centromeric region [GO:0000775]; condensed chromosome kinetochore [GO:0000777]
TOP2A	2.28	7.18E−26	7.60	–	Apoptotic chromosome condensation [GO:0030263]; cellular response to DNA damage stimulus [GO:0006974]; chromosome segregation [GO:0007059]	Chromosome, centromeric region [GO:0000775]; condensed chromosome [GO:0000793]; cytoplasm [GO:0005737]

Table 1 (continued)

Genes	CoMI	*p*-value[a]	Log2 (FC)[b]	Tissue specificity[c]	GO BP	GO CC
DLGAP5	2.20	9.00E−25	6.93	–	Mitotic chromosome movement towards spindle pole [GO:0007079]; positive regulation of mitotic metaphase/anaphase transition [GO:0045842]; positive regulation of transcription of Notch receptor target [GO:0007221]	Centriolar satellite [GO:0034451]; cytosol [GO:0005829]; mitochondrion [GO:0005739]
BIRC5	2.12	6.81E−24	6.88	–	Cell division [GO:0051301]; chromosome segregation [GO:0007059]; cytokine-mediated signaling pathway [GO:0019221]	Centriole [GO:0005814]; chromosome, centromeric region [GO:0000775]; chromosome passenger complex [GO:0032133]
NCAPG	2.11	4.91E−28	6.53	–	Cell division [GO:0051301]; mitotic chromosome condensation [GO:0007076]	Condensed chromosome [GO:0000793]; condensed chromosome, centromeric region [GO:0000779]; condensin complex [GO:0000796]
CDC45	1.99	9.54E−21	6.64	–	DNA replication [GO:0006260]; DNA replication checkpoint [GO:0000076]; DNA replication initiation [GO:0006270]	Centrosome [GO:0005813]; ciliary basal body [GO:0036064]; cytoplasm [GO:0005737]
AURKB	1.97	1.44E−23	6.66	–	Abscission [GO:0009838]; aging [GO:0007568]; anaphase-promoting complex-dependent catabolic process [GO:0031145]	Chromocenter [GO:0010369]; chromosome passenger complex [GO:0032133]; condensed chromosome, centromeric region [GO:0000779]
MELK	1.85	2.59E−23	6.80	–	Apoptotic process [GO:0006915]; cell population proliferation [GO:0008283]; G2/M transition of mitotic cell cycle [GO:0000086]	Cell cortex [GO:0005938]; cytoplasm [GO:0005737]; membrane [GO:0016020]
TROAP	1.83	1.20E−21	6.27	–	Cell adhesion [GO:0007155]	Cytoplasm [GO:0005737]
BUB1	1.72	1.92E−26	6.07	–	Apoptotic process [GO:0006915]; cell division [GO:0051301]; cell population proliferation [GO:0008283]	Condensed chromosome kinetochore [GO:0000777]; condensed nuclear chromosome kinetochore [GO:0000778]; condensed nuclear chromosome outer kinetochore [GO:0000942]
EPR1	1.72	3.69E−22	6.77	–	Cell surface receptor signaling pathway [GO:0007166]	Integral component of membrane [GO:0016021]
HJURP	1.71	3.55E−24	6.09	–	Cell cycle [GO:0007049]; CENP-A containing nucleosome assembly [GO:0034080]; chromosome segregation [GO:0007059]	Chromosome, centromeric region [GO:0000775]; condensed chromosome kinetochore [GO:0000777]; mitochondrion [GO:0005739]
FAM64A	1.67	9.65E−21	6.72	–	Cell cycle [GO:0007049]; cell division [GO:0051301]	Nucleolus [GO:0005730]; nucleoplasm [GO:0005654]

Table 1 (continued)

Genes	CoMI	*p*-value[a]	Log2 (FC)[b]	Tissue specificity[c]	GO BP	GO CC
KIF4A	1.64	9.11E−24	5.63	–	Anterograde axonal transport [GO:0008089]; antigen processing and presentation of exogenous peptide antigen via MHC class II [GO:0019886]; microtubule-based movement [GO:0007018]	Axon cytoplasm [GO:1904115]; chromosome [GO:0005694]; cytoplasm [GO:0005737]
ASF1B	1.63	1.00E−24	5.80	–	Blastocyst hatching [GO:0001835]; cell differentiation [GO:0030154]; DNA replication-dependent nucleosome assembly [GO:0006335]	Nuclear chromatin [GO:0000790]; nucleoplasm [GO:0005654]; protein-containing complex [GO:0032991]
NUSAP1	1.54	8.64E−24	5.63	–	Establishment of mitotic spindle localization [GO:0040001]; mitotic chromosome condensation [GO:0007076]; mitotic cytokinesis [GO:0000281]	Chromosome [GO:0005694]; cytoplasm [GO:0005737]; microtubule [GO:0005874]
CEP55	1.53	1.10E−26	5.42	–	Cranial skeletal system development [GO:1904888]; establishment of protein localization [GO:0045184]; midbody abscission [GO:0061952]	Centriolar satellite [GO:0034451]; centriole [GO:0005814]; centrosome [GO:0005813]
CENPA	1.52	5.29E−22	5.55	–	CENP-A containing nucleosome assembly [GO:0034080]; establishment of mitotic spindle orientation [GO:0000132]; kinetochore assembly [GO:0051382]	Chromosome, centromeric region [GO:0000775]; condensed chromosome inner kinetochore [GO:0000939]; condensed nuclear chromosome, centromeric region [GO:0000780]
SHOX2	1.48	6.10E−16	6.91	–	Cardiac atrium morphogenesis [GO:0003209]; cartilage development involved in endochondral bone morphogenesis [GO:0060351]; chondrocyte development [GO:0002063]	Nuclear chromatin [GO:0000790]
KIFC1	1.44	3.38E−19	5.72	–	Cell division [GO:0051301]; microtubule-based movement [GO:0007018]; mitotic metaphase plate congression [GO:0007080]	Early endosome [GO:0005769]; kinesin complex [GO:0005871]; membrane [GO:0016020]
KIAA0101	1.43	8.07E−21	5.95	–	Cellular response to DNA damage stimulus [GO:0006974]; centrosome cycle [GO:0007098]; DNA replication [GO:0006260]	Centrosome [GO:0005813]; nucleoplasm [GO:0005654]; nucleus [GO:0005634]

Table 1 (continued)

Genes	CoMI	*p*-value[a]	Log2 (FC)[b]	Tissue specificity[c]	GO BP	GO CC
KIF2C	1.33	3.31E−20	5.29	–	Antigen processing and presentation of exogenous peptide antigen via MHC class II [GO:0019886]; attachment of mitotic spindle microtubules to kinetochore [GO:0051315]; cell division [GO:0051301]	Centrosome [GO:0005813]; chromosome, centromeric region [GO:0000775]; condensed chromosome kinetochore [GO:0000777]
ATP8A2	1.31	8.43E−13	6.85	Brain	Aging [GO:0007568]; axonogenesis [GO:0007409]; detection of light stimulus involved in visual perception [GO:0050908]	Endosome [GO:0005768]; Golgi apparatus [GO:0005794]; integral component of membrane [GO:0016021]
CENPK	1.31	4.27E−21	5.26	–	CENP-A containing nucleosome assembly [GO:0034080]; kinetochore assembly [GO:0051382]; mitotic sister chromatid segregation [GO:0000070]	Condensed nuclear chromosome inner kinetochore [GO:0000941]; cytosol [GO:0005829]; nucleoplasm [GO:0005654]
FOXM1	1.26	7.18E−19	4.95	–	DNA damage response, signal transduction by p53 class mediator resulting in transcription of p21 class mediator [GO:0006978]; DNA repair [GO:0006281]; G2/M transition of mitotic cell cycle [GO:0000086]	Nuclear chromatin [GO:0000790]; nucleoplasm [GO:0005654]; nucleus [GO:0005634]
GTSE1	1.23	7.44E−19	5.23	–	DNA damage response, signal transduction by p53 class mediator resulting in cell cycle arrest [GO:0006977]; microtubule-based process [GO:0007017]; positive regulation of cell migration [GO:0030335]	Cytoplasmic microtubule [GO:0005881]; cytosol [GO:0005829]; membrane [GO:0016020]
IGFBP2	1.22	1.01E−14	5.58	–	Aging [GO:0007568]; cellular protein metabolic process [GO:0044267]; cellular response to hormone stimulus [GO:0032870]	Apical plasma membrane [GO:0016324]; cytoplasmic vesicle [GO:0031410]; extracellular exosome [GO:0070062]
KIF14	1.22	7.45E−18	5.20	–	Activation of protein kinase activity [GO:0032147]; cell division [GO:0051301]; cell proliferation in forebrain [GO:0021846]	Cytosol [GO:0005829]; Flemming body [GO:0090543]; kinesin complex [GO:0005871]
MKI67	1.22	1.18E−16	5.67	–	Cell cycle [GO:0007049]; cell population proliferation [GO:0008283]; regulation of chromatin organization [GO:1902275]	Chromosome [GO:0005694]; membrane [GO:0016020]; nuclear body [GO:0016604]

Table 1 (continued)

Genes	CoMI	*p*-value[a]	Log2 (FC)[b]	Tissue specificity[c]	GO BP	GO CC
F2R	1.21	9.11E−24	4.72	–	Activation of cysteine-type endopeptidase activity involved in apoptotic process [GO:0006919]; activation of MAPKK activity [GO:0000186]; anatomical structure morphogenesis [GO:0009653]	Caveola [GO:0005901]; cell surface [GO:0009986]; early endosome [GO:0005769]
FAM111B	1.19	1.50E−17	5.47	–	DNA replication [GO:0006260]	Chromatin [GO:0000785]; nucleus [GO:0005634]
PNMA5	1.18	2.62E−14	6.55	Brain	Positive regulation of apoptotic process [GO:0043065]	–
PRKCG	1.18	1.06E−11	7.41	Brain	Chemical synaptic transmission [GO:0007268]; chemosensory behavior [GO:0007635]; innervation [GO:0060384]	Calyx of Held [GO:0044305]; cell–cell junction [GO:0005911]; cytosol [GO:0005829]
MLF1IP	1.17	2.29E−19	4.91	–	CENP-A containing nucleosome assembly [GO:0034080]; chordate embryonic development [GO:0043009]; viral process [GO:0016032]	Centriolar satellite [GO:0034451]; condensed chromosome kinetochore [GO:0000777]; cytosol [GO:0005829]
CDCA2	1.17	5.25E−18	5.29	–	Cell cycle [GO:0007049]; cell division [GO:0051301]; chromosome segregation [GO:0007059]	Chromosome [GO:0005694]; cytosol [GO:0005829]; nucleoplasm [GO:0005654]
E2F8	1.16	7.40E−19	5.30	–	Cell cycle comprising mitosis without cytokinesis [GO:0033301]; cell population proliferation [GO:0008283]; chorionic trophoblast cell differentiation [GO:0060718]	Cytosol [GO:0005829]; nuclear chromatin [GO:0000790]; nucleolus [GO:0005730]
EZH2	1.15	2.37E−21	4.80	–	Cardiac muscle hypertrophy in response to stress [GO:0014898]; cellular response to hydrogen peroxide [GO:0070301]; cellular response to trichostatin A [GO:0035984]	Chromosome, telomeric region [GO:0000781]; cytoplasm [GO:0005737]; ESC/E(Z) complex [GO:0035098]
C1QL3	1.14	1.54E−11	6.92	Brain	Regulation of synapse organization [GO:0050807]	Collagen trimer [GO:0005581]; extracellular region [GO:0005576]
CDK1	1.14	2.71E−19	4.91	–	Activation of MAPK activity [GO:0000187]; anaphase-promoting complex-dependent catabolic process [GO:0031145]; animal organ regeneration [GO:0031100]	Centrosome [GO:0005813]; cyclin B1-CDK1 complex [GO:0097125]; cyclin-dependent protein kinase holoenzyme complex [GO:0000307]
BUB1B	1.13	9.46E−19	5.12	–	Anaphase-promoting complex-dependent catabolic process [GO:0031145]; apoptotic process [GO:0006915]; cell division [GO:0051301]	Anaphase-promoting complex [GO:0005680]; condensed chromosome kinetochore [GO:0000777]; condensed chromosome outer kinetochore [GO:0000940]

Table 1 (continued)

Genes	CoMI	*p*-value[a]	Log2 (FC)[b]	Tissue specificity[c]	GO BP	GO CC
KIF18A	1.13	1.24E−18	4.90	–	Antigen processing and presentation of exogenous peptide antigen via MHC class II [GO:0019886]; cellular response to estradiol stimulus [GO:0071392]; male meiotic nuclear division [GO:0007140]	Caveola [GO:0005901]; cytoplasm [GO:0005737]; cytosol [GO:0005829]
SST	1.12	3.64E−09	6.66	Brain	Cell–cell signaling [GO:0007267]; cell surface receptor signaling pathway [GO:0007166]; chemical synaptic transmission [GO:0007268]	Extracellular region [GO:0005576]; extracellular space [GO:0005615]; neuronal cell body [GO:0043025]
E2F2	1.11	2.48E−14	5.09	–	Cell cycle [GO:0007049]; intrinsic apoptotic signaling pathway by p53 class mediator [GO:0072332]; lens fiber cell apoptotic process [GO:1990086]	Nuclear chromatin [GO:0000790]; nucleoplasm [GO:0005654]; nucleus [GO:0005634]
RYR2	1.11	6.41E−12	7.23	–	Calcium ion transport [GO:0006816]; calcium ion transport into cytosol [GO:0060402]; calcium-mediated signaling [GO:0019722]	Calcium channel complex [GO:0034704]; cytoplasmic vesicle membrane [GO:0030659]; junctional sarcoplasmic reticulum membrane [GO:0014701]

[a] Calculated by T-test

[b] Absolute values of log2(fold change)

[c] RNA tissue specificity recorded in HPA database

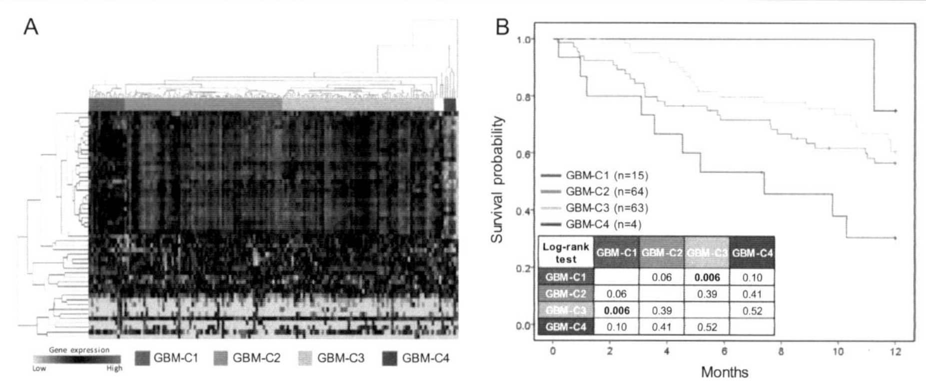

Fig. 4 Glioblastoma multiforme prognostic gene signature. **A** The hierarchal clustering of a 50-gene signature identified by CoMI in glioblastoma multiforme (GBM) tumor samples. Based on the gene expression patterns, the expression profile could be divided into four groups, including C1 (blue), C2 (green), C3 (orange), and C4 (purple). **B** The 12-month overall survival analysis in four groups of GBM patients. The inner table showed *p*-value of log-rank test compared between four groups of patients.

these approaches have shown to better define the prognosis of cancer patients, such as the PAM50 gene signature can use to classify breast tumors into four subtypes and to predict clinical prognosis [2, 3].

In this paper, we propose consensus mutual information (CoMI) to analyze omics data and to identify gene signatures. For multiple cancer omics data, we used CoMI to identify gene signatures, and those genes could reflect cancer-related signatures and have tissue-specific properties. In general, normal cells perform the function they are meant to perform, whereas cancer cells may not execute these functions. For example, cancerous thyroid cells may not produce thyroid hormone [20], and cancerous white blood cells are not functioning as they should [21]. On the other hand, the genes recorded in cancer hallmarks are often activated in cancer cells [22].

Here, we hypothesize that tissue-specific genes are often related to lost-of-tissue functions which are often down-regulated in the disease states. Our CoMI can identify these tissue-specific genes which are significant change (S_{Dist} score) and have consensus expressed values (S_{MI}) within cancer and normal samples. The changes between cancer and normal samples were quantified by considering global gene expression values for all genes in omics data. Based on our scoring function, CoMI identified the liver-specific genes, such as CYP1A2 (CoMI score = 0.99 and FC = − 7.76), CYP2C8 (CoMI score = 0.71 and FC = − 4.14), and CYP3A4 (CoMI score = 0.55 and FC = − 5.85), are the members of the cytochrome P450 family involving in Lipid metabolism (meta-z score = 2.74) as well as Xenobiotics biodegradation and metabolism (meta-z score = 3.72) in LIHC. Additionally, CoMI also identified gene SLC22A1 (CoMI score = 0.53 and FC = − 4.81) involving in liver unique functions, that is, bile secretion of Digestive system (meta-z score = 4.54). Conversely, MAP2K1, an essential component of the MAP kinase signal transduction pathway related to cancer hallmark, was not a tissue-specific gene. We found that MAP2K1 significantly changed in most cancers, but its expression values are not consistent (CoMI score = 0.20). These results show that our CoMI identified gene signatures not only involved in cancer-related progress, such as *Cell growth and death*, but also reflect tissue unique functions of *Nervous system* (meta-z score = 7.49) in GBM. For clinical prognosis, our identified 50-gene signature could stratify GBM patients into high- and low-risk groups, and could predict clinical outcomes with 12-month survival.

There were some limitations to the current study. Firstly, the omics data (i.e., NGS) is only collected from TCGA, and different sources and platforms, such as microarray, are needed to use to validate our method and identified gene signatures. Secondly, the results of this study are mostly based on bioinformatics analysis and predictions, and further experiments are needed to prove the tissue-specific properties of our identified gene signatures. Thirdly, although we identified prognostic gene signatures, those genes remain to be further explored in our future work.

Conclusion

In summary, we have proposed CoMI for analyzing omics data and discovering gene signatures. Our method accomplished the identification of genes and gene signatures, which have consistently and significantly changed between normal and disease states. Our results indicated that CoMI could identify gene signatures with tissue-specific properties for interested diseases, and is able to be applied to predict clinical prognosis.

Methods

To identify significantly and consistently expressed genes and gene signatures, we proposed a method, consensus mutual information (called CoMI), for analyzing omics data between normal and disease samples (Fig. 5A). We first calculated gene expression variations for each gene in a given omics data (Fig. 5B). For the evaluating of the consensus gene expression in normal and tissue states, we transferred the continuous gene expression to discrete integer symbols based on expression variations and intensities (Fig. 5C). Finally, we computed CoMI score for all genes, which can be used as a measure for significantly and consistently expressed genes and gene signatures between two states (Fig. 5D).

Omics and validation data

Here, we collected omics data (i.e. NGS) in LIHC, GBM, BLCA, BRCA, and COAD from The Cancer Genome Atlas (TCGA) databases [23]. These datasets contain 228 normal and 2315 tumor samples. To evaluate our method and compare to T-test and FC method, we collected the 1675 cancer-related genes derived from the Human Protein Atlas (HPA) database [18]. The tissue-specific genes were collected from HPA with protein annotation of tissue specificity in liver (409 genes), brain (1313 genes), urinary bladder (56 genes), breast (82 genes), or colon (147 genes). Finally,

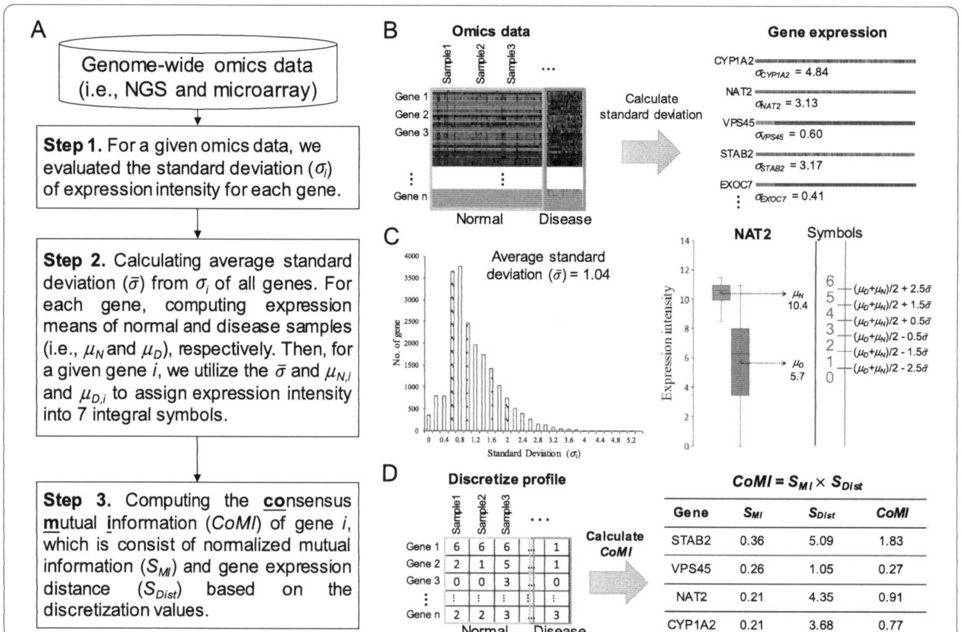

Fig. 5 Overview of consensus mutual information. **A** Flowchart describing the main procedure. **B** We firstly evaluated the standard deviation (σ_i) of gene expression intensity for each gene in a given omics data. **C** We then computing the average standard deviation ($\bar{\sigma}$) from σ_i of all genes, as well as expression means of normal and disease samples (i.e., μ_N and μ_D), respectively, in each gene. For a given gene i in sample j, we assigned expression intensity into 7 integral symbols by considering the $\bar{\sigma}$ and its μ_N and μ_D. **D** The gene expression values were converted to discretized integer symbol ranging from 0 to 6. The highly expressed genes were assigned to the highest symbol 6 and lowly expressed genes were assigned to the lowest symbol 0. (D) The calculation of consensus mutual information (CoMI) values of all genes

we calculated the p-values (i.e., T-test), fold changes, and SAM scores of all genes by using limma and samr R package [24, 25].

Discretization

To consider the readily quantifiable and significant expressions in the disease state of genes and gene signatures, we firstly evaluated the standard deviation (σ_i) of expression intensity of the gene i (Fig. 5B). Then, we calculated the average standard deviation ($\overline{\sigma}$) from σ_i of all genes (Fig. 5C). For each gene, computing expression means of normal (μ_N) and disease (μ_D) samples, respectively. For a given gene i in sample j, we utilize the average standard deviation ($\overline{\sigma}$), average expression ($\mu_{N,i}$ and $\mu_{D,i}$) to assign expression intensity ($EI_{i,j}$) into an integral symbol ($ES_{i,j}$) by using the following equations:

$$ES_{i,j} = \begin{cases} 6, & if EI_{i,j} > \frac{\mu_{D,i} + \mu_{N,i}}{2} + 2.5\overline{\sigma} \\ 5, & if \frac{\mu_{D,i} + \mu_{N,i}}{2} + 1.5\overline{\sigma} < EI_{i,j} \le \frac{\mu_{D,i} + \mu_{N,i}}{2} + 2.5\overline{\sigma} \\ 4, & if \frac{\mu_{D,i} + \mu_{N,i}}{2} + 0.5\overline{\sigma} < EI_{i,j} \le \frac{\mu_{D,i} + \mu_{N,i}}{2} + 1.5\overline{\sigma} \\ 3, & if \frac{\mu_{D,i} + \mu_{N,i}}{2} - 0.5\overline{\sigma} < EI_{i,j} \le \frac{\mu_{D,i} + \mu_{N,i}}{2} + 0.5\overline{\sigma} \\ 2, & if \frac{\mu_{D,i} + \mu_{N,i}}{2} - 1.5\overline{\sigma} < EI_{i,j} \le \frac{\mu_{D,i} + \mu_{N,i}}{2} - 0.5\overline{\sigma} \\ 1, & if \frac{\mu_{D,i} + \mu_{N,i}}{2} - 2.5\overline{\sigma} < EI_{i,j} \le \frac{\mu_{D,i} + \mu_{N,i}}{2} - 1.5\overline{\sigma} \\ 0, & if EI_{i,j} \le \frac{\mu_{D,i} + \mu_{N,i}}{2} - 2.5\overline{\sigma} \end{cases} \quad (1)$$

The gene expression values were converted to discretized integer symbols ranging from 0 to 6. The highly expressed gene was assigned to the symbol 6 and lowly expressed gene was assigned to the symbol 0. For instance, the $\overline{\sigma}$ is 1.04 in LIHC, and the μ_N and μ_D of the gene NAT2 are 10.4 and 5.7, respectively. The expression intensity (EI) of NAT2 is 10.8 in sample TCGA-DD-AAE3-01A-11R-A41C-07, which satisfies the $EI_{i,j} > \frac{\mu_{D,i}+\mu_{N,i}}{2} + 2.5\overline{\sigma}$ (i.e., 10.7), therefore, we assign the $ES_{NAT2,TCGA\text{-}DD\text{-}AAE3\text{-}01A\text{-}11R\text{-}A41C\text{-}07}$ value to 6.

CoMI: Consensus mutual information

For each gene i in a given omics data, we evaluated its consensus mutual information (CoMI) and to identify significantly and consistently expressed genes and gene signatures between normal and disease states (Fig. 5D). The CoMI of a gene is defined as:

$$CoMI = S_{MI} \times S_{Dist} \quad (2)$$

where S_{MI} is gene expression difference between two states using mutual information; S_{Dist} is the gene expression distance between two states by using mean distance. For the gene i, the S_{MI} is given as:

$$S_{MI} = \sum_{y=1}^{Y} \sum_{x=1}^{X} p(x,y) \log\left(\frac{p(x,y)}{p(x)p(y)}\right) \quad (3)$$

where $p(x,y)$ is the probability of gene i in symbol x and state y; $p(x)$ is the fraction of gene i in symbol x, and $p(y)$ is the fraction of samples in state y. Y is the number of states (here, Y = 2 for normal and disease states), and X is the number of symbols (here, X = 7). The S_{Dist} is given as:

$$S_{Dist} = \left(\frac{\sum_{n \in y1} ES_{i,n}}{N} - \frac{\sum_{d \in y2} ES_{i,d}}{D} \right) \tag{4}$$

where N and D are numbers of samples in the normal (y_1) and disease (y_2) states, respectively; $ES_{i,n}$ and $ES_{i,d}$ are the integral symbols of gene i at the sample for normal and disease states, respectively.

According to the equation of mutual information, the S_{MI} is related to the ratio between normal and cancer samples. S_{MI} is also related to the overlap of integer symbols between normal and cancer samples. For example, $S_{MI} = 1$, while *p(normal) = p(cancer)* and there is no overlap of integer symbols between normal and cancer samples; S_{MI} is from 0.3 to 0.5, while *p(normal) = p(cancer)* and half of the integer symbols in normal and cancer samples are the same; $S_{MI} = 1$, while *p(normal) = p(cancer)* and there is no overlap of integer symbols between normal and cancer samples; $S_{MI} = 0.44$, while 10 × *p(normal) = p(cancer)* and there is no overlap of integer symbols between normal and cancer samples; S_{MI} is from 0.3 to 0.4, while 10 × *p(normal) = p(cancer)* and half of the integer symbols in normal samples are the same as integer symbols in cancer samples; $S_{MI} = 0.28$, while 20 × *p(normal) = p(cancer)* and there is no overlap of integer symbols between normal and cancer samples. In our collected NGS data in LIHC, GBM, BLCA, BRCA, and COAD from TCGA databases, there are 228 normal and 2315 tumor samples and the expected maxima of S_{MI} might be 0.44. Here, we assumed that CoMI > 0.6 might be a suitable cut-off which could ensure only half of the integer symbols in normal samples are the same as integer symbols in cancer samples and the S_{Dist} is greater than 2 (Additional file 1: Fig. S7).

Moreover, we found that both S_{MI} and S_{Dist} could be used as good indexes to identify the genes with tissue-specific properties (Additional file 1: Fig. S8). S_{Dist} focuses on the distance between two states and is more related to the tissue-specific genes in our five data sets. In this study, we provide a scoring system that has a reliable discretizing method and consider both distances between two states and mutual information to identify gene signatures. Using S_{MI} and S_{Dist} could evaluate the distance, significantly and consistently expressed value of the gene between normal and disease states under the same scale.

Abbreviations

CoMI: Consensus mutual information; FC: Fold change; NGS: Next-generation sequencing; HPA: Human protein atlas; TCGA: The cancer genome atlas; LIHC: Liver hepatocellular carcinoma; GBM: Glioblastoma multiforme; BLCA: Bladder urothelial carcinoma; BRCA: Breast invasive carcinoma; COAD: Colon adenocarcinoma.

Supplementary Information

The online version contains supplementary material available at https://doi.org/10.1186/s12859-022-04682-2.

Additional file 1: Fig. S1. The correlations of the ranking of all genes were compared between CoMI, T-test, FC, and SAM in five cancer types. **Fig. S2**. The odds ratios of tissue-specific genes selected by CoMI and SAM in five cancer types, including LIHC, GBM, BRCA, and COAD. **Fig. S3**. Boxplot for the results of the ranking of tissue-specific genes identified by CoMI and T-test in five cancer types. **Fig. S4**. Boxplot for the results of the ranking of tissue-specific genes identified by CoMI and SAM in five cancer types. **Fig. S5**. Comparisons between CoMI and COPA for tissue-specific properties in five cancer types. **Fig. S6**. Kaplan–Meier plots representing patients stratified by the auto-select best cutoff of top-ranked 10 genes identified by CoMI in GBM. **Fig. S7.** The distributions of CoMI values with different S_{MI} and S_{Dist} values. **Fig. S8.** The relationship between S_{MI}, S_{Dist}, and tissue-specific genes.

Acknowledgements
We are thankful for the computing resources supported by the Center for Bioinformatics Research (CBR) by Ministry of Education ATU program and National Chiao Tung University, and the Center for Intelligent Drug Systems and Smart Bio-devices (IDS2B) of the Higher Education Sprout Project by the Ministry of Education (MOE) in Taiwan.

About this supplement
This article has been published as part of BMC Bioinformatics Volume 22 Supplement 10 2021: Selected articles from the 19th Asia Pacific Bioinformatics Conference (APBC 2021): bioinformatics.The full contents of the supplement are available at https://bmcbioinformatics.biomedcentral.com/articles/supplements/volume-22-supplement-10.

Author contributions
SHH analysed the omics data and wrote the manuscript; YSL designed the experiments; YCL and JYL provided contribution and discussion on biological meaning; YHC prepared the computing environments for R language and program code; JMY designed and supervised the study, provided the intellectual contribution, discussion and edited the manuscript. All authors have read and approved the final manuscript.

Funding
Publication costs are funded by the Ministry of Science and Technology (MOST 109-2634-F-009-021) in Taiwan, the Artificial Intelligence to Precision Health: Integrating Dynamic Physiological Signals and EMR to build a Medical Digital Twins Platform (MOST 109-2321-B-009-007), and the National Health Research Institutes (NHRI-EX109-10504PI). The funding bodies played no role in the design of the study and collection, analysis, and interpretation of data and in writing the manuscript.

Availability of data and materials
The omics data of LIHC, GBM, BLCA, BRCA, and COAD and clinical information were collected from TCGA cohort of Genomic Data Commons Data Portal database (https://portal.gdc.cancer.gov/). Kaplan–Meier analysis was performed by using R tool which could be obtained from R Project (https://www.r-project.org/).

Declarations

Ethics approval and consent to participate
Not applicable.

Consent to publish
Not applicable.

Competing interests
The authors declare that they have no conflict of interest.

Author details
[1]Institute of Bioinformatics and Systems Biology, National Yang Ming Chiao Tung University, Hsinchu 300193, Taiwan. [2]Department of Biological Science and Technology, National Yang Ming Chiao Tung University, Hsinchu 300193, Taiwan. [3]Center for Intelligent Drug Systems and Smart Bio-Devices, National Yang Ming Chiao Tung University, Hsinchu 300193, Taiwan. [4]Present Address: Graphen Inc., New York, NY 10110, USA.

Received: 8 April 2022 Accepted: 11 April 2022
Published online: 19 April 2022

References

1. Normanno N, De Luca A, Carotenuto P, Lamura L, Oliva I, D'Alessio A. Prognostic applications of gene expression signatures in breast cancer. Oncology. 2009;77(Suppl 1):2–8.
2. Parker JS, Mullins M, Cheang MC, Leung S, Voduc D, Vickery T, Davies S, Fauron C, He X, Hu Z, et al. Supervised risk predictor of breast cancer based on intrinsic subtypes. J Clin Oncol. 2009;27(8):1160–7.
3. Perou CM, Sorlie T, Eisen MB, van de Rijn M, Jeffrey SS, Rees CA, Pollack JR, Ross DT, Johnsen H, Akslen LA, et al. Molecular portraits of human breast tumours. Nature. 2000;406(6797):747–52.
4. Ryaboshapkina M, Hammar M. Tissue-specific genes as an underutilized resource in drug discovery. Sci Rep. 2019;9(1):7233.
5. Cui X, Churchill GA. Statistical tests for differential expression in cDNA microarray experiments. Genome Biol. 2003;4(4):210.
6. Jain N, Thatte J, Braciale T, Ley K, O'Connell M, Lee JK. Local-pooled-error test for identifying differentially expressed genes with a small number of replicated microarrays. Bioinformatics. 2003;19(15):1945–51.
7. Tusher VG, Tibshirani R, Chu G. Significance analysis of microarrays applied to the ionizing radiation response. Proc Natl Acad Sci U S A. 2001;98(9):5116–21.

8. Tomlins SA, Rhodes DR, Perner S, Dhanasekaran SM, Mehra R, Sun XW, Varambally S, Cao X, Tchinda J, Kuefer R, et al. Recurrent fusion of TMPRSS2 and ETS transcription factor genes in prostate cancer. Science. 2005;310(5748):644–8.
9. MacDonald JW, Ghosh D. COPA-cancer outlier profile analysis. Bioinformatics. 2006;22(23):2950–1.
10. Galitski T, Saldanha AJ, Styles CA, Lander ES, Fink GR. Ploidy regulation of gene expression. Science. 1999;285(5425):251–4.
11. Tibshirani R, Hastie T, Narasimhan B, Chu G. Diagnosis of multiple cancer types by shrunken centroids of gene expression. Proc Natl Acad Sci U S A. 2002;99(10):6567–72.
12. Gentles AJ, Newman AM, Liu CL, Bratman SV, Feng W, Kim D, Nair VS, Xu Y, Khuong A, Hoang CD, et al. The prognostic landscape of genes and infiltrating immune cells across human cancers. Nat Med. 2015;21(8):938–45.
13. Madu CO, Lu Y. Novel diagnostic biomarkers for prostate cancer. J Cancer. 2010;1:150–77.
14. Gaude E, Frezza C. Tissue-specific and convergent metabolic transformation of cancer correlates with metastatic potential and patient survival. Nat Commun. 2016;7:13041.
15. Sonawane AR, Platig J, Fagny M, Chen CY, Paulson JN, Lopes-Ramos CM, DeMeo DL, Quackenbush J, Glass K, Kuijjer ML. Understanding tissue-specific gene regulation. Cell Rep. 2017;21(4):1077–88.
16. Shlomi T, Cabili MN, Herrgard MJ, Palsson BO, Ruppin E. Network-based prediction of human tissue-specific metabolism. Nat Biotechnol. 2008;26(9):1003–10.
17. Yang WY, Rao PS, Luo YC, Lin HK, Huang SH, Yang JM, Yuh CH. Omics-based investigation of diet-induced obesity synergized with HBx, Src, and p53 mutation accelerating hepatocarcinogenesis in Zebrafish model. Cancers (Basel). 2019;11(12):1899.
18. Uhlen M, Fagerberg L, Hallstrom BM, Lindskog C, Oksvold P, Mardinoglu A, Sivertsson A, Kampf C, Sjostedt E, Asplund A, et al. Proteomics. Tissue-based map of the human proteome. Science. 2015;347(6220):1260419.
19. Clark PA, Iida M, Treisman DM, Kalluri H, Ezhilan S, Zorniak M, Wheeler DL, Kuo JS. Activation of multiple ERBB family receptors mediates glioblastoma cancer stem-like cell resistance to EGFR-targeted inhibition. Neoplasia. 2012;14(5):420–8.
20. Carter Y, Sippel RS, Chen H. Hypothyroidism after a cancer diagnosis: etiology, diagnosis, complications, and management. Oncologist. 2014;19(1):34–43.
21. Chandran R, Hakki M, Spurgeon S. Infections in leukemia, Sepsis Luciano Azevedo. IntechOpen; 2012.
22. Hanahan D, Weinberg RA. Hallmarks of cancer: the next generation. Cell. 2011;144(5):646–74.
23. Cancer Genome Atlas Research N, Weinstein JN, Collisson EA, Mills GB, Shaw KR, Ozenberger BA, Ellrott K, Shmulevich I, Sander C, Stuart JM. The cancer genome atlas pan-cancer analysis project. Nat Genet. 2013;45(10):1113–1120.
24. Ritchie ME, Phipson B, Wu D, Hu Y, Law CW, Shi W, Smyth GK. limma powers differential expression analyses for RNA-sequencing and microarray studies. Nucleic Acids Res. 2015;43(7):e47.
25. Li J, Tibshirani R. Finding consistent patterns: a nonparametric approach for identifying differential expression in RNA-Seq data. Stat Methods Med Res. 2013;22(5):519–36.

Publisher's Note

Springer Nature remains neutral with regard to jurisdictional claims in published maps and institutional affiliations.

Dey *et al. BMC Bioinformatics* (2021) 22:627
https://doi.org/10.1186/s12859-022-04704-z

BMC Bioinformatics

RESEARCH Open Access

Gene expression data classification using topology and machine learning models

Tamal K. Dey[1], Sayan Mandal[2] and Soham Mukherjee[1*]

From The 19th Asia Pacific Bioinformatics Conference (APBC 2021) Tainan, Taiwan. 3-5 February 2021

*Correspondence:
mukher26@purdue.edu

[1] Department of Computer Science, Purdue University, West Lafayette, IN, USA
Full list of author information is available at the end of the article

Abstract

Background: Interpretation of high-throughput gene expression data continues to require mathematical tools in data analysis that recognizes the shape of the data in high dimensions. Topological data analysis (TDA) has recently been successful in extracting robust features in several applications dealing with high dimensional constructs. In this work, we utilize some recent developments in TDA to curate gene expression data. Our work differs from the predecessors in two aspects: (1) Traditional TDA pipelines use topological signatures called barcodes to enhance feature vectors which are used for classification. In contrast, this work involves curating relevant features to obtain somewhat better representatives with the help of TDA. This representatives of the entire data facilitates better comprehension of the phenotype labels. (2) Most of the earlier works employ barcodes obtained using topological summaries as fingerprints for the data. Even though they are stable signatures, there exists no direct mapping between the data and said barcodes.

Results: The topology relevant curated data that we obtain provides an improvement in shallow learning as well as deep learning based supervised classifications. We further show that the representative cycles we compute have an unsupervised inclination towards phenotype labels. This work thus shows that topological signatures are able to comprehend gene expression levels and classify cohorts accordingly.

Conclusions: In this work, we engender representative persistent cycles to discern the gene expression data. These cycles allow us to directly procure genes entailed in similar processes.

Keywords: Topological data analysis, Gene expression, Persistent cycles, Neural network

Background

The rapid advances in genome-scale sequencing have dispensed a comprehensive list of genes for different organisms. These data gives us a broad scope to comprehend the developmental and functional processes of these organisms. Since the advent of DNA microarray, it is now possible to measure the expression levels of large number of genes simultaneously. This has made a holistic analysis possible for gene data using their expression levels. The stochastic nature of biological processes and associated noise acquired during the mining process pose a fundamental challenge in modelling a mathematical structure explaining these high dimensional data. We look into two problems in data analysis involving gene expressions that are of current research interest.

A genome-wide association study (GWAS) is a method to link a subset of genes to a particular disease or physical phenomenon in an organism. It has been especially important to identify specific gene subsets not only from a clinical perspective but also from a data science perspective as well. The assimilation of these subsets enable better phenotype identification and improve prediction of cohort status using machine learning based approach. Our definition of cohort follows its common usage in biology where *a cohort is a group of animals of the same species, identified by a common characteristic, which are studied over a period of time as part of a scientific or medical investigation*. In our case all cohorts for each experiment belong to the same taxa.For small or medium sized data sets, since the number of gene expression in a cohort profile is far greater than the number of sample cohorts, disease prediction using neural networks is challenging as these architectures largely succeed when the number of samples is much larger. It becomes important for these cases to identify a subset of genes whose expression levels reflect the phenotype of the cohorts.

In addition, it is often the case that some cohort have incorrect or uncorrelated data due to instrumental or manual error. Hence, their gene expressions may not reflect their phenotype class. We find in practice that the elimination of such instances leads to better prediction scores and performance. In this work, we use topological data analysis to investigate both of these issues. We identify cohorts which are topologically relevant (Section Topo-Curated Cohort). We show that the use of these cohorts to determine phenotypes instead of the entire set improves classification. Next, in Topo-Relevant Gene Expression section, we look into the classic GWAS problem mentioned above to identify a small subset of genes by using topological data analysis. We compare classification results obtained by using this reduced gene subsets against the one obtained by using full gene pool. The results for the receded gene profile yields better prediction rate.

Topological data analysis (TDA), loosely speaking, explains the shape of a data using topological structures. Topological properties can be thought to remain invariant under continuous deformation. For instance, given a donut made of clay, topologically its shape remains the same if we stretch, twist, or bend it but changes if we cut or glue it. The theory of Algebraic Topology lays the mathematical foundation formalising this idea. Persistent Homology is a method to derive topological structures from a given data. Topological signatures, particularly based on Persistent Homology, enjoy some nice theoretical properties including robustness and scale invariance. These features are global and more resilient to local perturbations. This has made TDA an exciting area in data analysis with encouraging results in medical imaging [1, 2], protein analysis [3, 4], and molecular architecture [5, 6] among others. In previous works it has been shown that genes sharing similar attributes tend to cluster in high dimensions [7, 8]. This is because protein encoding genes that are part of the same biological pathway or have similar functionality are corregulated. This ultimately leads such gene clusters to have similar expression profiles. The

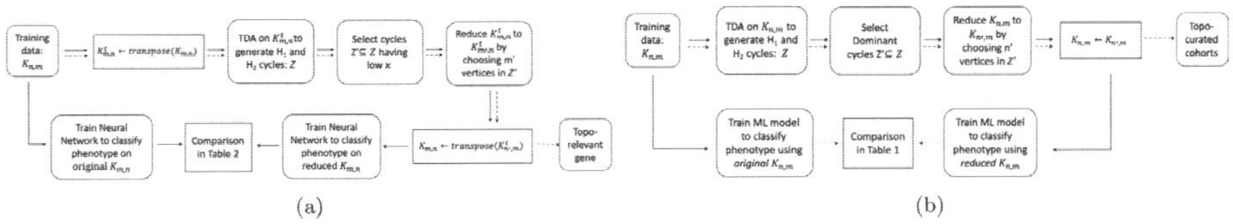

Fig. 1 **a** Flowchart for *topo-relevant gene* expression extraction. Refer to Section Topo-Relevant Gene Expression for details. **b** Flowchart for *topo-curated cohort* extraction. Refer to Section Topo-Curated Cohort for details. In both, bold lines show the path to take for training or testing large data. Dotted lines used in Fig. 2

property of clustering is essentially captured by the $zero^{th}$ order homology class in Persistent Homology (see next section). Motivated by these works, we are interested in finding if there exist relationships among similar genes in the higher order homology classes as well.

Traditional TDA pipelines use Persistent Homology to compute a set of intervals called barcodes which are used as topological feature in subsequent processing such as learning [3, 9–11]. While such barcodes provide robust topological signatures for the persistent features in data (such as tunnels, voids, loops, cycles etc.), their association to data is not immediately clear thus missing some crucial information. In effect, since these intervals represent homology classes, they contain the set of all loops around a topological hole. Thus using barcodes, it is hard to localize a feature, e.g., the shortest cycles or holes in a Persistent Homology class. This, in turn, hinders getting any direct mapping between the topological signatures and input cohorts or genes. So far there had been few studies addressing the problem of localizing persistent features and it has been shown that finding shortest cycles in given Persistent Homology classes is an NP-hard problem [12, 13]. However, taking advantage of the recent results in [12, 13], we are able to to compute good representative cycles for our application. These cycles capture definitive geometric features and provide a mapping between two domains of gene expression and topology.

In this paper we conduct two main experiments using the representative cycles: one to extract topologically relevant genes and the other to curate relevant cohorts. For these studies, some organisms were control units while others were either infected and/or injected with some antigen. The input consists of a matrix $\mathcal{K}$ which has n rows signifying the cohorts and their corresponding gene expressions in m columns. For obtaining and classifying topologically relevant (topo-relevant) genes, our experiment follow the pipeline in Fig. 1a whereas for determining curated cohorts, it follows the pipeline in Fig. 1b. For a large data set, we can trim out both insignificant cohorts and genes starting from the 'Training data $\mathcal{K}_{n,m}$'. This can be done following the pipeline in Fig. 2. We train our neural network architecture on the final curated dataset and thereby test against any unknown cohort. For our experiments, we work on gene expressions extracted from different organisms including Drosophila, *Mus muculus*, and *Homo sapiens*. We convert these data into a binary or multi-classification problem based on their phenotype and feed it into the pipeline. Our methodology and results have been listed in Computing Topological Signature of Gene-Expression Data section (Fig. 3).

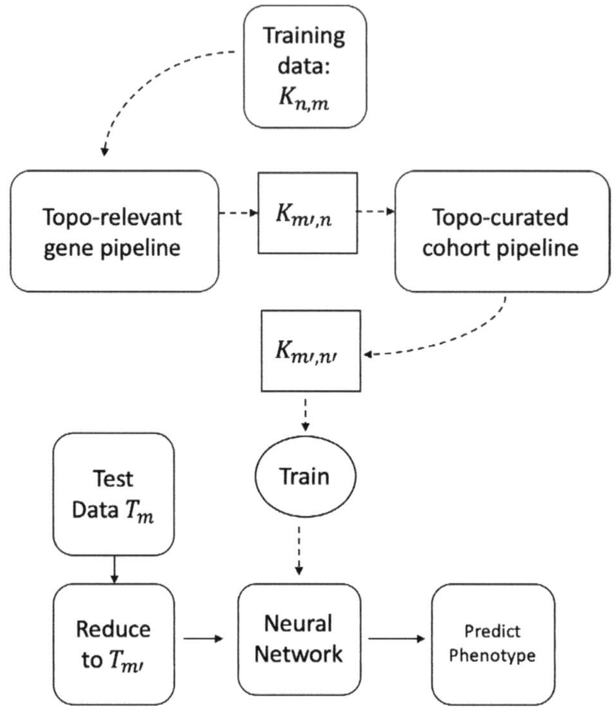

Fig. 2 Flowchart of proposed pipeline. For units *Topo-relevant gene pipeline* and *Topo-curated cohort pipeline* we follow the dotted lines in Fig. 1a and b respectively

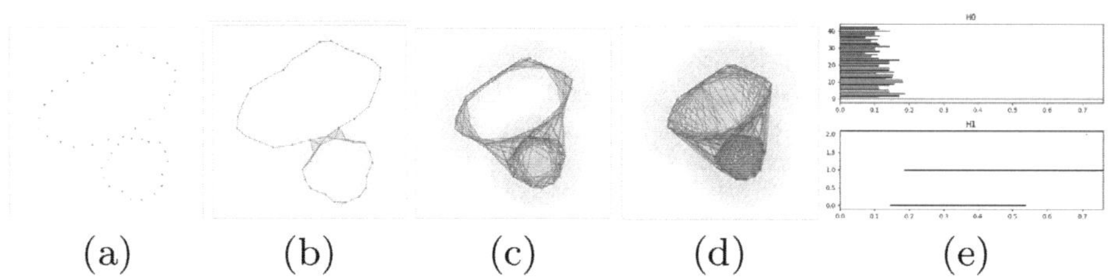

Fig. 3 Illustration of persistent homology. Here **e** represents the barcodes (in fig **e** Y-axis is No. of cycles and X-axis is radius r) associated with increasing radius from **a** to **d**. [Taken from [10] with authors' permission.]

Related works

In [14], gene expressions from peripheral blood data was used to build a model based on TDA network model and discrete Morse theory to look into routes of disease progression. These results on viruses suggest that Persistent Homology can be also used to study different forms of reticulate evolution. Topological structures have been used to analyse viruses by Emmet et al. [15]. They worked on influenza to show a bimodality of distribution of intervals in the persistent diagram. This bimodality indicates two scales of topological structure, corresponding to intra-subtype (involving one HA subtype) and inter-subtype (involving multiple HA subtypes) viral reassortments. Persistent Homology has also been used to identify DNA copy number aberrations [16]. Their experiments found a new amplification in 11q at the location of the progesterone receptor in the Luminal A subtype. Seemann et al. [17] used Persistent Homology to identify correlated patient samples in gene expression profiles. Their work focuses on the $\mathcal{H}_0$ homology class which is used to partition the point cloud. The paper by Nicolau et al. [18] identified a subgroup of breast cancers using topological data analysis in gene expressions.

Several works [19, 20] on use of machine learning techniques on gene expression profile have shown promising results. Kong et al. [21] used random forests to extract features for their Neural Network architecture. They investigate a problem similar to our *'Topo-relevant gene'* and the results show significant improvement. [22] analyzes gene expression data to classify cancer types. Different techniques of supervised learning are used to understand genes and classify cancer. The authors of [23] use machine learning to identify novel diagnostic and prognostic markers and therapeutic targets for soft tissue sarcomas. Their work shows overlap of three groups of tumors in their molecular profile.

Our contribution

We provide a technique based on persistent cycles (introduced in [12, 13]) to curate cohort data (datapoints) and gene expressions (features) (See Topo-Curated Cohort and Topo-Relevant Gene Expression sections). Through the experiments we show that these geometric structures, i.e. *cycles*, encode important information about the cohorts as choosing these *topo-curated cohorts* improve a classifier's accuracy. In a separate experiment, we demonstrate that choosing these *topo-curated gene expressions* provide a better classification. In a way, we provide empirical evidence that there is a one-to-one correspondence between topological features and important gene functionality.

Results

We now discuss the two ways to reduce the input $\mathcal{K}_{n,m}$ into $\mathcal{K}_{n'm'}$ where $n' \leq n$ and $m' \leq m$. The first section deals with finding pertinent cohorts, and the next with finding pertinent genes. In each subsection we describe the relevant procedure followed by results.

Topo-curated cohort

For our first proof of concept, we find a subset of cohorts who provide topologically relevant information for classification. The aim is to remove cohorts having either incorrect or uncorrelated data due to instrumental or manual error. Specifically, given $\mathcal{K}_{n,m}$, we would like to find $\mathcal{K}'_{n',m} \subseteq \mathcal{K}_{n,m}$ for $n \leq n'$ which improves classification odds for the cohorts. This subset of n' cohorts should therefore be topologically more relevant. We start by converting the matrix $\mathcal{K}_{n,m}$ into a point cloud. This point cloud has n points each of dimension m. Hence each cohort in the matrix is converted to an m-dimensional point where each dimension represents the expression level for each gene. We use Sparse Rips on the resulting point cloud to obtain a simplicial complex $\mathcal{K}$ and its filtration ($\mathcal{F}$) and apply the theory of Persistent Homology to obtain the set of finite intervals.

We consider the dataset $\mathbb{D}0$ having three phenotypes. We generate the longest 100 $\mathcal{H}_2$ cycles based on their interval length ($\delta - \beta$). For each cycle, we consider the constituent vertices and their corresponding phenotype labels ($\mathcal{X}$). We plot the count of $\mathcal{X}$ values in individual $\mathcal{H}_2$-cycles in Fig. 11a with the X, Y and Z axis representing $\mathcal{X} \in 0, 1,$ and 2 respectively. The black points indicate cycles where all vertices belong to a single phenotype. The red, green, and blue points indicate cycles having labels $(0,1)$, $(0,2)$ and $(1,2)$ respectively. The yellow points correspond to cycles having all three labels 0, 1, and 2. The takeaway from this plot is that, since most points are skewed towards some particular axis, most $\mathcal{H}_2$-cycles have constituent vertices who belong predominantly to some particular label in $\mathcal{X}$. Thus topological cycles in general are inclined towards some $\mathcal{X}$ labels without any supervision as they were not fed with the phenotype labels. Note that we added a small random noise to each point coordinate to illustrate multiplicity. Figure 11b plots similar values for the top 200 $\mathcal{H}_2$ cycles for dataset $\mathbb{D}1$. Since this dataset has two phenotypes, we get a $2d$ plot. The red labels denote cycles which have an equal constituent phenotypes, whereas blue and cyan represent skew, with blacks representing single labeled cycles as before. As is evident, most cycles exhibit a predominance in either $\mathcal{X} \in 0$ *or* 1. Based on the intuition of this plot, we define a cycle $\mathcal{Z}$ as a ***Dominant Cycle*** if, there exists a vertex set $U \subseteq Vert(\mathcal{Z})$[1] so that every vertex in U has the same label and $|U| \geq |Vert(\mathcal{Z})/2|$[2].

To illustrate the frequency of dominating cycles versus non-dominating ones, we plot the geodesic centers of the $\mathcal{H}_2$ cycles for $\mathbb{D}0$ by projecting them down to 2d using T-sNE (Fig. 4). Red vertices indicate non dominating cycles while each of the graded green points indicate the dominating ones. Clearly, most of the topology cycles are dominating and indicate a vote towards some phenotype class. The alpha values (denoted by the green bar at the right) indicates the ratio of the dominating phenotype in each cycle versus the other labels ($\mathcal{X}$). Hence, intuitively, more opaque a given point is, more is it dominated by a single class phenotype. Finally, we plot some of the individual dominating $\mathcal{H}_2$ cycles along with their phenotype labels in Fig. 5. Note that these points are part of the original D0 cohort point cloud and they were projected down to 3D using PCA.

Classification using machine learning

We work on several gene expression data extracted from different organisms. On each of these, we create a classification problem as described in the data section. For each dataset, we use the entire cohort list (irrespective of their phenotype) as an $(n \times m)$ dimensional point cloud. We generate the top 100 $\mathcal{H}_1$ and $\mathcal{H}_2$ cycles and select the *dominant cycles.* Next we select the vertices contained in these dominant cycles which form our new set of $n'(\leq n)$-cohorts. Taking gene expression for these n'-cohorts lets us form our new smaller matrix $\mathcal{K}'_{n',m}$. Thereafter, we train supervised classification models once using $\mathcal{K}_{n,m}$ and then again using $\mathcal{K}'_{n',m}$ and compare results for each. We use 10-fold cross validation by splitting the data randomly into 80−20% in each fold. For our classification models, we use two probability based classification models: Decision Tree and Naive Bayes.Note that we are interested in finding out whether the TDA pipeline can curate and retain the faithful representation of data. As a result we are comparing the performance of Decision Tree and Naive Bayes classifier on Topo-curated data. We do not report Support Vector Machine (SVM) result as its accuracy is too low to report. In general, probability based classification fared better than kernal based (SVM) techniques, hence we have

[1] $Vert(\mathcal{Z})$ denotes the constituent vertices on the cycle $\mathcal{Z}$.

[2] The modulo operator implies size of a set.

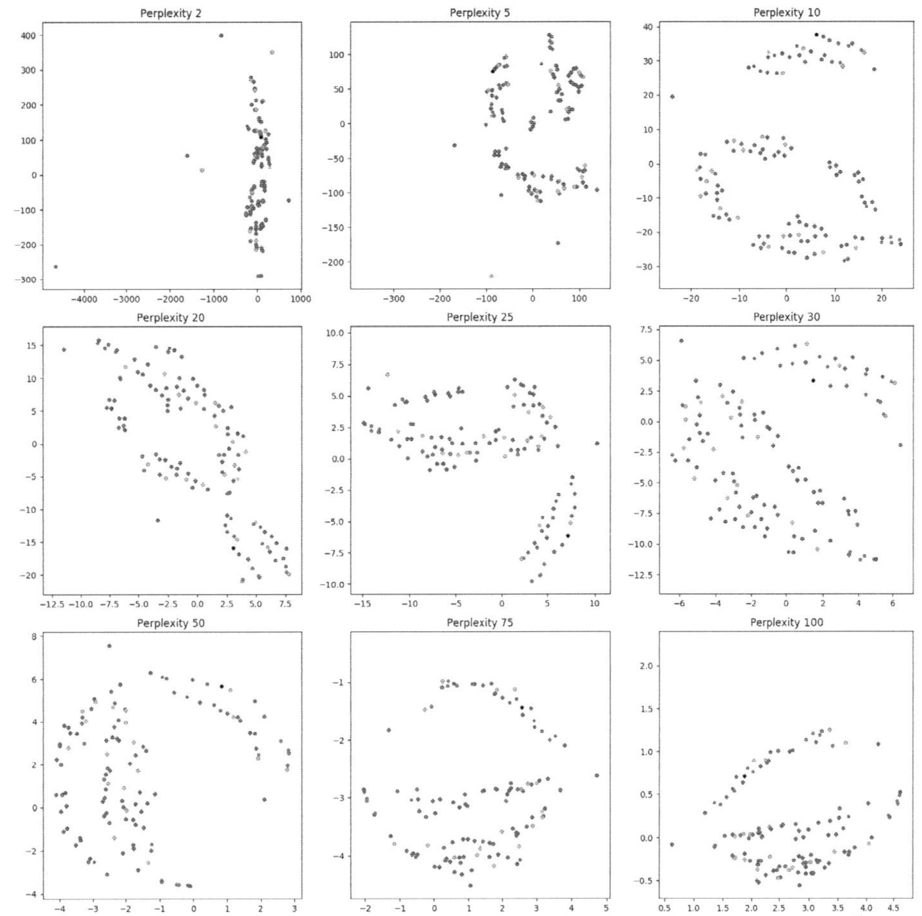

Fig. 4 Plot of geodesic centers for dominating cycles using t-SNE. Red vert: non dominating cycles. Graded green points: dominating cycles. Alpha values indicate ratio of dominating phenotype in each cycle versus the other labels.Range of perplexity values in the t-SNE plot indicate the uniformity of topological cycles

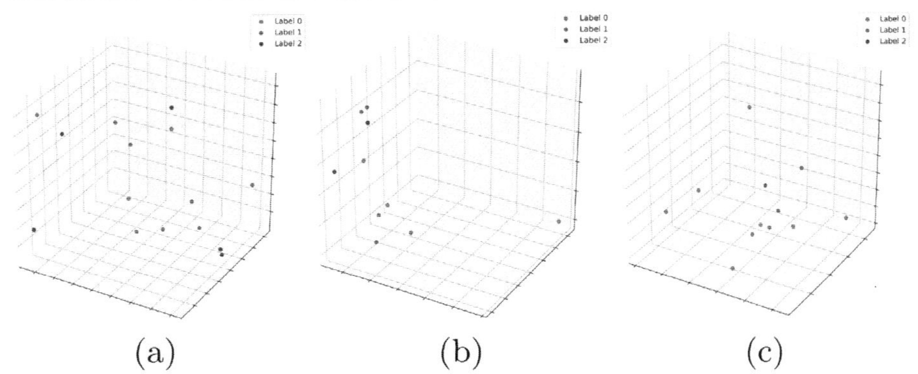

Fig. 5 Three figures plotting individual dominating cycles for gene dataset $\mathbb{D}0$. These cycles actually reside in m-dimensions and are projected down to 3D using Principal component analysis. The colors indicate cohort phenotype labels $\mathcal{X} \in 0, 1, 2$. **a** Cycle 1 spanned by 15 vertices with 3 vertices having label 0, 7 vertices having label 1 and 5 with label 2. **b** Cycle 2 spanned by 10 vertices with 5 vertices having label 0, 3 vertices having label 1 and 2 with label 2. **c** Cycle 3 spanned by 12 vertices with 9 vertices having label 0, 3 vertices having label 1

reported our results on the same. The average value of accuracy, precision, and recall for the 10-fold cross validation is reported in Table 4. The column 'FULL' represents training on $\mathcal{K}_{n,m}$ while $\mathcal{H}_1 + \mathcal{H}_2$ represent the union of n' topo-relevant cohorts obtained from the dominant cycles in either $\mathcal{H}_1$ or $\mathcal{H}_2$. We also get good classification statistics for the vertices in dominant cycles picked up only by $\mathcal{H}_2$ cycles only as reported in the same table. As is evident from the results, reduction in the number of cohorts leads to an increase in classification measures. Thus TDA is able to pick up cohorts who carry more decisive gene expression levels for their individual phenotype classes.

Topo-relevant gene expression

Our next problem is to reduce the matrix $\mathcal{K}_{n\times m}$ to $\mathcal{K}'$ of dimension $(n \times m')$ where $m' \ll m$. We use the persistent cycle descriptors $\mathcal{H}_1$ and $\mathcal{H}_2$ introduced in the previous section to extract $|m'|$ meaningful genes ($\mathcal{G}'$) such that $\mathcal{G}' \subset \mathcal{G}$. To this effect, we use the annotation of the gene set $\mathcal{G}$ based on their functional classification obtained from the *'Panther Classification System by Geneontology'* [24] and the *'NCBI Gene Data set'* [25]. Thus for each $g \in \mathcal{G}$, $\exists f : g \to R$, where R is a vector of functional attributes obtained from [24].

Once we obtain the representative cycles, we find the maximal cover of each cycle defined as follows:

Maximal cover of representative cycle (κ) For each gene expression $g \in \text{Vert}(\mathcal{Z})$ represented as vertices in a single representative cycle, we have a set of annotations $f(g)$. We select the minimum set consisting of at least one annotation for each $g \in Vert(\mathcal{Z})$. Let $\mathcal{S}$ be any set of annotations which contains at least one annotation for each $g \in \text{Vert}(\mathcal{Z})$. Thus,

$$\kappa = inf\{|\mathcal{S}| \mid \forall g \in \text{Vert}(\mathcal{Z}), \mathcal{S} \cap f(g) \neq \emptyset\}$$

The idea behind using κ is to get a sense of the functionality of the gene. A gene may be responsible for multiple processes described in the Panther and NCBI database. If κ is low or unity for a certain $\mathcal{Z}$, it probably indicates that the gene expressions involved in $\mathcal{Z}$ reflect the functionality captured by κ. This is illustrated in Fig. 6 where we plot some of the $\mathcal{H}_2$ cycles generated on $\mathcal{K}'$ with color annotated by their functionality. We use PCA as before to project the points down to 3-dimensions. The three figures illustrate three instances of different κ-values. Consider the example in Fig. 6a for getting the intuition behind κ. The six vertices representing genes in the $\mathcal{H}_2$ cycles have function annotations: {1: Localization, 2: Not annotated, 3: Metabolic process, Cellular process, 4: Metabolic process, Cellular process, Biological regulation, 5: Metabolic process, Cellular process, Localization, 6: Not Annotated}. Out of this the set: {Localization, Not annotated, and Metabolic process} covers all the vertices and hence κ is 3.

We choose $\mathcal{C}$ with low κ values and select their component genes as part of $\mathcal{G}'$. We can control the size of $\mathcal{G}'$ based on the value of κ.

For all our experiments, we run each architecture and obtain performance measures on $\mathcal{K}$ which contains the exhaustive set of m-genes. We re-run these experiments on our trimmed set $\mathcal{K}'$ containing $m'(\ll m)$ topologically significant genes. Note that we may use the topo-relevant cohort extraction to additionally reduce $\mathcal{K}'_{n'\times m}$ into $\mathcal{K}''_{n'\times m'}$. But since the public datasets we work on as our proof of concept have much less number of

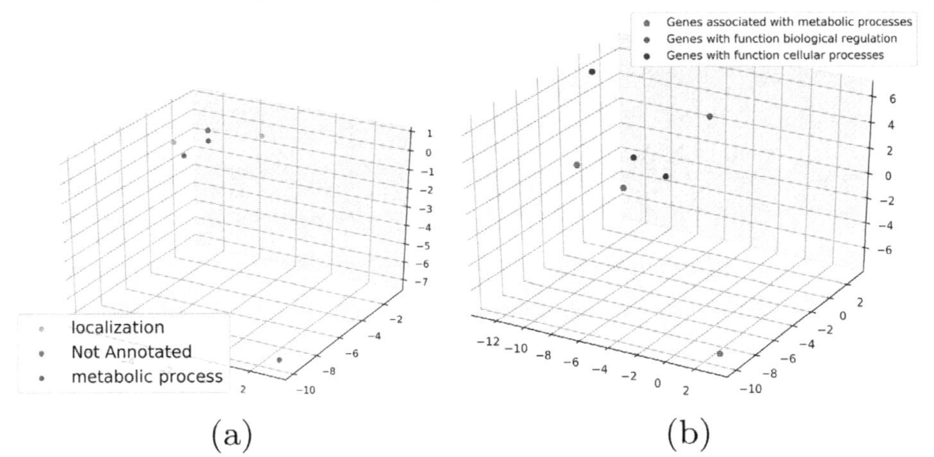

Fig. 6 Three 2-cycles for gene dataset $\mathbb{D}6$ with colors indicating gene function labels. Total number of colors indicate the κ value of the cycle. Note that a gene can be responsible for several functionalities. The legend in this plot takes into account only a single functionality which contributes towards the maximal cover of the representative cycle. **a** Low κ:3 **b** High κ:6

Table 1 Neural network result

Data Name	Human dengue (#:4415)		Human bone marrow (#: 469)		Human bowel disease (#: 1745)	
Method	$TP(\mathcal{Z})$	**Full**	$TP(\mathcal{Z})$	**Full**	$TP(\mathcal{Z})$	**Full**
# genes	1937	60619	5464	17258	1801	54715
Tr-Loss(e^{-2})	5.95	10.06	05.16	05.70	13.11	9.58
Tr-Acc	97.84	96.64	99.72	99.15	96.63	97.56
Tr-F1	97.86	96.48	99.72	99.15	96.60	97.55
Tr-Prec	97.86	96.48	99.72	99.15	96.60	97.55
Tr-Rec	97.86	96.48	99.72	99.15	96.60	97.55
Ts-Loss (e^{-2})	21.99	14.55	06.34	51.30	84.29	83.73
Ts-Acc	93.21	91.65	97.46	95.76	90.10	89.62
Ts-F1	92.26	90.67	96.95	95.74	90.34	89.66
Ts-Prec	93.48	90.67	96.95	95.74	90.34	89.66
Ts-Rec	93.48	90.67	96.95	95.74	90.34	89.66

The column $TP(\mathcal{Z})$ indicates the results on reduced gene set using topology. Full indicates results on the full gene set. Tr-Loss, Tr-Acc, Tr-F1, Tr-Prec,Tr-Rec is loss, accuracy, F1-score, precision, and recall on the training data. Whereas the prefix Ts- indicate the same on the test set

data to work with a Neural Net architecture, we do not trim the dataset. The results are in Table 1.

Neural network architecture

We use one dimensional convolutional neural network to perform experiments on gene-expression data. Our architecture is inspired by [21] who have managed to detect 'relevant' features of gene-expression data. The authors use a series of dense networks connected by activation functions. Since we provide some functional relevance among the genes, we sort them by their functionality and feed them to an additional convolutional layer on top of the model (Fig. 7). We start with this 1D-Convolutional Neural

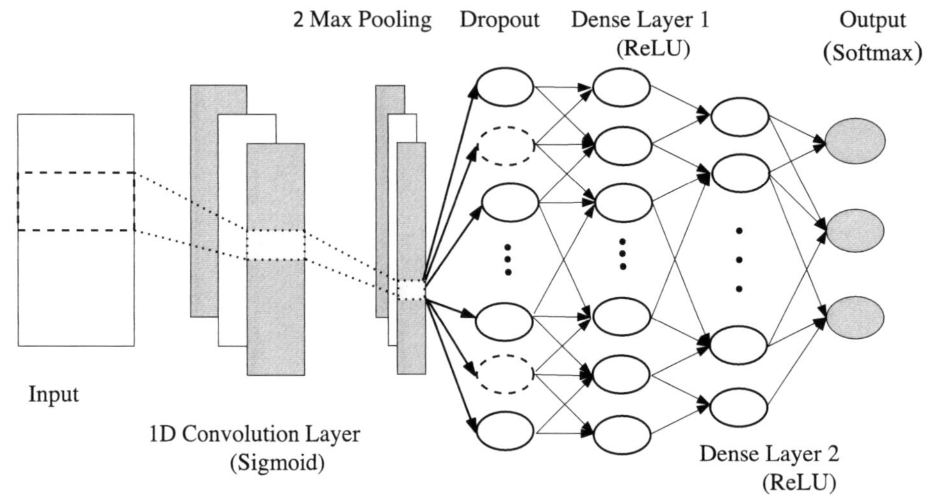

Fig. 7 Neural network architecture. Number of filters, filter size, dropout probability, number of units in dense layers are hyperparameters and are tuned for different datasets

Network (CNN) layer activated by the sigmoid function. Sigmoid is a traditional activation function which provides a smooth non linearity in the network and since the architecture is not too deep, we do not need to worry about its shortcomings like the vanishing gradient. This is followed by a max pooling of size 2 and subsequently a dropout layer. This layer is connected to two densely connected layers with decreasing sizes. These layers have ReLU (Rectified Linear Unit) as their activation function as used in the paper by [21]. In the end, we add a softmax activation layer to determine the final label of the data. The hyper-parameters of the network can be tuned using advanced hyperparameter optimization algorithm such as Bayesian Optimization. However, since this study is a proof of concept, and its purpose is to show the effectiveness of our feature selection, we fine tune them using manual observation.

Since the number of samples is still less for CNN, overfitting is an issue. Notice that, for this precise reason, we do not curate this data using the pipeline in Topo-Curated Cohort section. Dropout layers are added after each layer to further prevent overfitting and reduce high variance. We, however, do not initiate early stopping as those pipelines are not amenable to orthogonalization. Finally, the model is optimised using Adam (Adaptive Moment Optimizer) [26]. The dataset is split evenly into 80-20 and cross validated for 50 epochs. The neural network has been implemented in Python using Tensorflow and Keras. The results for our experiment on datasets $\mathbb{D}5, \mathbb{D}6$ and $\mathbb{D}7$ is shown in Table 1. The row *# genes* show that our architecture using vertices selected from topological cycles are less than 30% the size of the original gene pool. The results have, however, improved in all the cases. For experiments with Neural networks we follow the trend of the loss, accuracy and F1 score by plotting their value after every epoch in our algorithm. Figure 8 shows this result on dataset $\mathbb{D}7$. We see that the loss function on test data has been slightly higher but smoother than the full dataset. Despite this, using TDA the accuracy and F1 score has consistently performed better in every iteration for both the training and test data.

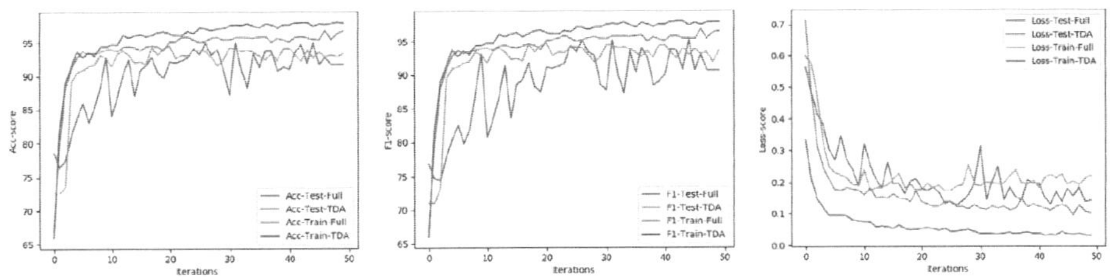

(a) (b) (c)

Fig. 8 Comparison of **a** Accuracy, **b** F1-score, **c** Loss function for 50 epochs. For the TDA curated data, red and yellow lines represent train and test scores respectively. For the full data, they are represented by the green and blue lines

Table 2 Comparison with standard outlier detection and unsupervised clustering methods

Dataset	Decision tree			Naive Bayes		
	Local outlier factor	DBSCAN	Topo-curated cohort	Local outlier factor	DBSCAN	Topo-curated cohort
Droso breeding	32.14	–	**79.34**	**57.14**	–	42.24
Droso parasitod	36.84	100	**83.50**	47.36	**66.67**	–
Mouse E Coli	30.91	21.05	**89.29**	40.00	21.05	**59.21**
Mouse prion	35.89	39.53	**61.62**	39.74	39.53	**57.84**
Mouse liver cancer	46.67	56.21	**72.95**	60.00	56.52	**72.32**
Human bowel disease	17.74	19.41	**51.09**	48.39	**53.40**	–

Bold values indicate the highest accuracy obtained for different classifiers and datasets under consideration

Discussions

Comparison with baseline

We compare our *Topo curated cohort* with the following standard outlier detection, unsupervised clustering methods.

- Local Outlier Factor
- Density-based spatial clustering of applications with noise [27] (DBSCAN)

It is noteworthy for the Droso breeding dataset DBSCAN fails to cluster and reports all the cohorts as outliers (Table 2). With DBSCAN as outlier detector and Decision Tree as a classifier, classifier's accuracy reaches upto 100% which is probably due to imbalance in the dataset and overfitting. In Table 2 we report the maximum accuracy obtained by our method, *i.e.* $max(Acc_{H1}, Acc_{H1+H2}, Acc_{H2})$ from Table 4. We then compare our *Topo-relevant gene expression* method with the following feature selection methods (See Table 3).

- Variance thresholding: Removes the low variance features that provide little information for modelling. [28–30].

Table 3 Comparison with standard feature selection methods

Dataset	Variance threshold	Select K-best	PCA	UMAP	Topo-relevant gene
Human dengue	92.26	93.41	74.21	85.39	**97.84**
Human bone marrow	95.37	92.25	76.50	**100.00**	99.72
Human bowel disease	61.62	61.69	61.62	61.63	**96.63**

The bold values indicate highest accuracy obtained for the feature selection methods

- Select K-best features: Selects k-features according to the highest scores. The scoring function used is F-value from *analysis of variance* (ANOVA). [29, 30]
- Principal Component Analysis (PCA): Although this is a dimensionality reduction technique and not a feature selection method we incorporate it because PCA is used widely while analyzing high-dimensional data such as gene expression [31].
- Uniform Manifold Approximation and Projection for Dimension Reduction (UMAP): UMAP is a manifold learning learning and TDA based dimensionality reduction technique. UMAP assumes the data to be uniformly distributed on a Riemannian manifold and finds a low dimensional embedding by building a 'fuzzy topological' structure on it [32].

We use the same neural net architecture described in Neural Network Architecture section and report the test-set accuracy after selecting the features with the aforementioned feature selection methods and compare with our *Topo-Relevant Gene Expression* procedure. We report the test-accuracy only because we have observed during the training that the datasets are prone to overfitting (Table 4).

Conclusions

We investigated into a topological technique to extract relevant cohorts and gene expressions so as to improve feature selections. Both our test cases show that the data follow topological alignment due to which the representative cycles covers the subset of vertices that are able to faithfully represent the data. As a result we are able to fit our training models better and reduce variance thereby getting better accuracy and f1-score. In future work, we will try to further tune our models so as to correlate the selected features with their functionality. For instance, there are cycles with low κ values that have unannotated genes as its constituent vertices. It would be interesting to study functionalities of such specific genes with the other genes in a cycle.

Methods

In this section we briefly describe the idea of Persistent Homology and their representative cycles. Since the idea is involved, the readers are directed to [33, 34] for more details on Persistent Homology. The representative cycles have been described in [12, 13]. Notice that, for curating cohorts, we convert the input $n \times m$ cohort-gene matrix to a point cloud of n points in m dimensions by treating each cohort as a point. Similarly, for curating gene expression we convert the transposed matrix to a point cloud of m points in n dimension with each gene as a single point. Our goal is to compute 'good' representative cycles in the Persistent Homology classes defined on a scaffold called 'filtration' built on top of these point clouds. Using these representative cycles, we identify the cohorts that are predominately present in a cycle and eliminate those which are not dominant in any of these cycles. Similar curation is done for gene expressions using the point cloud representing them.

Persistence signature of point cloud data

We start with a point cloud data in any n-dimensional Euclidean space. These will essentially be n-dimensional points describing individual gene expressions (or cohorts). To illustrate the theory of Persistent Homology, we consider a toy example of taking a set of points in two dimensions sampled uniformly from a two-hole structure (Fig. 3). We start growing n-balls around each point, increasing their radius r continually and tracking the behavior of the union of these growing balls. Starting from r=*0* (Fig. 3a), we notice that at some $r = r_1$ (Fig. 3b) both holes are prominent in the union of ball structure. Increasing r further (Fig. 3d), we fill the smaller hole followed by the larger ones. During the change in the structure of the union of balls due to increase in radius, the larger of the two holes *'persists'* for a larger range of r compared to the smaller one. Hence features that are more prominent are expected to persist for longer periods of increasing r. This exemplifies the basic intuition for topological persistence.

The persistence of holes are captured by a set of *birth-death* pairs (intervals) of homology classes that indicate at which value of r the class is born and where it dies. Each of these pairs is visualized using horizontal line segment known as a bar which together form the barcode [35] (Fig. 3e). The rank of a Persistent Homology group called the persistent Betti number captures the number of persistent features. This means, for the $zero^{th}$ homology group $\mathcal{H}_0$ consisting of $zero^{th}$ homology classes, $betti_0$ counts the number of connected components that arise in the filtration. For $\mathcal{H}_1$, $betti_1$ counts the number of *circular* holes (loops) being born as we proceed through the filtration. Similarly for $\mathcal{H}_2$, $betti_2$ gives a count of the number of surfaces enclosing three dimensional voids in the data. Thus, the short blue line segments of $betti_0$ in (Fig. 3e) correspond to the separate components that are joined to form one big connected component corresponding to the red line. The two long blue line segments of $betti_1$ correspond to the two holes in the structure, the largest representing the bigger hole.

For computational purposes, the growing sequence of the union of balls is converted to a growing sequence of triangulations, simplicial complexes in general, called a *filtration* (Fig. 9). The topological signatures are born when a series of say, edges (1-simplices), are connected to form a cycle and die when they are filled in with triangles (2-simplices). If we take the example in Fig. 9a, the theory of Persistent Homology suggests that in the filtration $\mathcal{F} = K_0 \rightarrow K_1 \rightarrow \ldots \rightarrow K_n = K$, the edges inserted in $\mathcal{K}_4$, $\mathcal{K}_5$ and $\mathcal{K}_6$ (1-simplex denoted σ_4^1, σ_5^1 and σ_6^1 respectively)are the creators as introduction of which create class of homology cycles. We can think the creators as the representative of the cycles. By convention when a triangle appears in the filtration it kills the youngest homology class and is denoted by pairing with creators. In $\mathcal{K}_7$ the triangle kills the cycle created by the edge that came in $\mathcal{K}_6$. So, it pairs with σ_6^1. In $\mathcal{K}_8$ the triangle pairs with σ_7^1 and the big hole (created by σ_4^1 and is the youngest creator unpaired) is filled up and destroyed by the last triangle (2-simplex denoted σ_9^2) inserted in $\mathcal{K}_9$. Thus σ_4^1 is paired with σ_9^2 for interval [4, 9). The

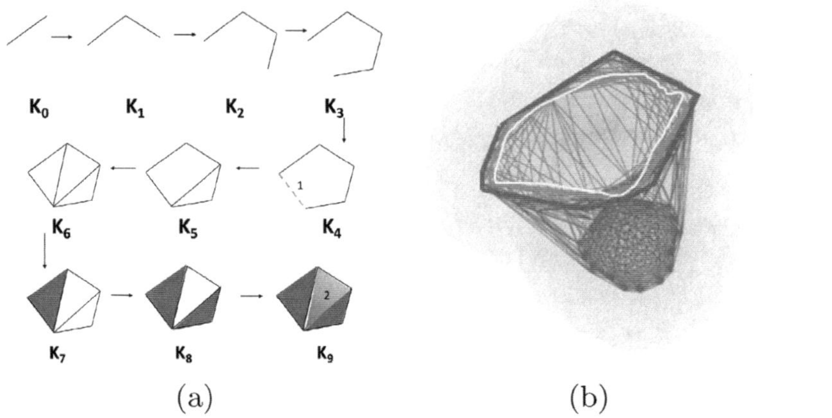

Fig. 9 **a** Filtration $\mathcal{F} = K_0, .., K_9$ explaining persistence pairing. The edge inserted in $\mathcal{K}_4$ dies when the green triangle in $\mathcal{K}_9$ appears. **b** Different $\mathcal{H}_1$ cycles for same homology class

problem with relying only on the barcodes is that they tell us when the classes are born or die given a filtration. But for each homology class, there can be several cycles in the same class (Fig. 9b). Ideally, we would like the tightest cycle (blue one) in the class to be a representative cycle for a given bar. However, it is shown in [12] that computing such cycles even for $\mathcal{H}_1$ is an NP-Hard problem. A follow up paper [13] shows that for dimensions ≥ 1, the problem remains NP-Hard. We therefore, use alternate polynomial time algorithms to build good representative $\mathcal{H}_1$ and $\mathcal{H}_2$ cycles given any barcode interval $[\beta, \delta)$. The first algorithm [12] computes a good but not necessarily the tightest representative cycles. The second algorithm [13] computes the tightest representative cycles but for a specific class of domains called pseudo-manifolds. We briefly describe these two algorithms.

Algorithm 1 Persistent 1-cycle computation

Input:
$\mathcal{K}$: Simplicial 2-complex
$\mathcal{F}$: filtration $\varnothing = K_0 \subseteq K_1 \subseteq \ldots \subseteq K_n = K$, and
$[\beta, \delta)$: finite interval in $\mathcal{F}$
Output: A persistent 1-cycle $\mathcal{C}_1$ for $[\beta, \delta)$
$G' \leftarrow \varnothing$
for $i \leftarrow 1, ..., \beta$ **do**
  if σ_i is creator & (it either gets paired after δ or never gets paired) **then**
    $c_i \leftarrow$ shortest cycle (S_h) containing σ_i in $\mathcal{K}_i$
    $\triangleright$ S_h: shortest path between boundary(σ_i) in $\mathcal{K}_{i-1} \bigcup \sigma_i$
    $G' \leftarrow G' \cup \{i\}$
  end if
end for
find a $G \subseteq G'$ s.t. $\sum_{g \in G} [c_g] = 0$ in $\mathcal{K}_\delta$
output $\sum_{g \in G} c_g$ as the persistent for $[\beta, \delta)$

Algo. 1 generates $\mathcal{H}_1$ cycles. The idea is briefly as follows: we know that at the birth time β of a 1-cycle (found by the persistence algorithm), an edge σ^1_β is inserted in $\mathcal{F}$ to form a cycle in $\mathcal{K}_\beta$. We hence check for the shortest path between the vertices of σ^1_β in $\mathcal{K}_{i-1}$ before σ^1_β is inserted. Since we know that at least one cycle *containing* σ^1_β is formed at $\mathcal{K}_\beta$, adding σ^1_β to this path gives us the shortest cycle at $\mathcal{K}_\beta$. At δ, we need to know which cycle belonging to the homology class has died. This can be a linear combination of any cycles still alive including the cycle found at $\mathcal{K}_\beta$. This is found using a strategy of annotations [4]. In fact, it is shown in the paper that the shortest cycle found at $\mathcal{K}_\beta$ is exactly the shortest cycle for the interval in most practical cases.

Algo. 2 is used to compute $\mathcal{H}_2$ cycles for an interval $[\beta, \delta)$ and can be extended to any $\mathcal{H}_n$. We first construct an undirected dual graph G for $\mathcal{K}$ where vertices of G are dual to 2-simplices of $\mathcal{K}$ and edges of G are dual to 1-simplices of $\mathcal{K}$. One dummy vertex termed as *infinite vertex* which does not correspond to any 2-simplices is added to G for graph edges dual to the boundary 1-simplices. We then build an undirected flow network on top of G where the source is the vertex dual to the death of an interval and the sink is the infinite vertex along with the set of vertices dual to those 2-simplices which are added to $\mathcal{F}$ after δ. If a 1-simplex is σ^1_β or added to $\mathcal{F}$ before σ^1_β, we let the capacity of its dual graph edge be its weight; otherwise, we let the capacity of its dual graph edge be $+\infty$. Finally, we compute a minimal cut of this flow network and return the 2-chain dual to the edges across the minimal cut as a minimal persistent cycle for the interval. The readers may consult the respective papers for $\mathcal{H}_1$-cycles [12] and $\mathcal{H}_2$-cycles [13] computations for more details.

Since computation of H_n-cycles is computationally expensive, especially in higher dimensions, we restrict ourselves with the computation of upto $\mathcal{H}_2$-cycles for our experiments. Most previous works on TDA had mainly included $\mathcal{H}_1$ intervals, with applications in gene expression being restricted to $\mathcal{H}_0$, so we hope to shed some new light into the problem even with this restricted setup.

Algorithm 2 Persistent 2-cycle computation

Input: $\mathcal{K}$: finite 2-weighted weak (3)-pseudomanifold $\mathcal{F}$: filtration $\mathcal{K}_0 \subseteq K_1 \subseteq \ldots \subseteq K_n$ of $\mathcal{K}$ $[\beta, \delta)$: finite interval of $D_2(\mathcal{F})$
Output: minimal persistent 2-cycle of $[\beta, \delta)$

procedure MINPERSCYCFIN($\mathcal{K}, \mathcal{F}, \beta, \delta$)
  ▷ Set up the complex K' being worked on
  $C^3 \leftarrow$ (3)-connected component of $\mathcal{K}$ containing $\sigma^{\mathcal{F}}_\delta$
  $\mathcal{K}' \leftarrow$ closure of the simplicial set C^3
  Construct dual graph
  $(G, \theta) \leftarrow$ DUALGRAPHFIN(K', d)
  assign capacity to G
  for $e \in E(G)$ **do**
    if $ind(\theta^{-1}(e)) \leq \beta$ **then**
      $c(e) \leftarrow w(\theta^{-1}(e))$
    else
      $c(e) \leftarrow +\infty$
    end if
  end for
  ▷ Set the source
  $s_1 \leftarrow \{\theta(\sigma^2_\delta)\}$
  ▷ Set the sink
  $s \leftarrow \{v \in V(G) | v \neq \phi, ind(\theta^{-1}(v)) > \delta\}$
  if ϕ in $V(G)$ **then**
    $s_2 \leftarrow s_2 \bigcup \phi$
  end if
  $(S*, T*) \leftarrow$ min cut of(G, s_1, s_2)
  return $\theta^{-1}(\xi(S*, T*))$
end procedure

Computing topological signature of gene-expression data

We work under the hypothesis that topological data analysis extracts relevant information sufficient for cohort classification. We note that topological feature extraction methods used in earlier works may not work in this setting. Traditionally, for many applications in bio science (say protein classification) and engineering, we find corresponding topological signatures using Persistent Homology for each sample (in this case cohorts or genes). These signatures are appended to the feature vectors. However, in this case, since each cohort is represented by a single 1D vector of gene expression levels, we are not able to find suitable signatures to append. This is why the algorithms we described in the previous section comes handy, as we will see in this section. We use our tools in two separate set of experiments. For algorithms 1 and 2, we need a simplicial complex $\mathcal{K}$, a filtration $\mathcal{F}$, and finite intervals. For all the studies in the paper, we use Sparse Rips [36] to obtain the simplicial complex $\mathcal{K}$ and its filtration ($\mathcal{F}$). We can apply the theory of Persistent Homology to obtain the set of all finite intervals. In addition, algorithm 2 requires a pseudo-manifold ($\widetilde{K}$) instead of a regular simplicial complex $\mathcal{K}$. For our case, this means that all triangles ($d = 2$-simplices) has at most two tetrahedrons ($d + 1 = 3$ -simplices) attached to it. We convert $\mathcal{K}$ into $\widetilde{K}$ by allowing at most two cofaces (tetrahedra) per triangle which appear first in the filtration:

- Add all $\sigma^{0...d}$ to $\widetilde{K}$:
- $\forall \sigma^{d} \in K$:
  - **Sort:** its co-faces $\mathcal{T} = \sigma^{d+1}$ by $\mathcal{F}(\sigma^{d+1})$
  - **If:** $|\mathcal{T}| \geq 2$, insert into $\widetilde{K}$, the first two σ^{d+1} in $\mathcal{T}$,
  - **Else:** insert $\mathcal{T}$ in $\widetilde{K}$

Dataset

We have a set of n cohorts ($\mathcal{C}$) each represented by the gene expression profile of m genes ($\mathcal{G}$). Thus our input is a matrix $\mathcal{K}$ of dimension ($n \times m$) where each $\mathcal{K}_{i,j}$ represents the jth gene of the ith cohort. In addition, we have $\mathcal{X} : \mathcal{C} \rightarrow I$, where I is the phenotype for the cohort. For instance, $\mathcal{X}(c) = 0$ may imply that c is healthy or control, whereas $\mathcal{X}(c) = 1$ may imply they are infected or treated with an antigen depending on the experiment. Throughout our experiments we will work on several datasets containing gene expression profile of different organisms [37]. We provide a brief description of these data (Table 5).

($\mathbb{D}$**0**): *Droso Breeding* In this data set, the *Drosophila melanogaster* larvae is bred on a *Aspergillus nidulans* infested breeding substrate. The phenotypes differ on the different breeding condition for the Drosophilas. We assign label 0 to control, label 1 to the Drosophilas bred on *Aspergillus nidulans* mutant laeA, and label 2 to both the Drosophilas bred on wild Aspergillus nidulans and sterigmatocystin. Note that in this experiment, mutating laeA from wild *Aspergillus nidulans* removes sterigmatocystin production. Hence, both the wild Aspergillus and the class with external sterigmatocystin should have similar gene expression profile.

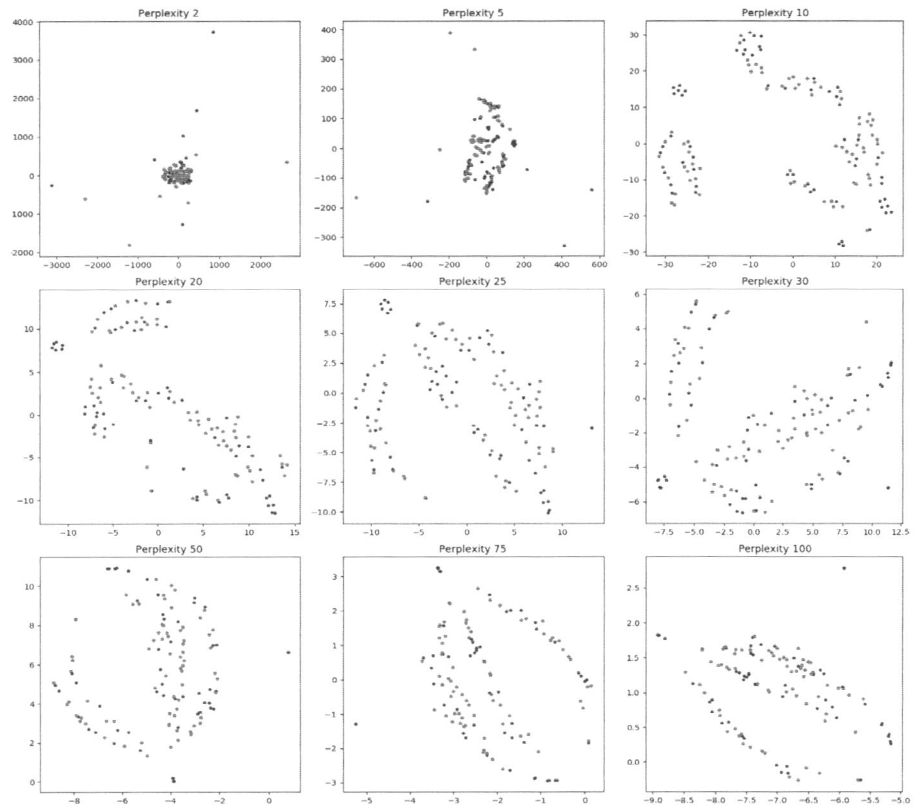

Fig. 10 t-SNE on entire cohort point cloud ($\mathbb{D}0$). Red vertices indicate cohorts included in top 100 $\mathcal{H}_2$ cycles whereas blue indicate otherwise. Note that different perplexity values for t-SNE indicate that the topological cycles are indeed uniform and not specific to a particular hyperparameter (perplexity) value

The experiments in the dataset website confirms this fact, as there is no change in any gene expression profile between these two classes. The number of cohorts in the database is 131. Link: https://www.ebi.ac.uk/gxa/experiments/E-MTAB-5344/Results.

($\mathbb{D}1$): *Droso Parasitod* The data contains the profile of Drosophila larvae after a parasitod attack. There are two labels on the phenotype, one for the control and the other for the cohorts under parasitod attack. Thus, we have a binary classification problem in this case. Total cohorts count is 89. Link to this dataset: https://www.ebi.ac.uk/gxa/experiments/E-MAXD-6/Results.

($\mathbb{D}2$): *Mouse Prion* This data has *Mus musculus* as the cohort. The experiment investigates into the effects of two different strains of the prion disease. The phenotypes are 'RML infected', '301V infected', and the healthy control which are assigned labels 0–2 respectively. Total cohort count is 321. Link:https://www.ebi.ac.uk/gxa/experiments/E-MTAB-76/Results

($\mathbb{D}3$): *Mouse Liver Cancer* This is again a binary classification problem of the *Mus muculus*. The two phenotypes are control type and liver cancer cohorts. We take healthy control as 0 and mice treated with carcenogenic compunds as 1. Total cohort count is 242. Link: https://www.ebi.ac.uk/arrayexpress/experiments/E-GEOD-18858/.

Table 4 Classification using topo-relevant cohort

EXPR		Decision tree			Naive Bayes classifier		
		FULL	H1+H2	H2	FULL	H1+H2	H2
Droso breeding	#	131	116	101	131	116	101
	Accuracy	0.714125	0.751768	0.793434	0.398146	0.412121	0.422444
	Precision	0.745417	0.815000	0.835000	0.389111	0.431756	0.451673
	Recall	0.712500	0.754167	0.795833	0.400000	0.416667	0.434478
Droso parasitod	#	89	85	51			
	Accuracy	0.792778	0.796667	0.811667	–	–	–
	Precision	0.817381	0.823571	0.859167	–	–	–
	Recall	0.792500	0.797500	0.825000	–	–	–
Mouse prion	#	321	292	168	321	292	146
	Accuracy	0.562310	0.616240	0.586843	0.555112	0.578489	0.576131
	Precision	0.562716	0.591471	0.543743	0.383462	0.378556	0.384572
	Recall	0.539712	0.564394	0.558267	0.415855	0.422354	0.423651
Mouse liver cancer	#	242	229	190	242	229	190
	Accuracy	0.682761	0.698934	0.729545	0.723232	0.723232	0.721404
	Precision	0.590716	0.579833	0.656051	0.444761	0.444761	0.412018
	Recall	0.573319	0.602582	0.641168	0.499837	0.499837	.506429
Mouse E.Coli	#	226	206	166	226	206	166
	Accuracy	0.880731	0.851794	0.892900	0.592770	0.592105	0.592105
	Precision	0.880541	0.853406	0.901481	0.604010	0.651101	0.652203
	Recall	0.868052	0.842963	0.891786	0.509841	0.511111	0.511111
Human bowel disease	#	1745	101	101			
	Accuracy	0.499698	0.510987	0.510987	–	–	–
	Precision	0.493808	0.509147	0.509147	–	–	–
	Recall	0.491258	0.501173	0.501173	–	–	–

Each of the data are explained in Dataset section. The # symbol indicates the size of each dataset. '–' in the table means the stats were too low: the relevant classifier was unable to classify the given data. The column 'FULL' represents training on the full dataset while $H_1 + H_2$ represent the union of n' topo-relevant cohorts obtained from the dominant cycles in either H_1 or H_2 whereas H_2 represents cohorts obtained from the dominant cycles in H_2

($\mathbb{D}4$): *Mouse EColi.* The three phenotypes in this dataset are the *Eschreichia coli*, Staphylococcus, and control. The total number of cohorts across all three phenotypes in 226. Link: https://www.ebi.ac.uk/gxa/experiments/E-ENAD-29/Results.

($\mathbb{D}5$): *Human Bowel Disease* A binary classification problem where the phenotype are from cohorts suffering Crohns Disease and placebo cases. This is a big dataset having gene expressions of 1745 human. Link: https://www.ebi.ac.uk/gxa/experiments/E-GEOD-100833/Results

($\mathbb{D}6$): *Human Bone Marrow* This data set contains gene expressions of patients having bone marrow failure and cytogeneic abnormalities along with healthy cohorts who serve as control. This dataset has 469 cohorts. Link: https://www.ebi.ac.uk/arrayexpress/experiments/E-GEOD-32719/.

($\mathbb{D}7$): *Human Dengue* This is yet another big dataset having two types of phenotypes where we have gene expression of Dengue patients versus cohort control. Cohort count for this dataset is 4415. Link: https://www.ncbi.nlm.nih.gov/geo/download/?acc=GSE116672.

Table 5 Cohort count in each phenotype. Details of the label assignment for each dataset can be found in its description

Dataset	Phenotype	Samples per class	Total Samples
Droso breeding	Control	32	131
	Aspergillus	34	
	Aspergillus + Sterigmatosystin	65	
Droso parasitod	Control	45	89
	Under parasitod attack	44	
Mouse Prion	RML infected	154	321
	301V infected	122	
	Control	45	
Mouse liver cancer	Control	63	242
	With liver cancer	179	
Mouse E.Coli	With E.Coli	64	226
	With staphylococcus	102	
	Control	60	
Human Bowel Disease	Crohn's disease	101	1745
	Placebo	1113	
	Control	531	
Human bone marrow	Bone marrow failure	391	469
	Control	78	
Human dengue	Dengue fever	3311	4415
	Controll	1104	

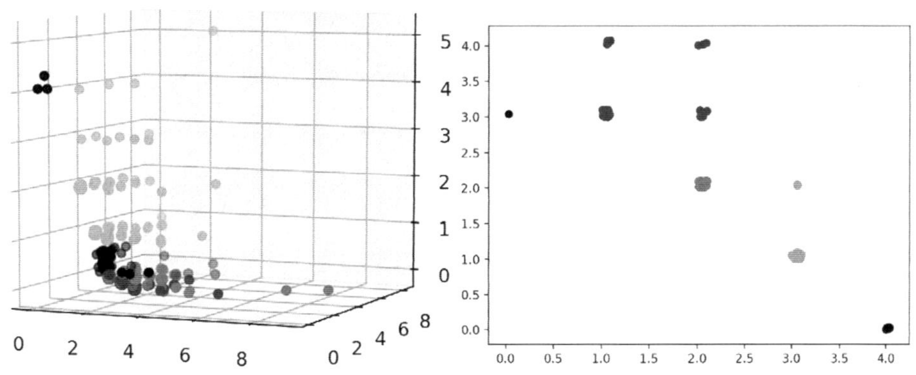

Fig. 11 **a** Count of vertex labels in individual $\mathcal{H}_2$-cycles for $\mathbb{D}0$. The red points indicate cycles having phenotype labels 0 and 1, blue indicates cycles with labels 1 and 2 whereas green (very few in the top 500 $\mathcal{H}_2$ cycles) indicates labels 0 and 2. **b** Count of vertex labels in individual $\mathcal{H}_2$-cycles for $\mathbb{D}1$. Red indicate cycles having equal phenotype labelled vertices. Blue and cyan indicate prevalence of label 0 and 1 respectively. In both the diagrams, black points indicate cycles having a single phenotype label

Since each data point reside in dimension > 3 we apply t-Distributed Stochastic Neighbor Embedding (t-SNE) on $\mathbb{D}0$ Drosophila dataset to obtain a 2D projection for visualization in Fig. 10. To get a sense of the distribution of topological cycles, we calculate the top 100 representative $\mathcal{H}_2$ cycles based on their interval length $(\delta - \beta)$. In Fig. 10, we color a cohort vertex red if it is contained in any of the top 100 $\mathcal{H}_2$ cycles. The cohorts not included are painted blue. This figure shows the uniform distribution of the topological cycles w.r.t the entire dataset (Fig. 11).

Abbreviations
TDA: Topological data analysis; ML: Machine learning; NCBI: National Center for Biotechnology Information; CNN: Convolutional neural network.

Acknowledgements
Not applicable.

About this supplement
This article has been published as part of BMC Bioinformatics Volume 22 Supplement 10 2021: Selected articles from the 19th Asia Pacific Bioinformatics Conference (APBC 2021): bioinformatics. The full contents of the supplement are available at https://bmcbioinformatics.biomedcentral.com/articles/supplements/volume-22-supplement-10.

Author contributions
TKD and SML designed the experiments. SML developed the software and SML, SME both carried out the experiments. SML played an integral role in writing the manuscript. SME generated necessary figures and tables for the manuscript. All authors have read and approved this manuscript for publication.

Funding
Publication cost was funded by National Science Foundation [Grant Nos. ccf-2049010 and ccf-1839252]. This sponsor had no role in the design of the study and collection, analysis, and interpretation of data or in writing this manuscript.

Data availability
The source code for computing persistent persistent 1-cycles can be found at https://github.com/Sayan-m90/PersLoop-viewer whereas persistent 2-cycles is hosted in https://github.com/Sayan-m90/Minimum-Persistent-Cycles. The Neural Network Architecture and the Probabilistic ML (Machine Learning) approaches on top of it can be built with any standard open source ML libraries. Experiments done to compare with the standard approaches can be found https://github.com/soham0209/Gene-Expression. The datasets are available publicly. Link of the datasets are provided in their respective description.

Declarations

Ethics approval and consent to participate
Not applicable since the gene expression data used in this study were publicly available, de-identified and obtained from the following source: Gene Expression Omnibus (GEO).

Consent for publication
Not applicable.

Competing interests
The authors declare that they have no competing interests.

Author details
[1]Department of Computer Science, Purdue University, West Lafayette, IN, USA. [2]Department of Computer Science and Engineering, The Ohio State University, Columbus, OH, USA.

Received: 28 April 2022 Accepted: 28 April 2022
Published online: 20 May 2022

References

1. Singh N, Couture HD, Marron JS, Perou C, Niethammer M. Topological descriptors of histology images. In: Wu G, Zhang D, Zhou L, editors. Machine learning in medical imaging. Cham: Springer; 2014. p. 231–9.
2. Gao M, Chen C, Zhang S, Qian Z, Metaxas D, Axel L. Segmenting the papillary muscles and the trabeculae from high resolution cardiac CT through restoration of topological handles. In: Gee JC, Joshi S, Pohl KM, Wells WM, Zöllei L, editors. Information processing in medical imaging. Berlin: Springer; 2013. p. 184–95.
3. Cang Z, Wei G-W. Topologynet: topology based deep convolutional and multi-task neural networks for biomolecular property predictions. PLoS Comput Biol. 2017;13(7):1–27.
4. Dey TK, Fan F, Wang Y. Computing topological persistence for simplicial maps. In: Proceedings of the thirtieth annual symposium on computational geometry. ACM; 2014. p. 345.
5. Pike JA, Khan AO, Pallini C, Thomas SG, Mund M, Ries J, Poulter NS, Styles IB. Topological data analysis quantifies biological nano-structure from single molecule localization microscopy. bioRxiv. 2018. https://doi.org/10.1101/400275.
6. Sauerwald N, Shen Y, Kingsford C. Topological data analysis reveals principles of chromosome structure throughout cellular differentiation. bioRxiv. 2019. https://doi.org/10.1101/540716.
7. Eisen MB, Spellman PT, Brown PO, Botstein D. Cluster analysis and display of genome-wide expression patterns. Proc Natl Acad Sci. 1998;95(25):14863–8.

8. Oyelade J, Isewon I, Oladipupo F, Aromolaran O, Uwoghiren E, Ameh F, Achas M, Adebiyi E. Clustering algorithms: their application to gene expression data. Bioinform Biol Insights. 2016;10:38316. https://doi.org/10.4137/BBI.S38316.
9. Dey TK, Mandal S. Protein classification with improved topological data analysis. In: WABI; 2018.
10. Mandal S, Guzmán-Sáenz A, Haiminen N, Basu S, Parida L. A topological data analysis approach on predicting phenotypes from gene expression data 2020.
11. Dey T, Mandal S, Varcho W. Improved image classification using topological persistence. In: Proceedings of the conference on vision, modeling and visualization; 2017. p. 161–8.
12. Dey TK, Hou T, Mandal S. Persistent 1-cycles: definition, computation, and its application. In: Marfil R, Calderón M, íaz del Río F, Real P, Bandera A, editors. Computational topology in image context. Cham: Springer; 2019. p. 123–36.
13. Dey TK, Hou T, Mandal S. Computing minimal persistent cycles: polynomial and hard cases. In: Proceedings of the thirty-first annual ACM-SIAM symposium on discrete algorithms. SODA '20. Society for Industrial and Applied Mathematics, USA; 2020. p. 2587–606.
14. Schofield JPR, Strazzeri FEA. A topological data analysis network model of asthma based on blood gene expression profiles. bioRxiv. 2019. https://doi.org/10.1101/516328.
15. Rabadan R, EDU C. Parametric inference using persistence diagrams: a case study in population genetics. Birth. 250:300.
16. Arsuaga J, Borrman T, Cavalcante R, Gonzalez G, Park C. Identification of copy number aberrations in breast cancer subtypes using persistence topology. Microarrays. 2015;4(3):339–69.
17. Seemann L, Shulman J, Gunaratne GH. A Robust topology-based algorithm for gene expression profiling. 2012. https://doi.org/10.5402/2012/381023. https://www.hindawi.com/journals/isrn/2012/381023/ Accessed 09 Mar 2020.
18. Nicolau M, Levine AJ, Carlsson G. Topology based data analysis identifies a subgroup of breast cancers with a unique mutational profile and excellent survival. Proc Natl Acad Sci. 2011;108(17):7265–70. https://doi.org/10.1073/pnas.1102826108.
19. Pirooznia M, Yang JY, Yang MQ, Deng Y. A comparative study of different machine learning methods on microarray gene expression data. BMC Genom. 2008;9(Suppl 1):13–13. https://doi.org/10.1186/1471-2164-9-S1-S13.
20. Tarek S, Elwahab RA, Shoman M. Gene expression based cancer classification. Egypt Inform J. 2017;18(3):151–9. https://doi.org/10.1016/j.eij.2016.12.001.
21. Kong Y, Yu T. A deep neural network model using random forest to extract feature representation for gene expression data classification. Sci Rep. 2018;8(1):16477. https://doi.org/10.1038/s41598-018-34833-6.
22. Hwang K-B, Cho D-Y, Park S-W, Kim S-D, Zhang B-T. Applying machine learning techniques to analysis of gene expression data: cancer diagnosis; 2002. p. 167–82.
23. van IJzendoorn DGP, Szuhai K, Briaire-de Bruijn IH, Kostine M, Kuijjer ML, Bovée JVMG. Machine learning analysis of gene expression data reveals novel diagnostic and prognostic biomarkers and identifies therapeutic targets for soft tissue sarcomas. PLoS Comput Biol. 2019;15(2):1006826. https://doi.org/10.1371/journal.pcbi.1006826.
24. Mi H, Muruganujan A, Ebert D, Huang X, Thomas PD. PANTHER version 14: more genomes, a new PANTHER GO-slim and improvements in enrichment analysis tools. Nucleic Acids Res. 2018;47(D1):419–26. https://doi.org/10.1093/nar/gky1038.
25. National Institutes of Health, U.: National Center for Biotechnology Information 1988. https://www.ncbi.nlm.nih.gov/gene.
26. Kingma DP, Adam BJ. A method for stochastic optimization. In: Bengio Y, LeCun Y, editors. 3rd International conference on learning representations, ICLR 2015, San Diego, CA, USA, May 7–9, 2015, conference track proceedings; 2015. arXiv:1412.6980
27. Schubert E, Sander J, Ester M, Kriegel HP, Xu X. DBSCAN revisited, revisited: why and how you should (still) use DBSCAN. ACM Trans Database Syst (TODS). 2017;42(3):1–21.
28. Kira K, Rendell LA, et al. The feature selection problem: traditional methods and a new algorithm. In: AAAI, vol 2; 1992. p. 129–34.
29. Guyon I, Elisseeff A. An introduction to variable and feature selection. J Mach Learn Res. 2003;3(Mar):1157–82.
30. Pedregosa F, Varoquaux G, Gramfort A, Michel V, Thirion B, Grisel O, Blondel M, Prettenhofer P, Weiss R, Dubourg V, Vanderplas J, Passos A, Cournapeau D, Brucher M, Perrot M, Duchesnay E. Scikit-learn: machine learning in Python. J Mach Learn Res. 2011;12:2825–30.
31. Shlens J. A tutorial on principal component analysis; 2014. arXiv preprint arXiv:1404.1100.
32. McInnes L, Healy J, Melville J. Umap: uniform manifold approximation and projection for dimension reduction; 2018. arXiv preprint arXiv:1802.03426.
33. Edelsbrunner H, Harer J. Computational topology: an introduction; 2010.
34. Zomorodian A, Carlsson G. Computing persistent homology. Discrete Comput Geom. 2005;33(2):249–74.
35. Carlsson G, Zomorodian A, Collins A, Guibas LJ. Persistence barcodes for shapes. Int J Shape Model. 2005;11(02):149–87.
36. Sheehy DR. Linear-size approximations to the vietoris-rips filtration. Discrete Comput Geom. 2013;49(4):778–96. https://doi.org/10.1007/s00454-013-9513-1.
37. Petryszak R, Keays M, Tang YA, Fonseca NA, Barrera E, Burdett T, Füllgrabe A, Fuentes AM-P, Jupp S, Koskinen S, Mannion O, Huerta L, Megy K, Snow C, Williams E, Barzine M, Hastings E, Weisser H, Wright J, Jaiswal P, Huber W, Choudhary J, Parkinson HE, Brazma A. Expression Atlas update-an integrated database of gene and protein expression in humans, animals and plants. Nucleic Acids Res. 2015;44(D1):746–52. https://doi.org/10.1093/nar/gkv1045.

Publisher's Note

AUTHOR INDEX

BioMed Central Ltd
236 Gray's Inn Road
London WC1X 8HB
UK

CURRAN ASSOCIATES INC.
proceedings
.com

ISBN 978-1-7138-5563-7

9 781713 855637